# About the author

**Paul Davidovits**, Professor of Chemistry at Boston College, was coawarded the prestigious R.W. Wood Prize from the Optical Society of America for his seminal work in optics. His contribution was foundational in the field of confocal microscopy, which allows engineers and biologists to produce optical sections through three-dimensional objects such as semiconductor circuits, living tissues, or a single cell. He has published more than 170 papers in physical chemistry and three books. He is a Fellow of the American Physical Society and of the American Association for Advancement of Science. The second edition of *Physics in Biology and Medicine* received the Alpha Sigma Nu Book Award in the Discipline of the Natural Sciences.

# Physics in biology and medicine sixth edition

A best-selling resource now in its sixth edition, Paul Davidovits' *Physics in Biology and Medicine* provides a high-quality and highly relevant physics grounding for students working toward careers in the medical and related professions. The text does not assume a prior background in physics but provides it as required. It discusses biological systems that can be analyzed quantitatively and demonstrates how advances in the life sciences have been aided by the knowledge of physical or engineering analysis techniques, with applications, practice, and illustrations throughout.

*Physics in Biology and Medicine*, sixth edition, includes new and revised material and corresponding exercises on many exciting developments in the field since the prior edition, including random walk, signal-to-noise, metabolism in connection with thermodynamics, and weight loss, and a detailed discussion of metamaterials.

This revised edition delivers helpful and engaging additions to the role and importance of physics in biology and medicine, including new coverage on metamaterials, metabolism, and environmental science. It is ideal for courses in biophysics, medical physics, and related subjects.

## Key features

- Provides practical information and techniques for building fundamental knowledge and applying physics and biology to the study of living systems.
- Includes numerous figures, examples, illustrative problems, and appendices, which provide convenient access to the important concepts of mechanics, electricity, and optics used in the text.
- Features new and revised coverage on metamaterials, random walk, and metabolism in connection with thermodynamics and weight loss.
- Offers online support, including a full solutions manual for qualified instructors and additional programming resources (PowerPoints) for students.

For additional information on the topics covered in the book, visit the instructor site: https://educate.elsevier.com/9780443215582

# Preface to the sixth edition

Until the mid-1800s, it was not clear to what extent the laws of physics and chemistry, which were formulated from the observed behavior of inanimate matter, could be applied to living matter. It was certainly evident that the laws were applicable on a large scale. Animals are clearly subject to the same laws of motion as inanimate objects. The question of applicability arose on a more basic level. Living organisms are very complex. Even a virus, which is one of the simplest biological organisms, consists of millions of interacting atoms. A cell, which is the basic building block of tissue, contains, on average, $10^{14}$ atoms. Living organisms exhibit properties not found in inanimate objects. They grow, reproduce, and decay. These phenomena are so different from the predictable properties of inanimate matter that many scientists in the early 19th century believed that different laws governed the structure and organization of molecules in living matter. Even the physical origin of organic molecules was in question. These molecules tend to be larger and more complex than molecules obtained from inorganic sources. It was thought that the large molecules found in living matter could be produced only by living organisms through a "vital force" that could not be explained by the existing laws of physics. This concept was disproved in 1828 when Friedrich Wöhler synthesized an organic substance, urea, from inorganic chemicals. Soon thereafter, many other organic molecules were synthesized without the intervention of biological organisms. Today most scientists believe that there is no special vital force residing in organic substances. Living organisms are governed by the laws of physics on all levels.

Much of the biological research over the past 100 years has been directed toward understanding living systems in terms of basic physical laws. This effort has yielded some significant successes. The atomic structure of many complex biological molecules has now been determined, and the role of these molecules within living systems has been described. It is now possible to explain the functioning of cells and many of their interactions with each other. Yet the work is far from complete. Even when the structure of a complex molecule is known, it is not possible at present to predict its function from its atomic structure. The mechanisms of cell nourishment, growth, reproduction, and communication are still understood only qualitatively. Many of the basic questions in biology remain unanswered. However, biological research has so far not revealed any areas where physical laws do not apply. The amazing properties of life seem to be achieved by the enormously complex organization in living systems.

The aim of this book is to relate some of the concepts in physics to living systems. In general, the text follows topics found in basic college physics texts. The discussion is organized into the following areas: solid mechanics, fluid mechanics, thermodynamics, sound, electricity, optics, and atomic and nuclear physics. A section on the growing field of nanotechnology in biology and medicine has been added to the later editions.

Each chapter contains a brief review of the background physics, but most of the text is devoted to the applications of physics to biology and medicine. No previous knowledge of biology is assumed. The biological systems to be discussed are described in as much detail as is necessary for the physical analysis. Whenever possible, the analysis is quantitative, requiring only basic algebra and trigonometry. The basic background concepts in mechanics, electricity, and optics are presented and reviewed in the appendices.

Many biological systems can be analyzed quantitatively. A few examples will illustrate the approach. Under the topic of mechanics, we calculate the forces exerted by muscles. We examine the maximum impact a body can sustain without injury. We calculate the height to which a person can jump, and we discuss the effect of an animal's size on the speed at which it can run. In our study of fluids, we examine quantitatively the circulation of blood in the body. The theory of fluids allows us also to calculate the role of diffusion in the functioning of cells and the effect of surface tension on the growth of plants in soil. Using the principles of electricity, we analyze quantitatively the conduction of impulses along the nervous system. Each section contains problems that explore and expand some of the concepts.

There are, of course, severe limits on the quantitative application of physics to biological systems. These limitations are discussed.

Many of the advances in the life sciences have been greatly aided by the application of the techniques of physics and engineering to the study of living systems. Some of these techniques are examined in the appropriate sections of the book.

This sixth edition of the book has been updated to include discussions of new topics in the application of physics in biology and medicine that have come to the forefront since the writing of the fifth edition of this book. The discussion of random walk has been extended to signal-to-noise calculations, which is a key parameter in obtaining accurate measurements in medical and biological studies. With overweight and obesity having become an important medical and societal issue, a section has been included in the new edition that examines physical exercise, calories, and weight loss from the perspective of thermodynamics and recent studies. A new chapter on metamaterials has been added to the book. The study of these novel materials with specifically designed properties provides a clear example of scientific collaborations in the fields of physics, engineering, biology, and medicine that are facilitating progress in all these important areas.

A word about units. Most physics and chemistry textbooks now use the MKS International System of Units (SI). In practice, however, a variety of units continue to be in use. For example, in the SI system, pressure is expressed in units of pascal ($N/m^2$). Both in common use and in the scientific literature, one often finds pressure also expressed in units of dynes/$cm^2$, Torr (mm Hg), psi, and atm. In this book, I have mostly used SI units. However, other units have also been used when common usage so dictated. In those cases, conversion factors have been provided either within the text or in a compilation at the end of Appendix A.

In the first edition of this book, I expressed my thanks to W. Chameides, M.D. Egger, L.K. Stark, and J. Taplitz for their help and encouragement. In the second edition, I thanked Prof. R.K. Hobbie and David Cinabro for their careful reading of the manuscript and helpful suggestions. In the fourth edition, I expressed my appreciation to Prof. Per Arne Rikvold for his careful reading of the text and his important comments. In the fifth edition I thanked Michelle Fisher, Editorial Project Manager at Elsevier/Academic Press, and Bharatwaj Varatharajan, Production Manager, for their help and involvement in the preparation of the fifth edition of the book. Here I want to thank Katey Birtcher, Publisher, STEM Education Content; Stephanie Cohen, Acquisitions Editor, Physical Science Textbooks; Mason Malloy, Editorial Project Manager, and Maria Shalini, Production Project Manager, all colleagues at Elsevier. They have provided encouragement and help and have greatly facilitated the publication of the sixth edition of this book.

**Paul Davidovits**

# Abbreviations

| | |
|---|---|
| μ | micron |
| μA | microampere |
| μV | microvolt |
| μV/m | microvolt per meter |
| A | ampere |
| Å | angstrom |
| atm | atmosphere |
| av. | average |
| BMI | body mass index |
| C | coulomb |
| c.g. | center of gravity |
| cal | calorie (gram calorie) |
| Cal | calorie (kilo calorie) |
| cm | centimeter |
| $cm^2$ | square centimeters |
| cos | cosine |
| cps | cycles per second |
| Cryo-EM | cryoelectron microscopy |
| CT | computerized tomography |
| dB | decibel |
| deg. | degree |
| diam | diameter |
| DLW | doubly labeled water |
| dyn | dyne |
| $dyn/cm^2$ | dynes per square centimeter |
| F | farad |
| f | frequency |
| F/m | farad/meters |
| ft | foot |
| ft/sec | feet per second |
| FUS | focused ultrasound surgery |
| g | gram |

| | |
|---|---|
| **h** | hour |
| **Hz** | hertz (cps) |
| **ICD** | implantable cardioverter-defibrillator |
| **in** | inch |
| **J** | joule |
| **KE** | kinetic energy |
| **kg** | kilogram |
| **km** | kilometer |
| **km/h** | kilometers per hour |
| **kph** | kilometers per hour |
| **lb** | pound |
| **lim** | limit |
| **L/min** | liters per minute |
| **m** | meter |
| **m/sec** | meters per second |
| **mA** | milliampere |
| **max** | maximum |
| **min** | minute |
| **mph** | miles per hour |
| **MRE** | magnetic resonance elastography |
| **MRI** | magnetic resonance imaging |
| **ms** | millisecond |
| **mV** | millivolt |
| **N** | Newton |
| **nm** | nanometer |
| **Nm** | newton meters |
| **NMR** | nuclear magnetic resonance |
| **PE** | potential energy |
| **psi** | pounds per square inch |
| **R** | Reynold's number |
| **rad** | radian |
| **RF** | radio frequency |
| **SAD** | seasonal affective disorder |
| **sec** | second |
| **sin** | sine |
| **SMT** | soil moisture tension |
| **SRR** | split ring resonator |
| **tan** | tangent |
| **TNF** | tumor necrosis factor |
| **V** | volt |
| **W** | watt |
| **Ω** | ohm |

# Contents

# Chapter 1

# Static forces

Mechanics is the branch of physics concerned with the effect of forces on the motion of bodies. It was the first branch of physics that was applied successfully to living systems, primarily to understanding the principles governing the movement of animals. Our present concepts of mechanics were formulated by Isaac Newton, whose major work on mechanics, *Principia Mathematica*, was published in 1687. The study of mechanics, however, began much earlier. It can be traced to the Greek philosophers of the 4th century BCE. The early Greeks, who were interested in both science and athletics, were also the first to apply physical principles to animal movements. Aristotle wrote, "The animal that moves makes its change of position by pressing against that which is beneath it. ... Runners run faster if they swing their arms for in extension of the arms there is a kind of leaning upon the hands and the wrist." Although some of the concepts proposed by the Greek philosophers were wrong, their search for general principles in nature marked the beginning of scientific thought.

After the decline of ancient Greece, the pursuit of all scientific work entered a period of lull that lasted until the Renaissance brought about a resurgence in many activities including science. During this period of revival, Leonardo da Vinci (1452–1519) made detailed observations of animal motions and muscle functions. Since da Vinci, hundreds of people have contributed to our understanding of animal motion in terms of mechanical principles. Their studies have been aided by improved analytic techniques and the development of instruments such as the photographic camera and electronic timers. Today the study of human motion is part of the disciplines of kinesiology, which studies human motion primarily as applied to athletic activities, and biomechanics, a broader area that is concerned not only with muscle movement but also with the physical behavior of bones and organs such as the lungs and the heart. The development of prosthetic devices such as artificial limbs and mechanical hearts is an active area of biomechanical research.

Mechanics, like every other subject in science, starts with a certain number of basic concepts and then supplies the rules by which they are interrelated. Appendix A summarizes the basic concepts in mechanics, providing a review rather than a thorough treatment of the subject. We will now begin our discussion of mechanics by examining static forces that act on the human body.

**Physics in Biology and Medicine.** https://doi.org/10.1016/B978-0-443-21558-2.00001-8

We will first discuss stability and equilibrium of the human body, and then we will calculate the forces exerted by the skeletal muscles on various parts of the body.

## 1.1 Equilibrium and stability

The Earth exerts an attractive force on the mass of an object; in fact, every small element of mass in the object is attracted by the Earth. The sum of these forces is the total weight of the body. This weight can be considered a force acting through a single point called the center of mass or center of gravity. As pointed out in Appendix A, a body is in static equilibrium if the vectorial sum of both the forces and the torques acting on the body is zero. If a body is unsupported, the force of gravity accelerates it, and the body is not in equilibrium. In order for a body to be in stable equilibrium, it must be properly supported.

The position of the center of mass with respect to the base of support determines whether the body is stable or not. A body is in stable equilibrium under the action of gravity if its center of mass is directly over its base of support (Fig. 1.1a, b). Under this condition, the reaction force at the base of support cancels the force of gravity and the torque produced by it. If the center of mass is outside the base, the torque produced by the weight tends to topple the body (Fig. 1.1c).

The wider the base on which the body rests, the more stable it is; that is, the more difficult it is to topple it. If the wide-based body in Fig. 1.1a is displaced, as shown in Fig. 1.2a, the torque produced by its weight tends to restore it to its original position ($F_r$ shown is the reaction force exerted by the surface on the body). The same amount of angular displacement of a narrow-based body results in a torque that will topple it (Fig. 1.2b). Similar considerations show that a body is more stable if its center of gravity is closer to its base.

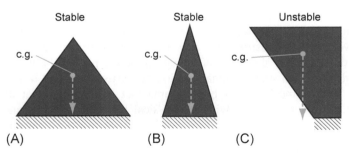

FIGURE 1.1 **Stability of bodies.** The position of the center of mass with respect to the base of support determines whether the body is stable or not. A body is in stable equilibrium under the action of gravity if its center of mass is directly over its base of support as in (A) and (B). Under this condition, the reaction force at the base of support cancels the force of gravity and the torque produced by it. The wider the base on which the body rests, the more stable it is; that is, the more difficult it is to topple it. If the center of mass is outside the base of support, the torque produced by the weight tends to topple the body (C).

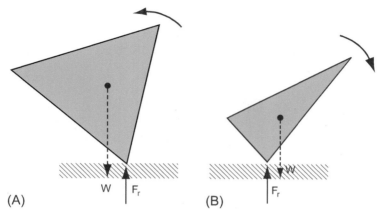

**FIGURE 1.2**  (A) Torque produced by the weight will restore the body to its original position. (B) Torque produced by the weight will topple the body. For simplicity, frictional forces that prevent the contact point from sliding are not shown. The frictional force horizontal to the left (A) and to the right (B). *(See Chapter 2 for discussion of frictional forces.)*

## 1.2 Equilibrium considerations for the human body

The center of gravity (c.g.) of an erect person with arms at the side is at approximately 56% of the person's height measured from the soles of the feet (Fig. 1.3). The center of gravity shifts as the person moves and bends. The act of balancing requires maintenance of the center of gravity above the feet. A person falls when his center of gravity is displaced beyond the position of the feet.

When carrying an uneven load, the body tends to compensate by bending and extending the limbs so as to shift the center of gravity back over the feet. For example, when a person carries a weight in one arm, the other arm swings away from the body and the torso bends away from the load (Fig. 1.4). This tendency of the body to compensate for uneven weight distribution often causes problems for people who have lost an arm, as the continuous compensatory bending of the torso can result in a permanent distortion of the spine. It is often recommended that amputees wear an artificial arm, even if they cannot use it, to restore balanced weight distribution.

## 1.3 Stability of the human body under the action of an external force

The body may of course be subject to forces other than the downward force of weight. Let us calculate the magnitude of the force applied to the shoulder that will topple a person standing at rigid attention. The assumed dimensions of the person are as shown in Fig. 1.5. In the absence of the force, the person is in stable equilibrium because his center of mass is above his feet, which are the base of support. The applied force $F_a$ tends to topple the body. When the

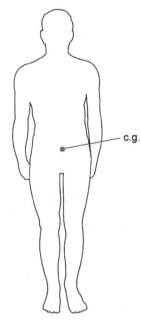

**FIGURE 1.3 Center of gravity for a person.**

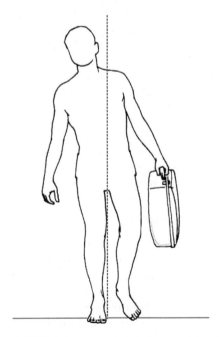

**FIGURE 1.4 A person carrying a weight.**

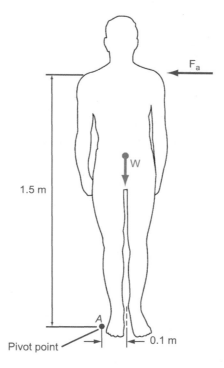

**FIGURE 1.5    A force applied to an erect person.**

person topples, he will do so by pivoting around point $A$—assuming that he does not slide. The counterclockwise torque $T_a$ about this point produced by the applied force is

$$T_a = F_a \times 1.5 \text{ m} \tag{1.1}$$

The opposite restoring torque $T_w$ due to the person's weight is

$$T_w = W \times 0.1 \text{ m} \tag{1.2}$$

Assuming that the mass m of the person is 70 kg, his weight W is

$$W = mg = 70 \times 9.8 = 686 \text{ newton (N)} \tag{1.3}$$

(Here g is the gravitational acceleration, which has the magnitude 9.8 m/ sec$^2$.) The restoring torque produced by the weight is therefore 68.6 newton-meter (Nm). The person is on the verge of toppling when the magnitudes of these two torques are just equal; that is, $T_a = T_w$ or

$$F_a \times 1.5 \text{ m} = 68.6 \text{ Nm} \tag{1.4}$$

Therefore, the force required to topple an erect person is

$$F_a = \frac{68.6}{1.5} = 45.7 \text{ N (10.3 lb)} \tag{1.5}$$

Actually, a person can withstand a much greater sideways force without losing balance by bending the torso in the direction opposite to the applied force (Fig. 1.6). This shifts the center of gravity away from the pivot point $A$, increasing the restoring torque produced by the weight of the body.

Stability against a toppling force is also increased by spreading the legs, as shown in Fig. 1.7 and discussed in Exercise 1.1.

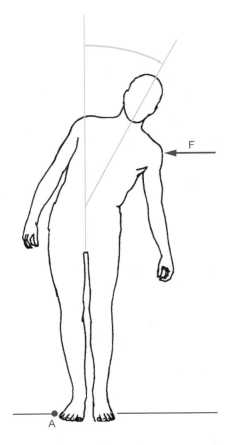

FIGURE 1.6 **Compensating for a side-pushing force.**

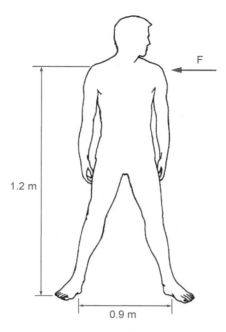

FIGURE 1.7    **Increased stability resulting from spreading the legs.**

## 1.4 Skeletal muscles

The skeletal muscles producing skeletal movements consist of many thousands of parallel fibers wrapped in a flexible sheath that narrows at both ends into tendons (Fig. 1.8). The tendons, which are made of strong tissue, grow into the bone and attach the muscle to the bone. Most muscles taper to a single tendon. But some muscles end in two or three tendons; these muscles are called, respectively, *biceps* and *triceps*. Each end of the muscle is attached to a different bone. In general, the two bones attached by muscles are free to move with respect to each other at the joints where they contact each other.

This arrangement of muscle and bone was noted by Leonardo da Vinci, who wrote, "The muscles always begin and end in the bones that touch one another, and they never begin and end on the same bone ..." He also stated, "It is the function of the muscles to pull and not to push except in the cases of the genital member and the tongue."

Da Vinci's observation about the pulling by muscles is correct. When fibers in the muscle receive an electrical stimulus from the nerve endings that are attached to them, they contract. This results in a shortening of the muscle and a corresponding pulling force on the two bones to which the muscle is attached.

There is a great variability in the pulling force that a given muscle can apply. The force of contraction at any time is determined by the number of individual

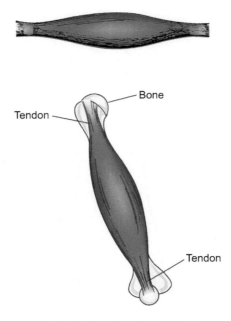

**FIGURE 1.8** **Drawing of a muscle.**

fibers that are contracting within the muscle. When an individual fiber receives an electrical stimulus, it tends to contract to its full ability. If a stronger pulling force is required, a larger number of fibers are stimulated to contract.

Experiments have shown that the maximum force a muscle is capable of exerting is proportional to its cross-section. From measurements, it has been estimated that a muscle can exert a force of about $7 \times 10^6$ dyn/cm$^2$ of its cross-sectional area $\left(7 \times 10^6 \text{ dyn/cm}^2 = 7 \times 10^5 \text{ Pa} = 102 \text{ lb/in}^2\right)$.

To compute the forces exerted by muscles, the various joints in the body can be conveniently analyzed in terms of levers. Such a representation implies some simplifying assumptions. We will assume that the tendons are connected to the bones at well-defined points and that the joints are frictionless.

Simplifications are often necessary to calculate the behavior of systems in the real world. Seldom are all the properties of the system known, and even when they are known, consideration of all the details is usually not necessary. Calculations are most often based on a model, which is assumed to be a good representation of the real situation.

## 1.5 Levers

A lever is a rigid bar free to rotate about a fixed point called the *fulcrum*. The position of the fulcrum is fixed so that it is not free to move with respect to the bar. Levers are used to lift loads in an advantageous way and to transfer movement from one point to another.

There are three classes of levers, as shown in Fig. 1.9. In a Class 1 lever, the fulcrum is located between the applied force and the load. A crowbar is an example of a Class 1 lever. In a Class 2 lever, the fulcrum is at one end of the bar; the force is applied to the other end; and the load is situated in between. A wheelbarrow is an example of a Class 2 lever. A Class 3 lever has the fulcrum at one end and the load at the other. The force is applied between the two ends. As we will see, many of the limb movements of animals are performed by Class 3 levers.

It can be shown from the conditions for equilibrium (see Appendix A) that, for all three types of levers, the force F required to balance a load of weight W is given by

$$F = \frac{Wd_1}{d_2} \tag{1.6}$$

where $d_1$ and $d_2$ are the lengths of the lever arms, as shown in Fig. 1.9 (see Exercise 1.2). If $d_1$ is less than $d_2$, the force required to balance a load is smaller than the load. The mechanical advantage M of the lever is defined as

$$M = \frac{W}{F} = \frac{d_2}{d_1} \tag{1.7}$$

Depending on the distances from the fulcrum, the mechanical advantage of a Class 1 lever can be greater or smaller than 1. By placing the load close to the fulcrum, with $d_1$ much smaller than $d_2$, a very large mechanical advantage can be obtained with a Class 1 lever. In a Class 2 lever, $d_1$ is always smaller than $d_2$; therefore, the mechanical advantage of a Class 2 lever is greater than 1. The situation is opposite in a Class 3 lever. Here $d_1$ is larger than $d_2$; therefore, the mechanical advantage is always less than 1.

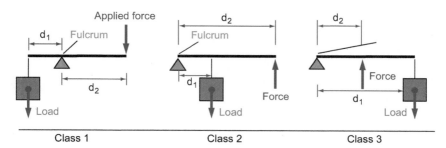

FIGURE 1.9   **The three classes of lever.**

A force slightly greater than is required to balance the load will lift it. As the point at which the force is applied moves through a distance $L_2$, the load moves a distance $L_1$ (see Fig. 1.10). The relationship between $L_1$ and $L_2$ (see Exercise 1.2) is given by

$$\frac{L_1}{L_2} = \frac{d_1}{d_2} \tag{1.8}$$

The ratio of velocities of these two points on a moving lever is likewise given by

$$\frac{v_1}{v_2} = \frac{d_1}{d_2} \tag{1.9}$$

Here $v_2$ is the velocity of the point where the force is applied, and $v_1$ is the velocity of the load. These relationships apply to all three classes of levers. Thus, it is evident that the excursion and velocity of the load are inversely proportional to the mechanical advantage.

## 1.6 The elbow

The two most important muscles producing elbow movement are the biceps and the triceps (Fig. 1.11). The contraction of the triceps causes an extension, or opening, of the elbow, while contraction of the biceps closes the elbow. In our analysis of the elbow, we will consider the action of only these two muscles. This is a simplification, as many other muscles also play a role in elbow movement. Some of them stabilize the joints at the shoulder as the elbow moves, and others stabilize the elbow itself.

Figure 1.12a shows a weight $W$ held in the hand with the elbow bent at a $100°$ angle. A simplified diagram of this arm position is shown in Fig. 1.12b.

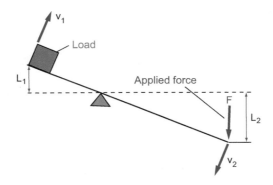

FIGURE 1.10  Motion of the lever arms in a Class 1 lever.

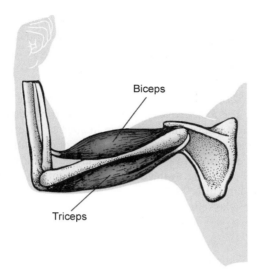

**FIGURE 1.11   The elbow.**

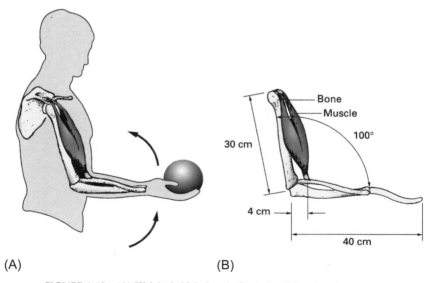

(A)                                    (B)

**FIGURE 1.12   (A) Weight held in hand. (B) A simplified drawing of (A).**

The dimensions shown in Fig. 1.12 are reasonable for a human arm, but they will, of course, vary from person to person. The weight pulls the arm downward. Therefore, the muscle force acting on the lower arm must be in the up direction. Accordingly, the prime active muscle is the biceps. The position of the upper arm is fixed at the shoulder by the action of the shoulder muscles. We will calculate, under the conditions of equilibrium, the pulling force $F_m$ exerted by the biceps muscle and the direction and magnitude of the reaction force $F_r$ at the fulcrum (the joint). The calculations will be performed by considering the arm position as a Class 3 lever, as shown in Fig. 1.13. The *x*- and *y*-axes are as shown in Fig. 1.13. The direction of the reaction force $F_r$ shown is a guess. The exact answer will be provided by the calculations.

In this problem we have three unknown quantities: The muscle force $F_m$, the reaction force at the fulcrum $F_r$, and the angle, or direction, of this force $\phi$. The angle $\theta$ of the muscle force can be calculated from trigonometric considerations, without recourse to the conditions of equilibrium. As is shown in Exercise 1.3, the angle $\theta$ is 72.6°.

For equilibrium, the sum of the x and y components of the forces must each be zero. From these conditions, we obtain the following:

$$x \text{ components of the forces}: F_m\cos\theta = F_r\cos\phi \qquad (1.10)$$

$$y \text{ components of the forces}: F_m\sin\theta = W + F_r\sin\phi \qquad (1.11)$$

These two equations alone are not sufficient to determine the three unknown quantities. The additional necessary equation is obtained from the torque conditions for equilibrium. In equilibrium, the torque about any point in Fig. 1.13 must be 0. For convenience, we will choose the fulcrum as the point for our torque balance.

The torque about the fulcrum must be 0. There are two torques about this point: A clockwise torque due to the weight and a counterclockwise torque due to the vertical y component of the muscle force. Since the reaction force $F_r$ acts at the fulcrum, it does not produce a torque about this point.

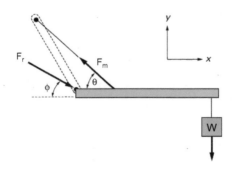

**FIGURE 1.13   Lever representation of Fig. 1.12.**

Using the dimensions shown in Fig. 1.12, we obtain

$$4 \text{ cm} \times F_m \sin\theta = 40 \text{ cm} \times W$$

$$F_m \sin\theta = 10W \tag{1.12}$$

Therefore, with $\theta = 72.6°$, the muscle force $F_m$ is

$$F_m = \frac{10W}{0.954} = 10.5W \tag{1.13}$$

With a 14-kg (31-lb) weight in hand, the force exerted by the muscle is

$$F_m = 10.5 \times 14 \times 9.8 = 1440 \text{ N} \text{ (325 lb)}$$

If we assume that the diameter of the biceps is 8 cm and that the muscle can produce a $7 \times 10^6$ dyn force for each square centimeter of area, the arm is capable of supporting a maximum of 334 N (75 lb) in the position shown in Fig. 1.13 (see Exercise 1.4).

The solutions of Eqs. 1.10 and 1.11 provide the magnitude and direction of the reaction force $F_r$. Assuming as before that the weight supported is 14 kg, these equations become

$$
\begin{aligned}
1440 \times \cos 72.6 &= F_r \cos\phi \\
1440 \times \sin 72.6 &= 14 \times 9.8 + F_r \sin\phi
\end{aligned}
\tag{1.14}
$$

or

$$
\begin{aligned}
F_r \cos\phi &= 430 \text{ N} \\
F_r \sin\phi &= 1240 \text{ N}
\end{aligned}
\tag{1.15}
$$

Squaring both equations, using $\cos^2\phi + \sin^2\phi = 1$ and adding them, we obtain

$$F_r^2 = 1.74 \times 10^6 \text{ N}^2$$

or

$$F_r = 1320 \text{ N} \text{ (298 lb)} \tag{1.16}$$

From Eqs. 1.14 and 1.15, the cotangent of the angle is

$$\cot\phi = \frac{430}{1240} = 0.347 \qquad (1.17)$$

and

$$\phi = 70.9°$$

Exercises 1.5, 1.6, and 1.7 present other similar aspects of biceps mechanics. In these calculations, we have omitted the weight of the arm itself, but this effect is considered in Exercise 1.8. The forces produced by the triceps muscle are examined in Exercise 1.9.

Our calculations show that the forces exerted on the joint and by the muscle are large. In fact, the force exerted by the muscle is much greater than the weight it holds up. This is the case with all the skeletal muscles in the body. They all apply forces by means of levers that have a mechanical advantage of less than 1. As mentioned earlier, this arrangement provides for greater speed of the limbs. A small change in the length of the muscle produces a relatively larger displacement of the limb extremities (see Exercise 1.10). It seems that nature prefers speed to strength. In fact, the speeds attainable at limb extremities are remarkable. A skilled pitcher can hurl a baseball at a speed in excess of 100 mph. Of course, this is also the speed of his hand at the point where he releases the ball.

## 1.7 The hip

Figure 1.14 shows the hip joint and its simplified lever representation, giving dimensions that are typical for a male body. The hip is stabilized in its socket by a group of muscles, which is represented in Fig. 1.14b as a single resultant force $F_m$. When a person stands erect, the angle of this force is about 71° with respect to the horizon. $W_L$ represents the combined weight of the leg, foot, and thigh. Typically, this weight is a fraction (0.185) of the total body weight W (i.e., $W_L = 0.185W$). The weight $W_L$ is assumed to act vertically downward at the midpoint of the limb.

We will now calculate the magnitude of the muscle force $F_m$ and the force $F_R$ at the hip joint when the person is standing erect on one foot as in a slow walk, as shown in Fig. 1.14. The force $W$ acting on the bottom of the lever is the reaction force of the ground on the foot of the person. This is the force that supports the weight of the body.

From equilibrium conditions, using the procedure outlined in Section 1.6, we obtain

$$F_m\cos71° - F_R\cos\theta = 0 \text{ (x components of the force} = 0) \qquad (1.18)$$

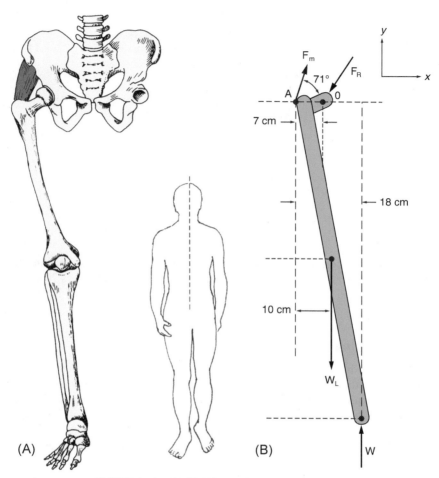

FIGURE 1.14   (A) The hip. (B) Lever representation.

$$F_m \sin 71° + W - W_L - F_R \sin\theta = 0 \text{ (y components of the force} = 0) \quad (1.19)$$

$$(F_R \sin\theta) \times 7 \text{ cm} + W_L \times 10 \text{ cm} - W \times 18 \text{ cm} = 0 \text{ (torque about point } A = 0)$$
$$(1.20)$$

Since $W_L = 0.185W$, from Eq. 1.20, we have

$$F_R \sin\theta = 2.31W$$

Using the result in Eq. 1.19, we obtain

$$F_m = \frac{1.50W}{\sin 71°} = 1.59W \tag{1.21}$$

From Eq. 1.18, we obtain

$$F_R \cos\theta = 1.59W \cos 71° = 0.52W$$

therefore,

$$\theta = \tan^{-1} 4.44 = 77.3°$$

and

$$F_R = 2.37W \tag{1.22}$$

This calculation shows that the force on the hip joint is nearly 2.5 times the weight of the person. Consider, for example, a person whose mass is 70 kg and weight is $9.8 \times 70 = 686$ N (154 lb). The force on the hip joint is 1625 N (366 lb).

### 1.7.1 Limping

Persons who have an injured hip limp by leaning toward the injured side as they step on that foot (Fig. 1.15). As a result, the center of gravity of the body shifts into a position more directly above the hip joint, decreasing the force on the injured area. Calculations for the case in Fig. 1.15 show that the muscle force $F_m = 0.47W$ and that the force on the hip joint is 1.28W (see Exercise 1.11). This is a significant reduction from the forces applied during a normal one-legged stance.

### 1.8 The back

When the trunk is bent forward, the spine pivots mainly on the fifth lumbar vertebra (Fig. 1.16a). We will analyze the forces involved when the trunk is bent at 60° from the vertical with the arms hanging freely. The lever model representing the situation is given in Fig. 1.16.

The pivot point $A$ is the fifth lumbar vertebra. The lever arm $AB$ represents the back. The weight of the trunk $W_1$ is uniformly distributed along the back; its effect can be represented by a weight suspended in the middle. The weight of the head and arms is represented by $W_2$ suspended at the end of the lever arm. The erector spinalis muscle, shown as the connection $D$-$C$ attached at a point two-thirds up the spine, maintains the position of the back. The angle

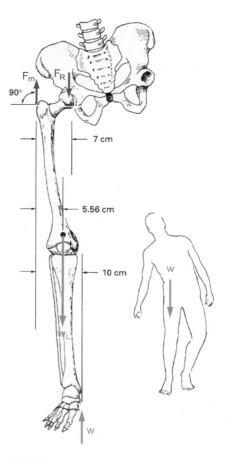

**FIGURE 1.15** **Walking on an injured hip.**

between the spine and this muscle is about 12°. For a 70-kg man, $W_1$ and $W_2$ are typically 320 N (72 lb) and 160 N (36 lb), respectively.

Solution to the problem is left as an exercise. It shows that just to hold up the body weight, the muscle must exert a force of 2000 N (450 lb) and the compressional force of the fifth lumbar vertebra is 2230 N (500 lb). If, in addition, the person holds a 20-kg weight in his hand, the force on the muscle is 3220 N (725 lb), and the compression of the vertebra is 3490 N (785 lb) (see Exercise 1.12).

This example indicates that large forces are exerted on the fifth lumbar vertebra. It is not surprising that backaches originate most frequently at this point. It is evident too that the position shown in the figure is not the recommended way of lifting a weight.

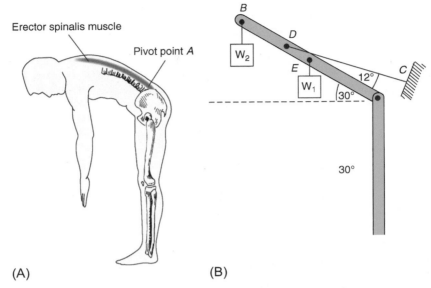

FIGURE 1.16 (A) The bent back. (B) Lever representation.

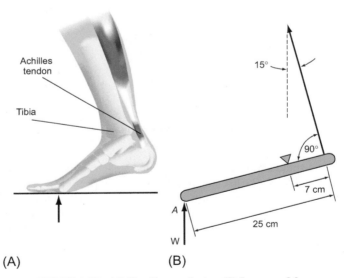

FIGURE 1.17 (A) Standing on tip-toe. (B) Lever model.

## 1.9 Standing tip-toe on one foot

The position of the foot when standing on tip-toe is shown in Fig. 1.17. The total weight of the person is supported by the reaction force at point *A*. This is a Class 1 lever with the fulcrum at the contact of the tibia. The balancing force is provided by the muscle connected to the heel by the Achilles tendon.

The dimensions and angles shown in Fig. 1.17b are reasonable values for this situation. Calculations show that while standing tip-toe on one foot, the compressional force on the tibia is 3.5W and the tension force on the Achilles tendon is 2.5 × W (see Exercise 1.13). Standing on tip-toe is a fairly strenuous position.

## 1.10 Dynamic aspects of posture

In our treatment of the human body, we have assumed that the forces exerted by the skeletal muscles are static. That is, they are constant in time. In fact, the human body (and bodies of all animals) is a dynamic system continually responding to stimuli generated internally and by the external environment. Because the center of gravity while standing erect is about half the height above the soles of the feet, even a slight displacement tends to topple the body. As has been demonstrated experimentally, the simple act of standing upright requires the body to be in a continual back-and-forth, left-right, swaying motion to maintain the center of gravity over the base of support. In a typical experiment designed to study this aspect of posture, the person is instructed to stand, feet together, as still as possible, on a platform that registers the forces applied by the soles of the feet (center of pressure). To compensate for the shifting center of gravity, this center of pressure is continually shifting by several centimeters over the area of the soles of the feet on a time scale of about half a second. Small back-and-forth perturbations of the center of mass (displacements less than about 1.5 cm) are compensated by ankle movements. Hip movements are required to compensate for larger displacements as well as for left-right perturbations.

The maintaining of balance in the process of walking requires a yet more complex series of compensating movements as the support for the center of gravity shifts from one foot to the other. Keeping the body upright is a highly complex task of the nervous system. The performance of this task is most remarkable when accidentally we slip and the center of gravity is momentarily displaced from the base of support. As is shown in Chapter 4, Exercise 4.9, without compensating movements, an erect human body that loses its balance will hit the floor in about 1 second. During this short time interval, the whole muscular system is called into action by the "righting reflex" to mobilize various parts of the body into motion so as to shift the center of mass back over the base of support. The body can perform amazing contortions in the process of restoring balance.

The nervous system obtains information required to maintain balance principally from three sources: Vision, the vestibular system situated in the inner ear that monitors movement and position of the head, and somatosensory system that monitors position and orientation of the various parts of the body.

With age, the efficiency of the functions required to keep a person upright decreases, resulting in an increasing number of injuries due to falls. In the United States, the number of accidental deaths per capita due to falls for persons above the age of 80 is about 60 times higher than for people below the age of 70.

Another aspect of the body dynamics is the interconnectedness of the musculoskeletal system. Through one path or another, all muscles and bones are connected to one another, and a change in muscle tension or limb position in one part of the body must be accompanied by a compensating change elsewhere. The system can be visualized as a complex tentlike structure. The bones act as the tent poles and the muscles as the ropes bringing into and balancing the body in the desired posture. The proper functioning of this type of a structure requires that the forces be appropriately distributed over all the bones and muscles. In a tent, when the forward-pulling ropes are tightened, the tension in the back ropes must be correspondingly increased; otherwise, the tent collapses in the forward direction. The musculoskeletal system operates in an analogous way. For example, excessive tightness, perhaps through overexertion, of the large muscles at the front of our legs will tend to pull the torso forward. To compensate for this forward pull, the muscles in the back must also tighten, often exerting excess force on the more delicate structures of the lower back. In this way, excess tension in one set of muscles may be reflected as pain in an entirely different part of the body.

## Exercises

**1.1.** (a) Explain why the stability of a person against a toppling force is increased by spreading the legs as shown in Fig. 1.7. (b) Calculate the force required to topple a person of mass = 70 kg, standing with his feet spread 0.9 m apart as shown in Fig. 1.7. Assume the person does not slide and the weight of the person is equally distributed on both feet.

**1.2.** Derive the relationships stated in Eqs. 1.6, 1.7, and 1.8.

**1.3.** Using trigonometry, calculate the angle $\theta$ in Fig. 1.13. The dimensions are specified in Fig. 1.12b.

**1.4.** Using the data provided in the text, calculate the maximum weight that the arm can support in the position shown in Fig. 1.12.

**1.5.** Calculate the force applied by the biceps and the reaction force ($F_r$) at the joint as a result of a 14-kg weight held in hand when the elbow is at (a) 160° and (b) 60°. Dimensions are as in Fig. 1.12.
Assume that the upper part of the arm remains fixed as in Fig. 1.12 and use calculations from Exercise 1.3. Note that under these conditions, the lower part of the arm is no longer horizontal.

**1.6.** Consider again Fig. 1.12. Now let the 14-kg weight hang from the middle of the lower arm (20 cm from the fulcrum). Calculate the biceps force and the reaction force at the joint.

**1.7.** Consider the situation when the arm in Fig. 1.13 supports two 14-kg weights, one held by the hand as in Fig. 1.13 and the other supported in the middle of the arm as in Exercise 1.6. (a) Calculate the force of the biceps muscle and the reaction force. (b) Are the forces calculated in part (a) the same as the sum of the forces produced when the weights are suspended individually?

**1.8.** Calculate the additional forces due to the weight of the arm itself in Fig. 1.13. Assume that the lower part of the arm has a mass of 2 kg and that its total weight can be considered to act at the middle of the lower arm, as in Exercise 1.6.

**1.9.** Estimate the dimensions of your own arm, and draw a lever model for the extension of the elbow by the triceps. Calculate the force of the triceps in a one-arm push-up in a hold position at an elbow angle of 100°.

**1.10.** Suppose that the biceps in Fig. 1.13 contracts 2 cm. What is the upward displacement of the weight? Suppose that the muscle contraction is uniform in time and occurs in an interval of 0.5 seconds. Compute the velocity of the point of attachment of the tendon to the bone and the velocity of the weight. Compare the ratio of the velocities to the mechanical advantage.

**1.11.** Calculate the forces in the limping situation shown in Fig. 1.15. At what angle does the force $F_R$ act?

**1.12.** (a) Calculate the force exerted by the muscle and the compression force on the fifth lumbar vertebra in Fig. 1.16. Use information provided in the text. (b) Repeat the calculations in part (a) for the case when the person shown in Fig. 1.16 holds a 20-kg weight in his hand.

**1.13.** Calculate the force on the tibia and on the Achilles tendon in Fig. 1.17.

Chapter 2

# Friction

If we examine the surface of any object, we observe that it is irregular. It has protrusions and valleys. Even surfaces that appear smooth to the eye show such irregularities under microscopic examination. When two surfaces are in contact, their irregularities intermesh, and as a result, there is a resistance to the sliding or moving of one surface on the other. This resistance is called *friction*. If one surface is to be moved with respect to another, a force has to be applied to overcome friction.

Consider a block resting on a surface as shown in Fig. 2.1. If we apply a force F to the block, it will tend to move. But the intermeshing of surfaces produces a frictional reaction force $F_f$ that opposes motion. In order to move the object along the surface, the applied force must overcome the frictional force. The magnitude of the frictional force depends on the nature of the surfaces; clearly, the rougher the surfaces, the greater the frictional force. The frictional property of surfaces is represented by the coefficient of friction $\mu$. The magnitude of the frictional force depends also on the force $F_n$ perpendicular to the surfaces that presses the surfaces together. The force $F_n$ includes the weight W of the block (W = mg; see Appendix A.7) plus any other force perpendicular to the surface. The magnitude of the force that presses the surfaces together determines to what extent the irregularities are intermeshed.

The frictional force $F_f$ is given by

$$F_f = \mu F_n \tag{2.1}$$

Distinction has to be made between the frictional force that acts on a moving object (called the kinetic frictional force) and the frictional force that acts on the object when it is stationary. The kinetic frictional force opposing motion of the object is obtained from Eq. 2.1 using the kinetic coefficient of friction $\mu_k$. In general, it takes a larger force to get the object moving against a frictional force than to keep it in motion. This is not surprising because in the stationary case, the irregularities of the two surfaces can settle more deeply into each other. The force that must be applied to an object to get it moving is again obtained from Eq. 2.1 but this time using the static coefficient of friction $\mu_s$. This is the magnitude of the maximum static frictional force.

The magnitude of the frictional force does not depend on the size of the contact area. If the surface contact area is increased, the force per unit area

Physics in Biology and Medicine. https://doi.org/10.1016/B978-0-443-21558-2.00002-X

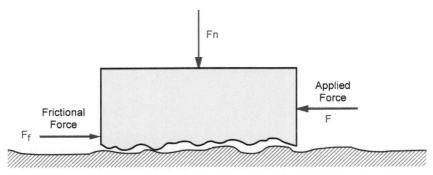

**FIGURE 2.1   Friction.** For simplicity, the reaction force that is equal and opposite to $F_n$ is not shown.

(pressure) is decreased, and this reduces the interpenetration of the irregularities. However, at the same time, the number of irregularities is proportionately increased. As a result, the total frictional force is unchanged. Coefficients of static and kinetic friction between some surfaces are shown in Table 2.1. As is evident, the coefficient of static friction for two given surfaces is somewhat larger than the coefficient of kinetic friction.

We have illustrated the concept of friction with surfaces sliding along each other, but frictional forces are encountered also in rolling (rolling friction) and in fluid flows (viscous friction). Rolling motion is not encountered in living systems, but viscous friction plays an important role in the flow of blood and other biological fluids.

Whereas sliding friction is independent of velocity, fluid friction has a strong velocity dependence. We will discuss this in Chapter 3.

Friction is everywhere around us. It is both a nuisance and an indispensable factor in the ability of animals to move. Without friction, an object that is pushed into motion would continue to move forever (Newton's first law,

**TABLE 2.1 Coefficients of friction, static ($\mu_s$), and kinetic ($\mu_k$).**

| Surfaces | $\mu_s$ | $\mu_k$ |
| --- | --- | --- |
| Leather on oak | 0.6 | 0.5 |
| Rubber on dry concrete | 0.9 | 0.7 |
| Steel on ice | 0.02 | 0.01 |
| Dry bone on bone | | 0.3 |
| Bone on joint, lubricated | 0.01 | 0.003 |

Appendix A). The slightest force would send us into eternal motion. It is the frictional force that dissipates kinetic energy into heat and eventually stops the object (see Exercise 2.1). Without friction, we could not walk, nor could we balance on an inclined plane (see Exercise 2.2). In both cases, friction provides the necessary reaction force. Friction also produces undesirable wear and tear and destructive heating of contact surfaces. Both nature and engineers attempt to maximize friction where it is necessary and minimize it where it is destructive. Friction is greatly reduced by introducing a fluid such as oil at the interface of two surfaces. The fluid fills the irregularities and therefore smooths out the surfaces. A natural example of such lubrication occurs in the joints of animals, which are lubricated by a fluid called the synovial fluid. This lubricant reduces the coefficient of friction by about a factor of 100. As is evident from Table 2.1, nature provides very efficient joint lubrication. The coefficient of friction here is significantly lower than for steel on ice.

We will illustrate the effects of friction with a few examples.

## 2.1 Standing at an incline

Referring to Fig. 2.2, let us calculate the angle of incline $\theta$ of an oak board on which a person of weight $W$ can stand without sliding down. Assume that she is wearing leather-soled shoes and that she is standing in a vertical position, as shown in the figure.

The force $F_n$ normal to the inclined surface is

$$F_n = W\cos\theta \tag{2.2}$$

The static frictional force $F_f$ is

$$F_f = \mu F_n = \mu_s W\cos\theta = 0.6W\cos\theta \tag{2.3}$$

The force parallel to the surface $F_p$, which tends to cause the sliding, is

$$F_p = W\sin\theta \tag{2.4}$$

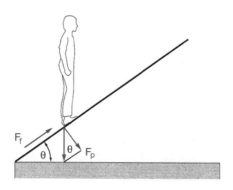

FIGURE 2.2  **Standing on an incline.**

The person will slide when the force $F_p$ is greater than the frictional force $F_f$; that is,

$$F_p > F_f \tag{2.5}$$

At the onset of sliding, these two forces are just equal; therefore,

$$F_f = F_p$$
$$0.6W\cos\theta = W\sin\theta \tag{2.6}$$

or

$$\frac{\sin\theta}{\cos\theta} = \tan\theta = 0.6$$

Therefore $\theta = 31°$.

## 2.2 Friction at the hip joint

We have shown in Chapter 1 that the forces acting on the joints are very large. When the joints are in motion, these large forces produce frictional wear, which could be damaging unless the joints are well lubricated. Frictional wear at the joints is greatly reduced by a smooth cartilage coating at the contact ends of the bone and by synovial fluid, which lubricates the contact areas. We will now examine the effect of lubrication on the hip joint in a person. When a person walks, the full weight of the body rests on one leg through most of each step. Because the center of gravity is not directly above the joint, the force on the joint is greater than the weight. Depending on the speed of walking, this force is about 2.4 times the weight (see Chapter 1). In each step, the joint rotates through about 60°. Since the radius of the joint is about 3 cm, the joint slides about 3 cm inside the socket during each step. The frictional force on the joint is

$$F_f = 2.4W\mu \tag{2.7}$$

The work expended in sliding the joint against this friction is the product of the frictional force and the distance over which the force acts (see Appendix A). Thus, the work expended during each step is

$$\text{Work} = F_f \times \text{distance} = 2.4\,W\mu(3\text{ cm}) = 7.2\,\mu W \text{ erg} \tag{2.8}$$

If the joint were not lubricated, the coefficient of friction ($\mu$) would be about 0.3. Under these conditions, the work expended would be

$$\text{Work} = 2.16 \times W \text{ erg} \tag{2.9}$$

This is a large amount of work to expend on each step. It is equivalent to lifting the full weight of the person by 2.16 cm. Furthermore, this work would be dissipated into heat energy, which would destroy the joint.

As it is, the joint is well lubricated, and the coefficient of friction is only 0.003. Therefore, the work expended in counteracting friction and the resultant heating of the joint are negligible. However, as we age, the joint cartilage begins to wear, efficiency of lubrication decreases, and the joints may become seriously damaged. Studies indicate that by the age of 70, about two-thirds of people have knee joint problems and about one-third have hip problems.

## 2.3 Spine fin of a catfish

Although in most cases good lubrication of bone-contact surfaces is essential, there are a few cases in nature where bone contacts are purposely unlubricated to increase friction. The catfish has such a joint connecting its dorsal spine fin to the rest of its skeleton (Fig. 2.3). Normally, the fin is folded flat against the body, but when the fish is attacked, the appropriate muscles pull the bone of the fin into a space provided in the underlying skeleton. Since the coefficient of friction between the fin bone and the skeleton is high, the frictional force tends to lock the fin in the up position. In order to remove the fin, a force must be applied in a predominantly vertical direction with respect to the underlying skeleton. The erect sharp fin discourages predators from eating the catfish.

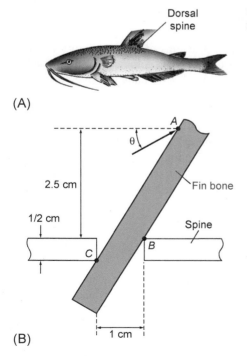

FIGURE 2.3    (A) Catfish. (B) Simplified representation of the spine in the catfish.

Figure 2.3b is a simplified representation of the spine and the protruding fin. The shaded block represents the movable fin bone, and the horizontal block is the spine holding the fin. Assume that a force F at an angle θ is applied at point *A* to dislodge the bone. The force is shown to act at point *A*, 2.5 cm above point *B*. The dimensions shown in the figure are to be used in the calculations required for Exercise 2.3. The applied force tips the bone, and as a result, reaction forces are set up at points *B* and *C*. The components of these forces normal to the fin-bone surface produce frictional forces that resist the removal of the bone. Calculation of some of the properties of the locking mechanism is left as an exercise.

## 2.4 Biotribology

Tribology is a field of engineering and physics that studies friction. Traditionally, most of the studies in this field focused on reducing friction in industrial equipment such as electric motors, generators, gears, and internal combustion engines. More recently, over the past 20 years or so, the field has expanded to study friction as it pertains to biology and medicine. This new subarea of tribology has been named "*biotribology*." We have already touched on this subject in the preceding two sections, where we described the effect of friction at the hip joint and the role of friction in the defensive positioning of the spine in the catfish. Here we examine briefly some of the other areas of this field.

Experiments in biotribology present problems of reproducibility and control principally, because the properties of a given biological material, such as skin or mucus, for example, change as a function of many variables such as the prevailing temperature, humidity, and time of day. The relevant materials also show wide variability from person to person. The properties of human skin that covers 1.5 to 2 m$^2$ of the body are significantly different at the various regions of the body and depend significantly on the age of the individual. For example, depending on the location, the roughness of human skin (the distance between the lowest and highest point at a given location) varies from 10 to 200 μm. In spite of these challenges, relevant issues in biotribology have been identified and progress has been made in this important field. We will briefly discuss three active areas of biotribology.

### 2.4.1 Oral biotribology

The principal components of the oral cavity (mouth) are teeth, tongue, and saliva. The composition of *saliva* and its role in the chemistry and tribology of the mouth are now well understood. Most of the saliva is produced by three pairs of salivary glands. Two of the glands are located on either side of the mouth below the jaw bone. The third pair of glands are located between the ear and the jaw. In addition, several smaller glands, distributed throughout the mouth, contribute to the flow of saliva. Human saliva consists of about 99% water and a highly complex mixture of proteins and ions that give the saliva its remarkable properties necessary for the proper functioning of the oral system.

Some of the proteins in saliva adhere strongly both to teeth and to soft tissue in the oral cavity. These adhered molecules form a substrate for the formation of a coating of saliva on oral surfaces that provides an excellent lubricant for these surfaces. The coefficient of friction of such a coated surface is about 0.02, similar to that of synovial fluid. The coefficient of friction of the surface-coated saliva is about two orders of magnitude lower than the uncoated surface or the surface coated with clear water. The low friction coating protects the teeth from excessive wear and also protects soft tissue from being bruised and damaged by hard foods. Another role of saliva is in the processing of food. In the process of chewing, the saliva is mixed with the food, making the mixture more slippery and promoting easy sliding down the esophagus into the stomach. A simple calculation shows that the thickness of the saliva layer on the oral surfaces is approximately 0.1 mm. (See Exercise 2.4.)

In addition to its tribological properties, saliva is also a mild antiseptic containing a range of compounds that have antibacterial and antiviral properties. Some fraction of these molecules also promotes blood clotting. This protective and healing nature of saliva is probably why wounded animals, including humans, respond instinctively to an injury by licking the wound. These attributes of saliva are likely the reason for the observation that in most cases wounds in the mouth heal faster than wounds on other parts of the body. The process of licking, of course, also clears the wound of debris that is likely to retard healing. As is pointed out in the literature, licking of wounds is not universally recommended. Licking may introduce into the wound bacteria and viruses that are immune to the disinfectant properties of the saliva. Such complications, however, seem to be rare.

Further, saliva is somewhat basic and functions as a buffer that keeps the acidity of the mouth near the neutral point ($pH \sim 7$). This is important because acid tends to leach out calcium and other minerals out of teeth, making them susceptible to cavities and other structural damage.

The normal wear of teeth is due to mastication (chewing). During normal wear, only about 40 $\mu$m of the tooth is worn off per year. A greater destructive effect on teeth is the increase in the consumption of soft drinks and sugar, which has contributed to an increase in cavities and tooth decay worldwide. Soft drinks are mostly acidic and therefore have a direct deleterious effect on teeth. The effect of sugar is indirect. The oral cavity contains a variety of bacteria that feed on sugar. Some of the bacteria are helpful in maintaining oral health. They control the population of other bacteria that cause oral diseases such as oral candida, for example. However, some bacteria feed on sugar and produce acids as byproducts that, as mentioned earlier, leach out the minerals that compose teeth. Here again, saliva has a significant role in repairing the damage. The inorganic ions in the saliva can remineralize, at least partially, the eroded tooth material. However, saliva cannot fully compensate for the dele-terious effects of a damaging diet.

### 2.4.2 Hair biotribology

Frictional forces act between *hair* fibers and between hair and objects in contact with hair such as a comb. In large measure, hair friction determines the esthetic properties of hair. Too much friction and the hair fibers stick together, giving the hair a stringy appearance. Not enough friction makes styling difficult, giving hair an unruly appearance. Millions of dollars annually are spent by the hair-products industry to study frictional properties of hair, and how it is affected by products such as shampoo, hair conditioner, and hair spray.

Measurement of hair friction coefficient is difficult, but techniques have been developed to perform such measurements. The effect on hair friction of environmental conditions such as natural hair oil, sweat, dirt, and hair dyes has been measured. These factors all tend to increase friction. Shampoos and hair conditioners are formulated to yield optimum hair properties as envisioned by the manufacturer.

Shampoo is designed to clean the hair and is essentially a soap. (See also Chapter 7.) The main cleaning agents in shampoos are surfactants. These are molecules that have hydrophilic (water-attracting) and a hydrophobic (water-repelling) components. The hydrophobic component attaches itself to hair oil, sweat, and other foreign components. The hydrophilic component at-taches to the rinsing water and washes away the dirt. This is essentially the way all soaps work. Hair conditioners contain positive ion surfactants and large fatty alcohol molecules that replace the natural washed-away oils that make the hair soft and adequately lubricated. Both shampoos and condi-tioners contain a variety of inactive ingredients, among them perfumes, foaming agents, preservatives, and coloring, designed to enhance the appeal of the products.

### 2.4.3 Biotribology of textiles

Friction is obviously present in areas of contact between skin and textiles. Under most conditions, such frictional contact does not cause problems. However, in some cases, repeated relative motion subject to friction can damage the skin. Most people have experienced an occasional blister, often on the heel or sole of the foot, caused by motion between the skin and an ill-fitting shoe—sock combination. The coefficient of friction increases when the skin and textile are wet, making skin damage more likely.

A simple blister is relatively easy to treat by improving the shoe—sock fit and cushioning the blister with padding. *Bedsores* (also called decubitus) are a much more serious condition found mostly in patients who require long periods of bed rest. They are caused by the pressure and friction at the areas of contact between the skin and the resting surface. Even though the patient appears to be at rest, ever-present tiny movements tend to frictionally abrade the skin. Pressure due to body weight reduces blood flow to the contact area, increasing the risk of serious ulceration and infection. Decubitus may also be a problem for people confined to wheelchairs, who likewise remain in a fixed pressure-bearing position for prolonged periods of time.

Several research groups are working on the development of new textiles for bedsheets and other applications involving skin contact that would reduce the friction on the skin of patients at risk of decubitus. Initially, five qualities are sought. (1) The material should readily wick away moisture so that the area of contact is kept as dry as possible. (2) While covering the same overall area, the material should present a smaller microscopic contact area. This can be achieved by increasing the free interfacial volume. (3) The penetration of the textile surface into the skin should be reduced. (4) The fibers of the material should be made more elastic, resulting in reduced abrasiveness. (5) Pressure on specific areas of the body should be reduced. In trial tests, several new materials have yielded promising improvements.

## Exercises

**2.1.** (a) Assume that a 50-kg skater, on level ice, has built up her speed to 30 km/h. How far will she coast before the sliding friction dissipates her energy? (Kinetic energy $= \frac{1}{2} mv^2$; see Appendix A.) (b) How does the distance of coasting depend on the mass of the skater?

**2.2.** Referring to Fig. 1.5, compute the coefficient of friction at which the tendency of the body to slide and the tendency to topple due to the applied force are equal.

**2.3.** (a) Referring to Fig. 2.3, assume that a dislodging force of 0.1 N is applied at $\theta = 20°$ and the angle between the fin bone and the spine is $45°$. Calculate the minimum value for the coefficient of friction between the bones to prevent dislodging of the bone. (b) Assuming that the coefficient of friction is 1.0, what is the value of the angle $\theta$ at which a force of 0.2 N will just dislodge the bone? What would this angle be if the bones were lubricated ($\mu = 0.01$)?

**2.4.** Calculate the thickness of the saliva layer on the inside of the oral cavity and the tongue using the following measured values (these are approximate values obtained by measuring the parameters for several people): Volume of oral cavity $\sim 40$ cm$^3$, and volume of tongue $\sim 30$ cm$^3$. In your calculation, assume the shape of both the oral cavity and the tongue is spherical.
Volume of residual saliva after swallowing all excess saliva $\sim 0.8$ mL $= 0.8$ cm$^3$. Assume that this amount of saliva coats evenly the tongue and the inside of the oral cavity. Neglect the saliva coating the teeth.

# Chapter 3

# Translational motion

In general, the motion of a body can be described in terms of *translational* and *rotational* motion. In pure translational motion, all parts of the body have the same velocity and acceleration (Fig. 3.1). In pure rotational motion, such as the rotation of a bar around a pivot, the rate of change in the angle θ is the same for all parts of the body (Fig. 3.2), but the velocity and acceleration along the body depend on the distance from the center of rotation. Many motions and movements encountered in nature are combinations of rotation and translation, as in the case of a body that rotates while falling. It is convenient, however, to discuss these motions separately. In this chapter, we discuss translation. Rotation is discussed in the following chapter.

The equations of translational motion for constant acceleration are presented In Appendix A and may be summarized as follows: In uniform acceleration, the final velocity (v) of an object that has been accelerated for a time $t$ is

$$v = v_0 + at \tag{3.1}$$

Here $v_0$ is the initial velocity of the object and a is the acceleration. Acceleration can therefore be expressed as

$$a = \frac{v - v_0}{t} \tag{3.2}$$

The average velocity during the time interval t is

$$v_{av} = \frac{v + v_0}{2} \tag{3.3}$$

The distance s traversed during this time is

$$s = v_{av}t \tag{3.4}$$

Using Eqs. 3.1 and 3.2, we obtain

$$s = v_0t = \frac{at^2}{2} \tag{3.5}$$

By substituting $t = (v - v_0)/a$ from Eq. 3.1 into Eq. 3.5, we obtain

Physics in Biology and Medicine. https://doi.org/10.1016/B978-0-443-21558-2.00003-1

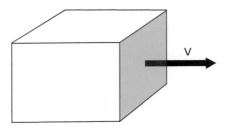

FIGURE 3.1  **Translational motion.**

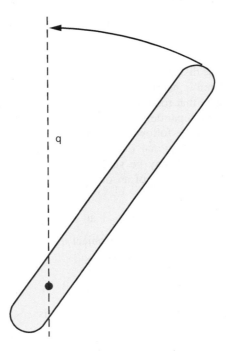

FIGURE 3.2  **Rotational motion.**

$$v^2 = v_0^2 + 2as \qquad (3.6)$$

Let us now apply these equations to some problems in the life sciences. Most of our calculations will relate to various aspects of jumping. Although in the process of jumping, the acceleration of the body is usually not constant, the assumption of constant acceleration is necessary to solve the problems without undue difficulties.

## 3.1 Vertical jump

Consider a simple vertical jump in which the jumper starts in a crouched position and then pushes off with her feet (Fig. 3.3).

We will calculate here the height H attained by the jumper. In the crouched position, at the start of the jump, the center of gravity is lowered by a distance c. During the act of jumping, the legs generate a force by pressing down on the surface. Although this force varies through the jump, we will assume that it has a constant average value F.

Because the feet of the jumper exert a force on the surface, an equal upward-directed force is exerted by the surface on the jumper (Newton's third law). Thus, there are two forces acting on the jumper: her weight (W), which is in the downward direction, and the reaction force (F), which is in the upward direction. The net upward force on the jumper is $F - W$ (see Fig. 3.4). This force acts on the jumper until her body is erect and her feet leave the ground. The upward force, therefore, acts on the jumper through a distance $c$ (see Fig. 3.3). The acceleration of the jumper in this stage of the jump (see Appendix A) is

$$a = \frac{F - W}{m} = \frac{F - W}{W/g} \qquad (3.7)$$

where W is the weight of the jumper and $g$ is the gravitational acceleration. A consideration of the forces acting on the Earth (Fig. 3.5) shows that an equal force accelerates the Earth in the opposite direction. However, the mass of the Earth is so large that its acceleration due to the jump is negligible.

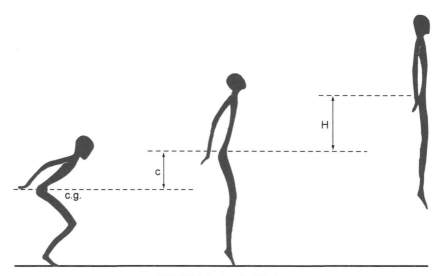

**FIGURE 3.3  Vertical jump.**

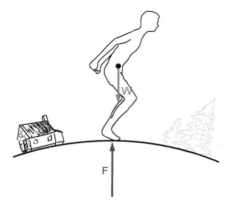

**FIGURE 3.4   Forces on the jumper.**

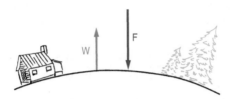

**FIGURE 3.5   Forces on the Earth.**

The acceleration shown in Eq. 3.7 takes place over a distance $c$. Therefore, the velocity v of the jumper at take-off, as given by Eq. 3.6 is

$$v^2 = v_0^2 + 2ac \tag{3.8}$$

Since the initial velocity at the start of the jump is zero (i.e., $v_0 = 0$), the take-off velocity is given by

$$v^2 = \frac{2\,(F - W)\,c}{W/g} \tag{3.9}$$

(Here we have substituted

$$a = \frac{F - W}{W/g}$$

into Eq. 3.8.)

After the body leaves the ground, the only force acting on it is the force of gravity W, which produces a downward acceleration $-g$ on the body. At the maximum height H, just before the body starts falling back to the ground, the velocity is zero. The initial velocity for this part of the jump is the take-off velocity v given by Eq. 3.9. Therefore, from Eq. 3.6, we obtain

$$0 = \frac{2\,(F - W)\,c}{W/g} - 2gH \tag{3.10}$$

From this, the height of the jump is

$$H = \frac{(F - W)\,c}{W} \tag{3.11}$$

The reaction force F is the sum of the weight W and the force $F_m$ generated by the leg muscles. That is $F = W + F_m$. Now let us estimate the numerical value for the height of the jump. Experiments have shown that in a good jump, a well-built person generates an average muscle force equal to the weight of the person (i.e., $F_m = W$). In that case, the height of jump is $H = c$. The distance c, which is the lowering of the center of gravity in the crouch, is proportional to the length of the legs. For an average person, this distance is about 60 cm, which is our estimate for the height of a vertical jump.

The height of a vertical jump can also be computed very simply from energy considerations. The work done on the body of the jumper by the force F during the jump is the product of the force F and the distance $c$ over which this force acts (see Appendix A). This work is converted to kinetic energy as the jumper is accelerated upward. At the full height of the jump H (before the jumper starts falling back to ground), the velocity of the jumper is zero. At this point, the kinetic energy is fully converted to potential energy as the center of mass of the jumper is raised to a height $(c + H)$. Therefore, from conservation of energy,

Work done on the body = Potential energy at maximum height

or

$$Fc = W(c + H) \tag{3.12}$$

From this equation, the height of the jump is, as before,

$$H = \frac{(F - W)c}{W} = \frac{(F_m + W - W)c}{W} = \frac{F_m c}{W}$$

Another aspect of the vertical jump is examined in Exercise 3.1.

## 3.2 Effect of gravity on the vertical jump

The weight of an object depends on the mass and size of the planet on which it is located. The gravitational constant of the moon, for example, is one-sixth that of the Earth; therefore, the weight of a given object on the moon is one-sixth its weight on the Earth.

From Eq. 3.11, the height of the jump on the Earth is

$$H = \frac{(F - W)c}{W} = \frac{F_m c}{W}$$

The force $F_m$ that accelerates the body upward depends on the strength of the leg muscles, and for a given person, this force is the same on the moon as on the Earth. Similarly, the lowering of the center of gravity c is unchanged with location. With $F'$ the reaction force on the moon $= F_m + W'$, the height of the jump on the moon $(H')$ is

$$H' = \frac{F_m c}{W'} \tag{3.13}$$

Here $W'$ is the weight of the person on the moon (i.e., $W' = W/6$). The ratio of the jumping heights at the two locations is

$$\frac{H'}{H} = \frac{W}{W'} \tag{3.14}$$

That is, if a person can jump to a height of 60 cm on Earth, that same person can jump up 3.6 m on the moon.

## 3.3 Running high jump

In the preceding sections, we calculated the height of a jump from a standing position and showed that the center of gravity could be raised about 60 cm. A considerably greater height can be attained by jumping from a running start. The current high jump record is about 2.5 m. The additional height is attained by using part of the kinetic energy of the run to raise the center of gravity off the ground. Let us calculate the height attainable in a running jump if the jumper could use all his/her initial kinetic energy $\left(\frac{1}{2}mv^2\right)$ to raise his/her body off the ground. If this energy were completely converted to potential energy by raising the center of gravity to a height H, then

$$MgH = \frac{1}{2}mv^2 \tag{3.15}$$

or

$$H = \frac{v^2}{2g}$$

To complete our estimate, we must consider two additional factors that increase the height of the jump. First, we should add the 0.6 m, which can be produced by the legs in the final push-off. Then we must remember that the center of gravity of a person is already about 1 m above the ground. With little extra effort, the jumper can alter the position of his body so that it is horizontal at its maximum height. This maneuver adds 1 m to the height of the bar he can clear. Thus, our final estimate for the maximum height of the running high jump is

$$H = \frac{v^2}{2g} + 1.6 \text{ m} \tag{3.16}$$

The maximum short distance speed of a good runner is about 10 m/sec. At this speed, our estimate for the maximum height of the jump from Eq. 3.16 is 6.7 m. This estimate is nearly three times the high jump record. Obviously, it is not possible for a jumper to convert all the kinetic energy of a full-speed run into potential energy.

In the unaided running high jump, only the force exerted by the feet is available to alter the direction of the running start. This limits the amount of kinetic energy that can be utilized to aid the jump. The situation is quite different in pole vaulting, where, with the aid of the pole, the jumper can in fact use most of the kinetic energy to raise his/her center of gravity. The 2023 men's pole-vaulting record is 6.22 m (20.41 ft), which is remarkably close to our estimate of 6.7 m. These figures would agree even more closely had we included in our estimate the fact that the jumper must retain some forward velocity to carry him/her over the bar.

## 3.4 Range of a projectile

A problem that is solved in most basic physics texts concerns a projectile launched at an angle $\theta$ and with initial velocity $v_0$. A solution is required for the range R, the distance at which the projectile hits the Earth (see Fig. 3.6). It is shown that the range is

$$R = \frac{v_0^2 \sin 2\theta}{g} \tag{3.17}$$

For a given initial velocity, the range is maximum when $\sin 2\theta = 1$ or $\theta = 45°$. In other words, a maximum range is obtained when the projectile is launched at a 45° angle. In that case, the range is

$$R_{\max} = \frac{v_0^2}{g} \tag{3.18}$$

Using this result, we will estimate the distance attainable in broad jumping.

## 3.5 Standing broad jump

When the jumper projects himself into the broad jump from a stationary crouching position (Fig. 3.7), his acceleration is determined by the resultant of

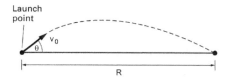

FIGURE 3.6  **Projectile.**

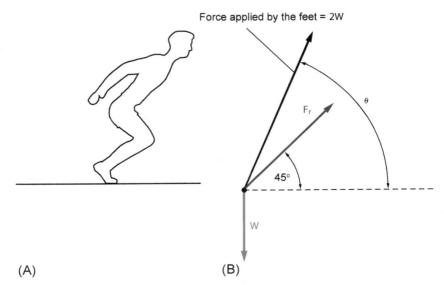

Force applied by the feet = 2W

(A)

(B)

**FIGURE 3.7** (A) The standing broad jump. (B) The associated forces.

two forces: the downward force of gravity, which is simply equal to his weight W, and the force generated by the feet, which he can apply in any direction. In order to maximize the distance of the jump, the launching velocity and therefore also the resultant force should be directed at a 45° angle.

We will assume as before that a jumper can generate with his feet a force equal to twice the body weight. The magnitude of the resultant force ($F_r$) and the angle $\theta$ at which the legs must apply the force to the body are obtained from the following considerations.

The horizontal and vertical components of the resultant force (see Fig. 3.7) are, respectively,

$$\text{Horizontal component of } F_r: F_r\cos45° = 2W\cos\theta \qquad (3.19)$$

and

$$\text{Vertical component of } F_r: F_r\sin45° = 2W\sin\theta - W \qquad (3.20)$$

Here we have two equations that can be solved to yield the two unknown quantities $F_r$ and $\theta$ (see Exercise 3.3). The magnitude of the force $F_r$ is

$$F_r = 1.16W$$

The optimum angle $\theta$ at which the legs apply the force 2W is $\theta = 65.8°$. We will again assume that the force that launches the jumper is applied over a distance of 60 cm, which is the extent of the crouching position. The acceleration produced by the resultant force is

$$a = \frac{F_r}{m} = \frac{1.16W}{W/g} = 1.16g$$

The launching velocity v of the jumper is therefore $v = \sqrt{2as}$. With s = 60 cm, the velocity is 3.70 m/sec. The distance (R) of the standing broad jump is, from Eq. 3.18,

$$R = \frac{v^2}{g} = \frac{13.7}{9.8} = 1.4 \text{ m}$$

The range of the jump can be significantly increased by swinging both the legs and the arms in the direction of the jump, which results in an increase in the forward momentum of the body. Other aspects of the standing broad jump are presented in Exercises 3.4 and 3.5.

## 3.6 Running broad jump (long jump)

Let us assume that a jumper launches the jump at full speed of 10 m/sec. The push-off force (2W) generated by the legs provides the vertical component of the launching velocity. From this force, we have to subtract the weight (W) of the jumper. The acceleration produced by the net force is

$$a = \frac{2W - W}{m} = \frac{W}{W/g} = g$$

If the push-off force acts on the jumper over a distance of 60 cm (the extent of the crouch) and if it is directed entirely in the vertical y direction, the vertical component of the velocity $v_y$ during the jump is given by

$$v_y^2 = 2as = 2 \times g \times 0.6 = 11.8 \text{ m}^2/\text{sec}^2$$

$$v_y = 3.44 \text{ m/sec}$$

Since the horizontal component of the launching velocity $v_x$ is the running velocity, the magnitude of the launching velocity is

$$v = \sqrt{v_x^2 + v_y^2} = 10.6 \text{ m/sec}$$

The launch angle θ is

$$\theta = \tan^{-1}\frac{v_y}{v_x} = \tan^{-1}\frac{3.44}{10} = 19°$$

From Eq. 3.17, the range R of the jump is

$$R = \frac{v^2 \sin2\theta}{g} = \frac{112.4 \sin38°}{g} = 7.06 \text{ m}$$

Considering the approximations used in these calculations, this estimate is in reasonable agreement with the current (2023) world record set in the 1990's, of about 9 m for men and 7.5 m for women.

## 3.7 Motion through air

We have so far neglected the effect of air resistance on the motion of objects, but we know from experience that this is not a negligible effect. When an object moves through air, the air molecules have to be pushed out of its way. The resulting reaction force pushes back on the body and retards its motion—this is the source of fluid friction in air. We can deduce some of the properties of air friction by sticking our hand outside a moving car. Clearly, the greater the velocity with respect to the air, the larger the resistive force. By rotating our hand, we observe that the force is greater when the palms face the direction of motion. We therefore conclude that the resistive force increases with the velocity and the surface area in the direction of motion. It has been found that the force due to air resistance $F_a$ can be expressed approximately as

$$F_a = CAv^2 \tag{3.21}$$

where v is the velocity of the object with respect to the air, A is the area facing the direction of motion, and C is the coefficient of air friction. The coefficient C depends somewhat on the shape of the object. In our calculations, we will use the value $= 0.88 \text{ kg}/\text{m}^3$.

Because of air resistance, there are two forces acting on a falling body: the downward force of gravity W and the upward force of air resistance. From Newton's second law (see Appendix A), we find that the equation of motion in this case is

$$W - F_a = ma \tag{3.22}$$

When the body begins to fall, its velocity is zero and the only force acting on it is the weight; but as the body gains speed, the force of air resistance grows, and the net accelerating force on the body decreases. If the body falls from a sufficiently great height, the velocity reaches a magnitude such that the force due to air resistance is equal to the weight. Past this point, the body is no longer accelerated and continues to fall at a constant velocity, called the *terminal velocity* $v_t$. Because the force on the body in Eq. 3.22 is not constant, the solution of this equation cannot be obtained by simple algebraic techniques. However, the terminal velocity can be obtained without difficulty. At the terminal velocity, the downward force of gravity is canceled by the upward force of air resistance, and the net acceleration of the body is zero. That is,

$$W - F_a = 0 \tag{3.23}$$

or

$$F_a = W$$

From Eq. 3.21, the terminal velocity is therefore given by

$$v_t = \sqrt{\frac{W}{CA}} \qquad (3.24)$$

From this equation, the terminal velocity of a falling person with mass 70 kg and an effective area of 0.2 m$^2$ is

$$v_t = \sqrt{\frac{W}{CA}} = \sqrt{\frac{70 \times 9.8}{0.88 \times 0.2}} = 62.4 \text{ m/sec (140 mph)}$$

The terminal velocity of different-sized objects that have a similar density and shape is proportional to the square root of the linear size of the objects. This can be seen from the following argument. The weight of an object is proportional to the volume, which is in turn proportional to the cube of the linear dimension L of the object,

$$W \propto L^3$$

The area is proportional to L$^2$. Therefore, from Eq. 3.24, the terminal velocity is proportional to $\sqrt{L}$, as shown here:

$$v_t \propto \sqrt{\frac{W}{A}} = \sqrt{\frac{L^3}{L^2}} = \sqrt{L}$$

This result has interesting implications for the ability of animals to survive a fall. With proper training, a person can jump from a height of about 10 m without sustaining serious injury. From this height, a person hits the ground at a speed of

$$v = \sqrt{2gs} = 14 \text{ m/sec (46 ft/sec)}$$

Let us assume that this is the speed with which any animal can hit the ground without injury. At this speed, the force of air resistance on an animal the size of a man is negligible compared to the weight. But a small animal is slowed down considerably by air friction at this speed. A speed of 8.6 m/sec is the terminal velocity of a 1-cm bug (see Exercise 3.6). Such a small creature can drop from any height without injury. Miners often encounter mice in deep coal mines but seldom rats. A simple calculation shows that a mouse can fall down a 100-m mine shaft without severe injury. However, such a fall will kill a rat.

Air friction has an important effect on the speed of falling raindrops and hailstones. Without air friction, a 1-cm diameter hailstone, for example, falling from a height of 1000 m, would hit the Earth at a speed of about 140 m/sec. At such speeds, the hailstone would certainly injure anyone on whom it fell. As it is, air friction slows the hailstone to a safe terminal velocity of about 8.3 m/sec (see Exercise 3.8).

## 3.8 Energy consumed in physical activity

Animals do work by means of muscular movement. The energy required to perform the work is obtained from the chemical energy in the food eaten by the animal. In general, only a small fraction of the energy consumed by the muscles is converted to work. For example, in bicycling at a rate of one leg extension per second, the efficiency of the muscles is 20%. In other words, only one-fifth of the chemical energy consumed by the muscle is converted to work. The rest is dissipated as heat. The energy consumed per unit time during a given activity is called the metabolic rate.

Muscle efficiency depends on the type of work and on the muscles involved. In most cases, the efficiency of the muscles in converting caloric food energy to work is less than 20%. However, in our subsequent calculations, we will assume a 20% muscular efficiency.

We will calculate the amount of energy consumed by a 70-kg person jumping up 60 cm for 10 minutes at a rate of one jump per second. The external mechanical work performed by the leg muscles in each jump is

$$\text{Weight} \times \text{Height of jump} = 70 \text{ kg} \times 9.8 \times 0.6 = 411 \text{ J}$$

The total muscle work during the 10 minutes of jumping is

$$411 \times 600 \text{ jumps} = 24.7 \times 10^4 \text{ J}$$

If we assume a muscle efficiency of 20%, then in the act of jumping, the body consumes

$$24.7 \times 10^4 \times 5 = 1.23 \times 10^6 \text{ J} = 294 \times 10^3 \text{ cal} = 294 \text{ kcal}$$

This is about the energy content in two doughnuts.

In a similar vein, A. H. Cromer (see Bibliography) calculates the metabolic rate while running. In the calculation, it is assumed that most of the work done in running is due to the leg muscles accelerating each leg to the running speed v, and then decelerating it to 0 velocity as one leg is brought to rest and the other leg is accelerated. The work in accelerating the leg of mass m is $\frac{1}{2}mv^2$. The work done in the deceleration is also $\frac{1}{2}mv^2$. Therefore, the total amount of work done during each stride is $mv^2$. As is shown in Exercise 3.9, typically, a 70-kg person (leg mass 10 kg) running at 3 m/sec (9-min mile) with a muscle efficiency of 20% and step length of 1 m expends 1350 J/sec or 1160 kcal/h. This is in good agreement with measurements. The energy required to overcome air resistance in running is calculated in Exercise 3.10.

In connection with the energy consumption during physical activity, we should note the difference between work and muscular effort. Work is defined as the product of force and the distance over which the force acts (see Appendix A). When a person pushes against a fixed wall, his/her muscles are not performing any external work because the wall does not move. Yet it is evident that considerable energy is used in the act of pushing. All the energy is expended in the body to keep the muscles balanced in the tension necessary for the act of pushing.

## Exercises

**3.1.** Experiments show that the duration of upward acceleration in the standing vertical jump is about 0.2 sec. Calculate the power generated in a 60-cm jump by a 70-kg jumper assuming that c = H, as in the text.

**3.2.** A 70-kg astronaut is loaded so heavily with equipment that on Earth, he can jump only to a height of 10 cm. How high can he jump on the Moon? (Use the assumptions related to Eq. 3.11 in the text. As in the text, assume that the force generated by the legs is twice the unloaded weight of the person and the gravitational constant on the moon is 1/6 that on Earth.)

**3.3.** Solve Eqs. 3.19 and 3.20 for the two unknowns $F_r$ and $\theta$.

**3.4.** What is the time period in the standing broad jump during which the jumper is in the air? Assume that the conditions of the jump are as described in the text.

**3.5.** Consider a person on the moon who launches herself into a standing broad jump at 45°. The average force generated during launching is, as stated in the text, F = 2W, and the distance over which this force acts is 60 cm. The gravitational constant on the moon is 1/6 that on earth. Compute (a) the range of the jump; (b) the maximum height of the jump; and (c) the duration of the jump.

**3.6.** Calculate the terminal velocity of a 1-cm bug. Assume that the density of the bug is 1 g/cm³ and that the bug is spherical in shape with a diameter of 1 cm. Assume further that the area of the bug subject to air friction is $\pi r^2$.

**3.7.** Calculate the radius of a parachute that will slow a 70-kg parachutist to a terminal velocity of 14 m/sec.

**3.8.** Calculate the terminal velocity of (a) 1-cm diameter hailstone and (b) a 4-cm diameter hailstone. Density of ice is 0.92 gm/cm³. Assume that the area subject to air friction is $\pi r^2$.

**3.9.** Using the approach discussed in the text, calculate the energy expended per second by a person running at 3 m/sec (9-min mile) with a muscle efficiency of 20%. Assume that the leg mass m = 10 kg and the step length is 1 m.

**3.10.** Compute the power necessary to overcome air resistance when running at 4.5 m/sec (6-min mile) against a 30 km/h wind. (Use data in the text and assume area of person facing the wind is 0.2 m².)

Chapter 4

# Angular motion

As was stated in Chapter 3, most natural movements of animals consist of both
linear and angular motion. In this chapter, we will analyze some aspects of
angular motion contained in the movement of animals. The basic equations
and definitions of angular motion used in this chapter are reviewed in
Appendix A.

## 4.1 Forces on a curved path

The simplest angular motion is one in which the body moves along a curved
path at a constant angular velocity, such as when a runner travels along a
circular path or an automobile rounds a curve. The usual problem here is to
calculate the centrifugal forces and determine their effect on the motion of the
object.

A common problem solved in many basic physics texts requires determination of the maximum speed at which an automobile can round a curve
without skidding. We will solve this problem because it leads naturally to an
analysis of running. Consider a car of weight W moving on a curved level road
that has a radius of curvature R. The centrifugal force $F_c$ exerted on the
moving car (see Appendix A) is

$$F_c = \frac{mv^2}{R} = \frac{Wv^2}{gR} \tag{4.1}$$

For the car to remain on the curved path, a centripetal force must be
provided by the frictional force between the road and the tires. The car begins
to skid on the curve when the centrifugal force is greater than the frictional
force.

When the car is on the verge of skidding, the centrifugal force is just equal
to the frictional force; that is,

$$\frac{Wv^2}{gR} = \mu W \tag{4.2}$$

Here $\mu$ is the *coefficient of friction* between the tires and the road surface.
From Eq. 4.2, the maximum velocity $v_{max}$ without skidding is

$$v_{max} = \sqrt{\mu g R} \qquad (4.3)$$

Safe speed on a curved path may be increased by banking the road along the curve. If the road is properly banked, skidding may be prevented without recourse to frictional forces. Fig. 4.1 shows a car rounding a curve banked at an angle $\theta$. In the absence of friction, the reaction force $F_n$ acting on the car must be perpendicular to the road surface. The vertical component of this force supports the weight of the car. That is,

$$F_n \cos\theta = W \qquad (4.4)$$

To prevent skidding on a frictionless surface, the total centripetal force must be provided by the horizontal component of $F_n$; that is,

$$F_n \sin\theta = \frac{W v^2}{gR} \qquad (4.5)$$

where R is the radius of road curvature.

The angle $\theta$ for the road bank is obtained by taking the ratio of Eqs. 4.4 and 4.5. This yields

$$\tan\theta = \frac{v^2}{gR} \qquad (4.6)$$

## 4.2 A Runner on a curved track

A runner on a circular track is subject to the same type of forces described in discussion of the automobile. As the runner rounds the curve, she leans toward the center of rotation (Fig. 4.2a). The reason for this position can be understood from an analysis of the forces acting on the runner. Her foot, as it makes contact with the ground, is subject to the two forces, shown in Fig. 4.2b: An upward force W, which supports her weight, and a centripetal reaction force $F_{cp}$, which counteracts the centrifugal force. The resultant force $F_r$ acts on the runner at an angle $\theta$ with respect to the vertical axis.

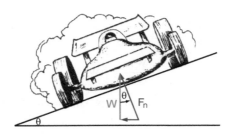

FIGURE 4.1   Banked curve.

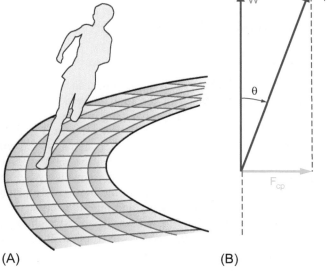

**(A)** **(B)**

**FIGURE 4.2** (A) Runner on a curved track. (B) Forces acting on the foot of the runner.

If the runner were to round the curve remaining perpendicular to the surface, this resultant force would not pass through her center of gravity and an unbalancing torque would be applied on the runner (see Exercise 4.1). If the runner adjusts her position by leaning at an angle $\theta$ toward the center of rotation, the resultant force $F_r$ passes through her center of gravity and the unbalancing torque is eliminated.

The angle $\theta$ is obtained from the following relationships (see Fig. 4.2b):

$$F_r \sin\theta = F_{cp} = \frac{Wv^2}{gR} \tag{4.7}$$

and

$$F_r \cos\theta = W \tag{4.8}$$

Therefore

$$\tan\theta = \frac{v^2}{gR} \tag{4.9}$$

The proper angle for a speed of 6.7 m/sec (this is a 4-min mile) on a 15-m-radius track is

$$\tan\theta = \frac{(6.7)^2}{9.8 \times 15} = 0.305$$

$$\theta = 17°$$

No conscious effort is required to lean into the curve. The body auto-matically balances itself at the proper angle. Other aspects of centrifugal force are examined in Exercises 4.2, 4.3, and 4.4.

## 4.3 Pendulum

Since the limbs of animals are pivoted at the joints, the swinging motion of animals is basically angular. Many of the limb movements in walking and running can be analyzed in terms of the swinging movement of a *pendulum*.

Whenever a system can store energy in two (or more) modes and the energy can flow readily from one mode to another such a system will exhibit resonance. That is, the energy exchange between the modes proceeds at a well-defined rate designated by the resonance frequency $f_{res.}$, sometimes called the natural frequency of the system. In the motion of a pendulum the exchange occurs between potential energy and kinetic energy.

The simple pendulum shown in Fig. 4.3 consists of a weight attached to a string, the other end of which is attached to a fixed point. If the pendulum is displaced a distance $A$ from the center position and then released, it will swing back and forth under the force of gravity. Such a back-and-forth movement is called a *simple harmonic motion*. The number of times the pendulum swings back and forth per second is the resonance frequency ($f_{res}$). The time for completing one cycle of the motion (i.e., from A to A′ and back to A) is called the *period* (T). Frequency and period are inversely related; that is, $T = 1/f$. As is shown in most college physics text books and also in several internet presentations, (see for example Morgan, 1969). If the angle of displacement is small, the period and the resonant frequency $f_{res}$ are given by

$$T = \frac{1}{f} = 2\pi\sqrt{\frac{\ell}{g}} \qquad (4.10)$$

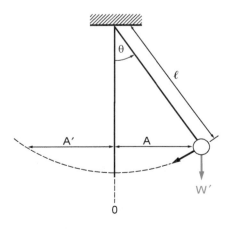

**FIGURE 4.3  The simple pendulum.**

where g is the gravitational acceleration and $\ell$ is the length of the pendulum arm. Although this expression for T is derived for a small-angle swing, it is a good approximation even for a relatively wide swing. For example, when the swing is through 120° (60° in each direction), the period is only 7% longer than predicted by Eq. 4.10.

As the pendulum swings, there is continuous interchange between potential and kinetic energy. At the extreme of the swing, the pendulum is momentarily stationary. Here its energy is entirely in the form of potential energy. At this point, the pendulum, subject to acceleration due to the force of gravity, starts its return toward the center. The acceleration is tangential to the path of the swing and is at a maximum when the pendulum begins to return toward the center. The maximum tangential acceleration $a_{max}$ at this point is given by

$$a_{max} = \frac{4\pi^2 A}{T^2} \tag{4.11}$$

As the pendulum is accelerated toward the center, its velocity increases, and the potential energy is converted to kinetic energy. The velocity of the pendulum is at its maximum when the pendulum passes the center position (0). At this point, the energy is entirely in the form of kinetic energy, and the velocity $(v_{max})$ here is given by

$$v_{max} = \frac{2\pi A}{T} \tag{4.12}$$

An ideal pendulum, that is one without friction, once set in motion would continue to swing endlessly. However, frictional losses dissipate the kinetic and to a lesser extent potential energy in the swinging pendulum. As a result if energy is not added into the motion of the pendulum, the swinging will eventually halt.

An important feature of resonant system is the ease of coupling energy at the resonant frequency into a resonant system. A well-known example is the back and forth motion of a playground swing that can be analyzed as a pendulum. With very little effort the amplitude of the swing can be increased if the pushing force at the back of the swing is at its resonant frequency.

## 4.4 Walking

Some aspects of walking can be analyzed in terms of the simple harmonic motion of a pendulum. The motion of one foot in each step can be considered approximately a half-cycle of a simple harmonic motion (Fig. 4.4). Assume that a person walks at a rate of 120 steps/min (2 steps/sec) and that each step is 90 cm long. In the process of walking, each foot rests on the ground for 0.5 seconds and then swings forward 180 cm and comes to rest again 90 cm ahead of the other foot. Since the forward swing takes 0.5 seconds, the full period of the harmonic motion is 1 second. The speed of walking v is

$$v = 90 \text{ cm} \times 2 \text{ steps/sec} = 1.8 \text{ m/sec (4 mph)}$$

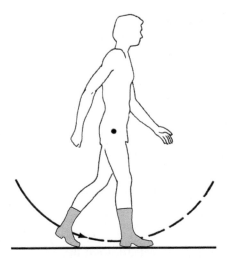

**FIGURE 4.4** **Walking.**

The maximum velocity of the swinging foot $v_{max}$ is, from Eq. 4.12,

$$v_{max} = \frac{2\pi A}{T} = \frac{6.28 \times 90}{1} = 5.65 \text{ m/sec } (12.6 \text{ mph})$$

Thus, at its maximum velocity, the foot moves about three times faster than the body. The maximum acceleration is

$$a_{max} = \frac{4\pi^2 A}{T^2} = 35.4 \text{ m/sec}^2$$

This is 3.6 times the acceleration of gravity. These formulas can also be applied to running (see Exercise 4.5).

## 4.5 Physical pendulum

The simple pendulum shown in Fig. 4.3 is not an adequate representation of the swinging leg because it assumes that the total mass is located at the end of the pendulum while the pendulum arm itself is weightless. A more realistic model is the physical pendulum, which takes into account the distribution of weight along the swinging object (see Fig. 4.5). It can be shown (see Morgan, 1969[1]) that under the force of gravity, the period of oscillation T for a physical pendulum is

$$T = 2\pi \sqrt{\frac{I}{Wr}} \tag{4.13}$$

---

1. References to the bibliography are given in square brackets.

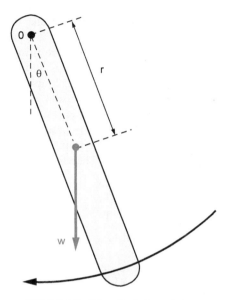

FIGURE 4.5 **The physical pendulum.**

Here I is the moment of inertia of the pendulum around the pivot point O (see Appendix A); W is the total weight of the pendulum, and r is the distance of the center of gravity from the pivot point. (The expression for the period in Eq. 4.13 is again strictly correct only for small angular displacement.)

## 4.6 Speed of walking and running

In the analysis of walking and running, the leg may be regarded as a physical pendulum with a moment of inertia of a thin rod pivoted at one end. The moment of inertia I for the leg (see Appendix A) is, therefore,

$$I = \frac{m\ell^2}{3} = \frac{W}{g}\frac{\ell^2}{3} \tag{4.14}$$

where W is the weight of the leg and $\ell$ is its length. If we assume that the center of mass of the leg is at its middle $\left(r = \frac{1}{2}\ell\right)$, the period of oscillation is

$$T = 2\pi\sqrt{\frac{I}{Wr}} = 2\pi\sqrt{\frac{(W/g)(\ell^2/3)}{W\ell/2}} = 2\pi\sqrt{\frac{2}{3}\frac{\ell}{g}} \tag{4.15}$$

For a 90-cm-long leg, the period is 1.6 seconds.

Because each step in the act of walking can be regarded as a half-swing of a simple harmonic motion, the number of steps per second is simply the inverse of the half-period. In a most effortless walk, the legs swing at their natural frequency, and the time for one step is $T/2$. Walking faster or slower requires additional muscular exertion and is more tiring. In Exercise 4.6, we calculate that for a person with 90-cm-long legs and 90-cm step length, the most effortless walking speed is 1.13 m/sec (2.53 mph). Similar considerations apply to the swinging of the arms (see Exercise 4.7).

We can now deduce the effect of the walker's size on the speed of walking. The speed of walking is proportional to the product of the number of steps taken in a given time and the length of the step. The size of the step is in turn proportional to the length of the leg $\ell$. Therefore, the speed of walking v is proportionally related as follows:

$$v \propto \frac{1}{T} \times \ell \tag{4.16}$$

But because $1/T$ is proportional to $\sqrt{1/\ell}$ (see Eq. 4.15)

$$v \propto \frac{1}{\sqrt{\ell}} \ell = \sqrt{\ell} \tag{4.17}$$

Thus, the speed of the natural walk of a person increases as the square root of the length of his/her legs. The same considerations apply to all animals: The natural walk of a small animal is slower than that of a large animal.

The situation is different when a person (or any other animal) runs at full speed. Whereas in a natural walk the swing torque is produced primarily by gravity, in a fast run the torque is produced mostly by the muscles. Using some reasonable assumptions, we can show that similarly built animals can run at the same maximum speed, regardless of differences in leg size.

We assume that the length of the leg muscles is proportional to the length of the leg ($\ell$) and that the area of the leg muscles is proportional to $\ell^2$. The mass of the leg is proportional to $\ell^3$. In other words, if one animal has a leg twice as long as that of another animal, the area of its muscle is four times as large and the mass of its leg is eight times as large.

The maximum force that a muscle can produce $F_m$ is proportional to the cross-sectional area of the muscle. The maximum torque $L_{max}$ produced by the muscle is proportional to the product of the force and the length of the leg; that is,

$$L_{max} = F_m \ell \propto \ell^3$$

The expression in the equation for the period of oscillation is applicable to a pendulum swinging under the force of gravity. In general, the period of oscillation for a physical pendulum under the action of a torque with maximum value of $L_{max}$ is given by

$$T = 2\pi \sqrt{\frac{I}{L_{max}}} \tag{4.18}$$

Because the mass of the leg is proportional to $\ell^3$, the moment of inertia (from Eq. 4.14) is proportional to $\ell^5$. Therefore, the period of oscillation in this case is proportional to $\ell$ as shown

$$T \propto \sqrt{\frac{\ell^5}{\ell^3}} = \ell$$

The maximum speed of running $v_{max}$ is again proportional to the product of the number of steps per second and the length of the step. Because the length of the step is proportional to the length of the leg, we have

$$v_{max} \propto \frac{1}{T}\ell \propto \frac{1}{\ell} \times \ell = 1$$

This shows that the maximum speed of running is independent of the leg size, which is in accordance with observation: A fox, for example, can run at about the same speed as a horse.

Eqs. 4.10 and 4.15 illustrate another clearly observed aspect of running. When a person runs at a slow pace, the arms are straight, as in walking. However, as the speed of running (that is, the number of steps in a given interval) increases, the elbows naturally assume a bent position. In this way, the effective length $(\ell)$ of the pendulum decreases. This in turn increases the natural frequency of the arm, bringing it into closer synchrony with the increased frequency of steps.

## 4.7 Energy expended in running

In Chapter 3, we obtained the energy expended in running by calculating the energy needed to accelerate the leg to the speed of the run and then decelerating it to rest. Here we will use the physical pendulum as a model for the swinging leg to compute this same quantity. We will assume that in running, the legs swing only at the hips. This model is, of course, not strictly correct because in running, the legs swing not only at the hips but also at the knees. We will now outline a method for calculating the energy expended in swinging the legs.

During each step of the run, the leg is accelerated to a maximum angular velocity $\omega_{max}$. In our pendulum model, this maximum angular velocity is reached as the foot swings past the vertical position 0 (see Fig. 4.6).

The rotational kinetic energy at this point is the energy provided by the leg muscles in each step of the run. This maximum rotational energy $E_r$ (see Appendix A) is

$$E_r = \frac{1}{2}I\omega_{max}^2$$

Here I is the moment of inertia of the leg. The angular velocity $\omega_{max}$ is obtained as follows. From the rate of running, we can compute the period of oscillation T for the leg modeled as a pendulum. Using this value for the

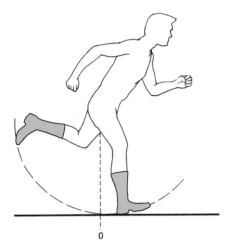

FIGURE 4.6 **Running.**

period, we calculate from Eq. 4.12 the maximum linear velocity $v_{max}$ of the foot. The angular velocity (see Appendix A) is then

$$\omega_{max} = \frac{v_{max}}{\ell}$$

where $\ell$ is the length of the leg. In computing the period T, we must note that the number of steps per second each leg executes is one-half of the total number of steps per second. In Exercise 4.8, it is shown that, based on the physical pendulum model for running, the amount of work done during each step is $1.6\, mv^2$. In Chapter 3, using different considerations, the amount of work done during each step was obtained as $mv^2$. Considering that both approaches are approximate, the agreement is certainly acceptable.

In calculating the energy requirements of walking and running, we assumed that the kinetic energy imparted to the leg is fully (frictionally) dissipated as the motion of the limb is halted within each step cycle. In fact, a significant part of the kinetic energy imparted to the limbs during each step cycle is stored as potential energy and is converted to kinetic energy during the following part of the gait cycle, as in the motion of an oscillating pendulum or a vibrating spring. The assumption of full energy dissipation at each step results in an overestimate of the energy requirements for walking and running. This energy overestimate is balanced by the underestimate due to the neglecting of movement of the center of mass up and down during walking and running, as is discussed in following Sections 4.8 and 4.9.

## 4.8 Alternate perspectives on walking and running

In Sections 4.4 to 4.7 we presented relatively simple models of walking and running. More detailed and accurate descriptions can be found in various technical journals. (For a review of the field, see the article by Novacheck (1998), Bibliography). However, the basic approach in the various methods of analysis is similar in that the highly complex interactive musculoskeletal system involved in walking and/or running is represented by a simplified structure that is amenable to mathematical analysis.

In our treatment of walking and running we considered only the pendulum-like motion of the legs. A more detailed treatment considers also the motion of the center of mass. A way to model the center of mass motion in walking is to consider the motion of the center of mass during the course of a step. Consider the start of the step when both feet are on the ground, with one foot ahead of the other. At this point, the center of mass is between the two feet and is at its lowest position (see Fig. 4.7). We chose left to be the rear foot. The step begins with the rear foot leaving the ground and swinging forward. The center of mass is at its highest point when the swinging foot is in line with the stationary foot. As the swinging foot passes the stationary foot, it becomes the forward foot and the step is completed with the two feet once again on the ground with the right foot now in the rear. The center of mass trajectory is an arc, as depicted in Fig. 4.7.

In the sequence of the step described in the figure, the center of mass is alternately behind and then in front of the point of the single-foot contact with the ground as the free leg swings forward. That is, when the rear left foot starts swinging forward, it is of course off the ground, and the center of mass is behind the supporting right foot. During this part of the step, the center of mass is swinging toward the stationary right foot and its kinetic energy is converted to potential energy (as in the upward swing of a pendulum, the supporting foot being the fulcrum). After the left foot passes the stationary right foot, the center of mass shifts forward of the right foot and accelerates as the potential energy is

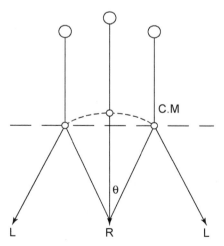

**FIGURE 4.7** Center of mass trajectory in the course of one step.

converted to kinetic energy (downward swing of the pendulum). With a step length of 90 cm and the center of mass (with feet together) 1 m above the ground, the center of mass is raised 11 cm during each swinging cycle of the step (see Exercise 4.10). This is an upper limit because in this simplified treatment it is assumed that the legs remain straight throughout the step.

For the above case, the work done in raising the center of mass by 0.11 m is 0.11 m × W (in units of Joule) during each step. Because the body in the process of walking is not a perfect pendulum, only part of this potential energy is converted back into kinetic energy. To reduce the energy expenditure, the body seeks adjustments to minimize the up-and-down movement of the center of mass (see Section 4.9).

A clear distinction is made between walking and running. During walking, at one point in the step cycle, both feet are in contact with the ground. In running, there is an interval during the step when both feet are off the ground. During a walking step, the center of mass trajectory is similar to that of an inverted swinging pendulum with the fulcrum at the point where the two feet pass one another (Fig. 4.7). Running can be compared to a person on a pogo stick as if bouncing from one leg to another.

In Fig. 4.8 we show the energy required per meter of distance covered as a function of speed, for walking and for running. (This figure is based on the work of Alexander (1968) as presented by Novacheck (1998).) The minimum energy consumed per distance traveled (as shown in Fig. 4.8) occurs at a walking speed of about 1.3 m/sec (2.9 mph). This is only 13% higher than the

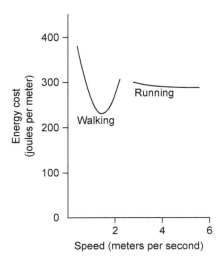

**FIGURE 4.8   Fraction of body weight for various parts of the body.** Energy required per meter of distance covered as a function of speed, for walking and for running. *(From Novacheck TF. The biomechanics of running. Gait Posture 1998;7:77–95.)*

most effortless walking speed of 1.13 m/sec we calculated on the basis of the swinging-leg pendulum model in Section 4.6. This close agreement is somewhat fortuitous, brought about by canceling effect of the approximations entailed in the calculation as discussed in Section 4.7. As shown in the figure, the energy consumed per distance traveled increases at both lower and higher walking speeds. For speeds typically less than 2 m/sec ( ∼4.5 mph), walking is more efficient than running (that is, requiring less energy per distance traveled). Past this speed, most people will spontaneously break into a run consuming less energy.

As shown in Fig. 4.8, the energy cost of running is about 300 J/m. As discussed in Section 3.8 (Exercise 3.9), the energy required to run at 3 m/sec (9-min mile) was calculated to be 1350 J/sec, that is, 450 J/m. Considering the approximate nature of the calculations and the difference in the methods, the agreement between the two numbers is again remarkably good.

## 4.9 Carrying loads

Carrying a load requires energy. Measurements have shown that for most humans, as well as animals such as dogs, horses, and rats, the energy expended at a given walking speed increases directly with the weight of the load being carried. Specifically, carrying a load that is 50% of the body weight increases energy consumption by 50%. For most people, this added energy expenditure is the same whether they carry the load on their backs or on their heads.

Recent studies have been focused on the extraordinary load-carrying abilities of women in certain areas of East Africa who can walk with relative ease carrying large loads balanced on their heads. Quantitative measurements have shown that women from the Luo and Kikuyu tribes can carry loads up to about 20% of their body weight without any measurable increase in their energy consumption. Past this weight, the energy consumption increases in proportion to the weight carried minus the 20%. That is, carrying a load of 50% of the body weight increases their energy consumption by 30% (50%−20%). The experimenters of Heglund et al. suggest that the high load-carrying efficiency of these women is due to greater efficiency of exchange from gravitational potential energy to kinetic energy in the pendulum-like movement of the center of mass discussed in Section 4.8. In other words their walking more closely approximates a swinging pendulum. What specific aspect of the movement or training brings about these enhanced load-carrying abilities is not yet understood.

## Exercises

Some of the problems in this chapter require a knowledge of the weight of human limbs. Use Table 4.1 to compute these weights.

**TABLE 4.1 Fraction of body weight for various parts of the body.**

|  | Fraction of body weight |
| --- | --- |
| Head and neck | 0.07 |
| Trunk | 0.43 |
| Upper arms | 0.07 |
| Forearms and hands | 0.06 |
| Thighs | 0.23 |
| Legs and feet | 0.14 |
| Total | 1.00 |

From Cooper JM, Glassow RB. Kinesiology. 3rd ed. St. Louis, MO: The C. V. Mosby Co.; 1972. p. 174.

**4.1.** Explain why a runner is subject to a torque if she rounds a curve maintaining a vertical position.

**4.2.** In the act of walking, the arms swing back and forth through an angle of 45° each second. Using the following data, and the data in Table 4.1, calculate the average force on the shoulder due to the centrifugal force. The mass of the person is 70 kg, and the length of the arm is 90 cm. Assume that the total mass of the arms is located at the midpoint of the arm.

**4.3.** Consider the carnival ride in which the riders stand against the wall inside a large cylinder. As the cylinder rotates, the floor of the cylinder drops and the passengers are pressed against the wall by the centrifugal force. Assuming that the coefficient of friction between a rider and the cylinder wall is 0.5 and that the radius of the cylinder is 5 m, what are the minimum angular velocity and the corresponding linear velocity of the cylinder that will hold the rider firmly against the wall?

**4.4.** If a person stands on a rotating pedestal with his arms loose, the arms will rise toward a horizontal position. (a) Explain the reason for this phenomenon. (b) Calculate the angular (rotational) velocity of the pedestal for the angle of the arm to be at 60° with respect to the horizontal. What is the corresponding number of revolutions per minute? Assume that the length of the arm is 90 cm and the center of mass is at mid-length.

**4.5.** Calculate the maximum velocity and acceleration of the foot of a runner who does a 100-m dash in 10 seconds. Assume that the length of a step is

1 m, that the length of the leg is 90 cm, and that the center of mass is at midlength.

**4.6.** What is the most effortless walking speed for a person with 90-cm-long legs if the length of each step is 90 cm?

**4.7.** While walking, the arms swing under the force of gravity. Compute the period of the swing. How does this period compare with the period of the leg swing? Assume arm length of 90 cm.

**4.8.** Using the physical pendulum model for running described in the text, derive an expression for the amount of work done during each step.

**4.9.** Compute the length of time for an erect human body without compensating movements to hit the floor once it loses its balance. Assume that the falling body behaves as a physical pendulum pivoted on the floor within the period given by Eq. 4.13. The full length of the body is 2 m.

**4.10.** Calculate the distance the center of mass is raised in the course of one step with parameters and assumptions as discussed in Section 4.8. (Refer to Fig. 4.7.)

# Chapter 5

# Elasticity and strength of materials

So far, we have considered the effect of forces only on the motion of a body. We will now examine the effect of forces on the shape of the body. When a force is applied to a body, the shape and size of the body change. Depending on how the force is applied, the body may be stretched, compressed, bent, or twisted. *Elasticity* is the property of a body that tends to return the body to its original shape after the force is removed. If the applied force is sufficiently large, however, the body is distorted beyond its elastic limit, and the original shape is not restored after removal of the force. A still larger force will rupture the body. We will review briefly the theory of deformation and then examine the damaging effects of forces on bones and tissue.

## 5.1 Longitudinal stretch and compression

Let us consider the effect of a stretching force F applied to a bar (Fig. 5.1). The applied force is transmitted to every part of the body, and it tends to pull the material apart. This force, however, is resisted by the cohesive force that holds the material together. The material breaks when the applied force exceeds the cohesive force. If the force in Fig. 5.1 is reversed, the bar is compressed, and its length is reduced. Similar considerations show that initially the compression is elastic, but a sufficiently large force will produce permanent deformation and then breakage.

Stress S is the internal force per unit area acting on the material; it is defined as[1]

$$S \equiv \frac{F}{A} \tag{5.1}$$

Here F is the applied force and $A$ is the area on which the force is applied.

The force applied to the bar in Fig. 5.1 causes the bar to elongate by an amount $\Delta \ell$. The fractional change in length $\Delta \ell / \ell$ is called the *longitudinal strain* $S_t$; that is,

---

1. The $\equiv$ symbol is read "defined as."

**Physics in Biology and Medicine. https://doi.org/10.1016/B978-0-443-21558-2.00005-5**

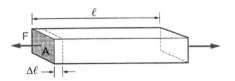

**FIGURE 5.1** Stretching of a bar due to an applied force.

**TABLE 5.1** Young's modulus and rupture strength for some materials.

| Material | Young's modulus (dyn/cm$^2$)* | Rupture strength (dyn/cm$^2$) |
|---|---|---|
| Steel | $200 \times 10^{10}$ | $450 \times 10^7$ |
| Aluminum | $69 \times 10^{10}$ | $62 \times 10^7$ |
| Bone | $14 \times 10^{10}$ | $100 \times 10^7$ compression |
| | | $83 \times 10^7$ stretch |
| | | $27.5 \times 10^7$ twist |
| Tendon | | $68.9 \times 10^7$ stretch |
| Muscle | | $0.55 \times 10^7$ stretch |

*dyn/cm$^2$ = $10^{-1}$ Pa (pascal).

$$S_t \equiv \frac{\Delta \ell}{\ell} \qquad (5.2)$$

Here $\ell$ is the length of the bar and $\Delta \ell$ is the change in the length due to the applied force. If reversed, the force in Fig. 5.1 will compress the bar instead of stretching it. (Stress and strain remain defined as before.) In 1676, Robert Hooke observed that while the body remains elastic, the ratio of stress to strain is constant (Hooke's law); that is,

$$\frac{S}{S_t} = Y \qquad (5.3)$$

The constant of proportionality $Y$ is called *Young's modulus*. Young's modulus has been measured for many materials, some of which are listed in Table 5.1. The breaking or rupture strength of these materials is also shown.

## 5.2 A spring

Two energy modes are present in a vibrating spring; the potential energy due to compression or stretching of the spring, and the kinetic energy of the vibrating mass associated with the spring. Potential and kinetic energy exchange throughout a cycle of vibration at a regular rate given by the resonance frequency ($f_{res}$) as

$$f_{res} = \frac{1}{2\pi}\sqrt{\frac{K}{m}} \tag{5.4}$$

Here $K$ is the spring constant and $m$ is the mass associated with the spring. This and the following expressions in this are derived in most college physics text books as well as in several internet presentations (see for example Morgan, J., 1969).

As will be shown a useful analogy can be drawn between a spring and the elastic properties of materials. Consider the spring shown in Fig. 5.2.

The force F required to stretch (or compress) the spring is directly proportional to the amount of stretch; that is,

$$F = -K\Delta\ell \tag{5.5}$$

The constant of proportionality K is called the *spring constant*. The negative sign indicates that the force is in the direction opposite to $\Delta\ell$.

A stretched (or compressed) spring contains potential energy; that is, work can be done by the stretched spring when the stretching force is removed. The energy E stored in the spring (see Morgan, 1969) is given by

$$E = \frac{1}{2}K(\Delta\ell)^2 \tag{5.6}$$

An elastic body under stress is analogous to a spring with a spring constant $YA/\ell$. This can be seen by expanding Eq. 5.3 as

$$\frac{S}{S_t} = \frac{F/A}{\Delta\ell/\ell} = Y$$

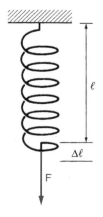

**FIGURE 5.2  A stretched spring.**

By re-arranging this equation we obtain an expression for force F as

$$F = \frac{YA}{\ell}\Delta\ell \tag{5.7}$$

This equation is identical to the equation for a spring with a spring constant

$$K = \frac{YA}{\ell} \tag{5.8}$$

By analogy with the spring (see Eq. 5.5), the amount of energy stored in a stretched or compressed body is

$$E = \frac{1}{2}\frac{YA}{\ell}(\Delta\ell)^2 \tag{5.9}$$

## 5.3 Bone fracture: Energy considerations

Knowledge of the maximum energy that parts of the body can safely absorb allows us to estimate the possibility of injury under various circumstances. We shall first calculate the amount of energy required to break a bone of area A and length $\ell$. Assume that the bone remains elastic until fracture. Let us designate the breaking stress of the bone as $S_B$ (see Fig. 5.3). The corresponding force $F_B$ that will fracture the bone is, from Eq. 5.7,

$$F_B = S_B A = \frac{YA}{\ell}\Delta\ell \tag{5.10}$$

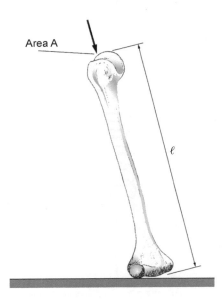

Area A

$\ell$

**FIGURE 5.3** Compression of a bone.

The compression $\Delta\ell$ at the breaking point is, therefore,

$$\Delta\ell = \frac{S_B \ell}{Y} \tag{5.11}$$

From Eq. 5.9, the energy stored in the compressed bone at the point of fracture is

$$E = \frac{1}{2}\frac{YA}{\ell}(\Delta\ell)^2 \tag{5.12}$$

Substituting for $\Delta\ell = S_B\ell/Y$, we obtain

$$E = \frac{1}{2}\frac{A\ell\, S_B^2}{Y} \tag{5.13}$$

As an example, consider the fracture of two leg bones that have a combined length of about 90 cm and an average area of about 6 cm$^2$. From Table 5.1, the breaking stress $S_B$ is $10^9$dyn/cm$^2$, and Young's modulus for the bone is $14 \times 10^{10}$ dyn/cm$^2$. The total energy absorbed by the bones of one leg at the point of compressive fracture is, from Eq. 5.13,

$$E = \frac{1}{2}\frac{6 \times 90 \times 10^{18}}{14 \times 10^{10}} = 19.25 \times 10^8 \text{ erg} = 192.5 \text{ J}$$

The combined energy in the two legs is twice this value, or 385 J. This is the amount of energy in the impact of a 70-kg person jumping from a height of 56 cm (1.8 ft), given by the product mgh. (Here m is the mass of the person, g is the gravitational acceleration, and h is the height.) If all this energy is absorbed by the leg bones, they may fracture.

It is certainly possible to jump safely from a height considerably greater than 56 cm if, on landing, the joints of the body bend and the energy of the fall is redistributed to reduce the chance of fracture. The calculation does, however, point out the possibility of injury in a fall from even a small height. Similar considerations can be used to calculate the possibility of bone fracture in running (see Exercise 5.1).

## 5.4 Impulsive forces

In a sudden collision, a large force is exerted for a short period of time on the colliding object. The general characteristic of such a collision force as a function of time is shown in Fig. 5.4. The force starts at zero, increases to some maximum value, and then decreases to zero again. The time interval $t_2 - t_1 = \Delta t$ during which the force acts on the body is the duration of the collision. Such a short-duration force is called an *impulsive force*.

Because the collision takes place in a short period of time, it is usually difficult to determine the exact magnitude of the force during the collision.

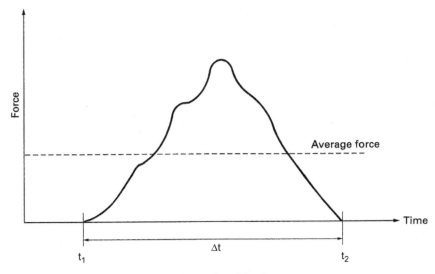

**FIGURE 5.4  Impulsive force.**

However, it is relatively easy to calculate the average value of the impulsive force $F_{av}$. It can be obtained simply from the relationship between force and momentum given in Appendix A; that is,

$$F_{av} = \frac{mv_f - mv_i}{\Delta t} \qquad (5.14)$$

Here $mv_i$ is the initial momentum of the object and $mv_f$ is the final momentum after the collision. For example, if the duration of a collision is $6 \times 10^{-3}$ seconds and the change in momentum is 2 kg m/sec, the average force that acted during the collision is

$$F_{av} = \frac{2 \text{ kg m/sec}}{6 \times 10^{-3} \text{ sec}} = 3.3 \times 10^2 \text{ N}$$

Note that, for a given momentum change, the magnitude of the impulsive force is inversely proportional to the collision time; that is, the collision force is larger in a fast collision than in a slower collision.

## 5.5 Fracture due to a fall: Impulsive force considerations

In the preceding section, we calculated the injurious effects of collisions from energy considerations. Similar calculations can be performed using the concept of impulsive force. The magnitude of the force that causes the damage is computed from Eq. 5.14. The change in momentum due to the collision is usually easy to calculate, but the duration of the collision $\Delta t$ is difficult to determine precisely. It depends on the type of collision. If the colliding objects are hard, the collision time is very short, a few milliseconds. If one of the objects is soft and yields during the collision, the duration of the collision is lengthened, and as a result, the impulsive force is reduced. Thus, falling into soft sand is less damaging than falling on a hard concrete surface.

When a person falls from a height h, his/her velocity on impact with the ground, neglecting air friction (see Eq. 3.6), is

$$v = \sqrt{2gh} \tag{5.15}$$

The momentum on impact is

$$mv = m\sqrt{2gh} = W\sqrt{\frac{2h}{g}} \tag{5.16}$$

After the impact, the body is at rest, and its momentum is therefore zero $(mv_f = 0)$. The change in momentum is

$$mv_i - mv_f = W\sqrt{\frac{2h}{g}} \tag{5.17}$$

The average impact force, from Eq. 5.14, is

$$F = \frac{W}{\Delta t}\sqrt{\frac{2h}{g}} = \frac{m}{\Delta t}\sqrt{2gh} \tag{5.18}$$

Now comes the difficult part of the problem: Estimate of the collision duration. If the impact surface is hard, such as concrete, and if the person falls with his/her joints rigidly locked, the collision time is estimated to be about $10^{-2}$ seconds. The collision time is considerably longer if the person bends his/her knees or falls on a soft surface.

From Table 5.1, the force per unit area that may cause a bone fracture is $10^9 \ \mathrm{dyn/cm^2}$. If the person falls flat on his/her heels, the area of impact may be about 2 $\mathrm{cm^2}$. Therefore, the force $F_B$ that will cause fracture is

$$F_B = 2 \ \mathrm{cm^2} \times 10^9 \ \mathrm{dyn/cm^2} = 2 \times 10^9 \ \mathrm{dyn} \ (4.3 \times 10^3 \ \mathrm{lb})$$

From Eq. 5.18, the height h of fall that will produce such an impulsive force is given by

$$h = \frac{1}{2g}\left(\frac{F\Delta t}{m}\right)^2 \tag{5.19}$$

For a man with a mass of 70 kg, the height of the jump that will generate a fracturing average impact force (assuming $\Delta t = 10^{-2}$ seconds) is given by

$$h = \frac{1}{2g}\left(\frac{F\Delta t}{m}\right)^2 = \frac{1}{2 \times 980}\left(\frac{2 \times 10^9 \times 10^{-2}}{70 \times 10^3}\right)^2 = 41.6 \text{ cm } (1.37 \text{ ft})$$

This is close to the result that we obtained from energy considerations. Note, however, that the assumption of a 2 cm$^2$ impact area is reasonable but somewhat arbitrary. The area may be smaller or larger depending on the nature of the landing; furthermore, we have assumed that the person lands with legs rigidly straight. Exercises 5.2 and 5.3 provide further examples of calculating the injurious effect of impulsive forces.

## 5.6 Airbags: Inflating collision protection devices

The impact force may also be calculated from the distance the center of mass of the body travels during the collision under the action of the impulsive force. This is illustrated by examining the inflatable safety device used in automobiles (see Fig. 5.5). An inflatable bag is located in the dashboard of the car. In a collision, the bag expands suddenly and cushions the impact of the passenger. The forward motion of the passenger must be stopped in about 30 cm of motion if contact with the hard surfaces of the car is to be avoided. The average deceleration (see Eq. 3.6) is given by

$$a = \frac{v^2}{2s} \tag{5.20}$$

where v is the initial velocity of the automobile (and the passenger) and s is the distance over which the deceleration occurs. The average force that produces the deceleration is

$$F = ma = \frac{mv^2}{2s} \tag{5.21}$$

where m is the mass of the passenger.

For a 70-kg person with a 30-cm allowed stopping distance, the average force is

$$F = \frac{70 \times 10^3 v^2}{2 \times 30} = 1.17 \times 10^3 \times v^2 \text{ dyn}$$

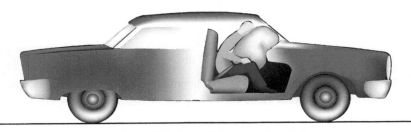

FIGURE 5.5   **Inflating collision protective device.**

At an impact velocity of 70 km/h $\left( = 1.94 \times \frac{10^3 \text{ cm}}{\text{sec}} = 43.5 \text{ mph} \right)$, the average stopping force applied to the person is $4.42 \times 10^9$ dyn. If this force is uniformly distributed over a $1000 \text{ cm}^2$ area of the passenger's body, the applied force per $\text{cm}^2$ is $4.42 \times 10^6$ dyn. This is just below the estimated strength of body tissue.

The necessary stopping force increases as the square of the velocity. At a 105-km impact speed, the average stopping force is about $10^{10}$ dyn and the force per $\text{cm}^2$ is $10^7$ dyn. Such a force would probably injure the passenger.

In the design of this safety system, the possibility has been considered that the bag may be triggered during normal driving. If the bag were to remain expanded, it would impede the ability of the driver to control the vehicle; therefore, the bag is designed to remain expanded for only the short time necessary to cushion the collision. (For an estimate of this period, see Exercise 5.4.)

## 5.7 Whiplash injury

Neck bones are rather delicate and can be fractured by even a moderate force. Fortunately, the neck muscles are relatively strong and are capable of absorbing a considerable amount of energy. If, however, the impact is sudden, as in a rear-end collision, the body is accelerated in the forward direction by the back of the seat, and the unsupported neck is then suddenly yanked back at full speed. Here the muscles do not respond fast enough and all the energy is absorbed by the neck bones, causing the well-known whiplash injury (see Fig. 5.6). The whiplash injury is described quantitatively in Exercise 5.5.

## 5.8 Falling from great height

There have been reports of people who jumped out of airplanes with parachutes that failed to open and yet survived because they landed on soft snow. It was found in these cases that the body made about a 1-m-deep depression in the surface of the snow on impact. The credibility of these reports can be verified by calculating the impact force that acts on the body during the landing. It is shown

FIGURE 5.6 Whiplash.

in Exercise 5.6 that if the decelerating impact force acts over a distance of about 1 m, the average value of this force remains below the magnitude for serious injury even at the terminal falling velocity of 62.5 m/sec (140 mph).

## 5.9 Osteoarthritis and exercise

In the preceding sections of this chapter, we discussed possible damaging effects of large impulsive forces. In the normal course of daily activities, our bodies are subject mostly to smaller repetitive forces such as the impact of feet with the ground in walking and running. A still not fully resolved question is to what extent are such smaller repetitive forces, particularly those encountered in exercise and sport, damaging. Osteoarthritis is the commonly suspected damage resulting from such repetitive impact.

Osteoarthritis is a joint disease characterized by a degenerative wearing out of the components of the joint, among them the synovial membrane and cartilage tissue. As a result of such wear and tear the joint loses flexibility and strength accompanied by pain and stiffness. Eventually, the underlying bone may also start eroding. Osteoarthritis is a major cause of disability at an older age. Knees are the most commonly affected joint. After the age of 65, about 60% of men and 75% of women are to some extent affected by this condition.

Over the past several years, a number of studies have been conducted to determine the link between exercise and osteoarthritis. The emerging conclusion is that joint injury is most strongly correlated with subsequent development of osteoarthritis. Most likely this is the reason why people engaged in high-impact injury-prone sports are at a significantly greater risk of osteoarthritis. Further, there appears to be little risk associated with recreational running 20 to 40 km a week ($\sim$13 to 25 miles).

It is not surprising that an injured joint is more likely to be subsequently subject to wear and tear. As shown in Chapter 2, Table 2.1, the coefficient of kinetic friction ($\mu_k$) of an intact joint is about 0.003. The coefficient of friction for unlubricated bones is a hundred times higher. A joint injury usually compromises, to some extent, the lubricating ability of the joint leading to increased frictional wear and osteoarthritis. This simple picture would lead one to expect that the progress of osteoarthritis would be more rapid in the joints of people who are regular runners than in a control group of nonrunners. Yet this does not appear to be the case. Osteoarthritis seems to progress at about the same rate in both groups, indicating that the joints possess some ability to self-repair. These conclusions remain tentative and are subject to further study.

## 5.10 Three-dimensional (3-D) printing a new technique for shaping materials

*Three-dimensional (3-D) printing* is a technique for producing three-dimensional objects from (usually computer-generated) two-dimensional

(2-D) images. The method was developed in the 1980s and is now being employed in the manufacturing of complex industrial equipment and components and with increasing frequency and success, in medical procedures and biology experiments.

The principle of 3-D printing is based on the technology of ink jet printing developed in the 1950s for the purpose of printing 2-D computer-generated text or images onto paper or other materials. By the 1980s, ink jet printers were relatively inexpensive and widely used.

Several ink jet printer designs are in current use. The purpose of the present discussion is to provide a simplified description of the operating principles of the device without presenting the often-complex design details. A simplified schematic of the printer is shown in Fig. 5.7. Ink is pumped into a small reservoir that has an exit nozzle typically 20 μm in diameter. A piezoelectric crystal is attached to the back of the reservoir. The exiting ink jet breaks up into droplets at a rate governed by the frequency of the vibrating crystal. Typically, 100,000 droplets are formed per second, each usually 50 μm in diameter. The droplets move out of the nozzle at a speed typically about 20 m/ sec. The number of droplets, droplet diameter, and speed are well controlled and are a function of the printer design.

As the droplets pass out of the nozzle, they are electrically charged. The amount of charge on the droplet is controlled droplet by droplet by the computer-generated signal. The droplet charge is governed by image to be printed. The charged droplets pass between two plates held at a high voltage, typically a few thousand volts. In their passage between the high-voltage plates, the droplets are deflected with the trajectory of the droplets governed by the amount of charge on the droplet. Approximately 500 droplets impact per inch of printing. This droplet-charging feature of the printer controls the

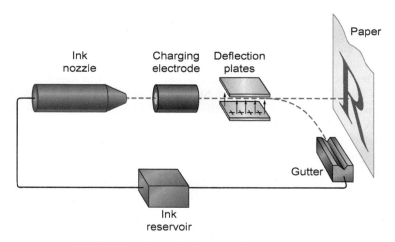

**FIGURE 5.7   Schematic diagram of an inkjet printer.**

fine details of the image, for example, a portion of a single letter. At the same time, the horizontal placement of the droplets is governed by the computer-controlled side-to-side motion of the ink jet assembly along a horizontal guide. The vertical positioning is determined by the motion of the paper (or other material) by accurately positioned rollers. When a blank space is required, the charged droplets are deflected into a collecting tube that recirculates the unused ink. The combination of small droplets and the absorbing properties of paper results in the fast drying of the ink so that the printed page emerges dry.

Printing a 3-D object begins by generating a layer-by-layer representation of the object to be printed. This is done using some version the several available computer design programs. In other words, we need to have a computer-generated slice-by-slice (say going from bottom to top) digital image of the object. In typical applications, slices are 0.1 mm thick, although for special applications the cuts can be an order of magnitude thinner. The simplest version of the 3-D printer resembles an ink jet printer except that the printing is done on a horizontal substrate and, instead of ink, an acrylic liquid (or some other photopolymer) is used. (See Exercise 5.10.) When illuminated by an ultraviolet light (usually produced by a laser), the liquid hardens into a substance similar to Plexiglas. Once the layer hardens, the next layer is printed. Layer by layer, the object is built up to completion. 3-D printing is a slow process. A full 3-D reproduction may take a few minutes to several hours depending on the complexity of the object. In the newer versions of 3-D printers, a variety of materials can be used including some metals as well as several biocompatible materials.

In a 2014 *New Yorker* article, Jerome Groopman describes one of the early applications of 3-D printing to solve a medical problem. In February of 2012, a 3-month-old boy was brought to the University of Michigan Children's Hospital in Ann Arbor with a rare condition wherein the tissue in a portion of his bronchia, the air passageway into his lungs, was so weak that it frequently collapsed endangering the child's life. A conventional surgical repair of the weak tissue was deemed too risky because of the child's small size. The child had to be placed on a ventilator. A 3-D printing method was suggested. A CT scan of the affected area was taken. Such scans, as discussed in Section 16.7, produce X-ray images of a thin section by a thin section of the organ examined. With 3-D printing, techniques using a strong flexible biocompatible material similar to one used in surgical sutures, these sectional images were used to construct an exact 3-D replica of the weakened bronchial passage. This support was inserted into the child's brachia preventing its collapse and making it possible to take the child off the ventilator. The properties of the material were such that it expanded as the child grew. With time, the material of the prosthesis dissolved and healthy tissue replaced it. Similar techniques are now being used to replace or repair heart valves, reconstruct segments of damaged human skulls, and repair a variety of malformations. To aid in the surgical operations, 3-D printing is often used to produce exact models of organs to be operated on, including the brain, ahead of complex surgery.

A major challenge in the field of 3-D printing is the reproduction of a working organ such as the heart or a kidney. With existing technology, it is relatively simple to grow macroscopic quantities of specific organ cells. Using 3-D printing techniques with the suspended cells as the "ink," it is now possible to print an object in the shape of a kidney. However, such an object could not be self-sustaining or perform the task of a kidney. Kidney or any other organ requires a network of capillaries that distribute blood through the whole volume of the organ providing nourishment to the cells and removing waste from them. Further, an organ like the kidney requires a network of nephrons that regulates the complex composition of the blood. The normal kidney contains about a million nephrons. The 3-D printer would therefore be required to print within the kidney cells, the complex network of vascular and nephron connections. Such a task is highly complex and at this point far from the state of the art of 3-D printing. The workers in this field are optimistic and predict that this can be accomplished within the next 10 years.

## 5.11 Joint replacement

Hip and knee *joint replacements* are now a common medical procedure. About 700,000 knee and 300,000 hip replacement operations are performed annually in the United States. This ratio is typical worldwide. About twice as many knee replacements as hip replacements are performed everywhere such procedures are offered. Ankle and shoulder joint replacements, while not as frequent, are also becoming routine. Currently, about 50,000 shoulder joint and 30,000 ankle joint replacements are performed annually in the United States. Joint replacements are done usually to relieve pain, and correct motion limitations due to joint damage caused by disease, wear, and injuries. Successful joint replacements required the development of appropriate materials for the artificial joints and the attainment of surgical skills to insert these implements into the body.

### 5.11.1 Hip replacement

Surgical procedures attempting to correct hip joint disorders date to the late 1700s. Between the years 1800s and the 1950s, a number of surgeons experimented with a range of joint replacement techniques, developing new methodologies, and gaining experience with materials that could be adapted for artificial joints. Progress in the field was incremental but steady. However, till the work of John Charnley, joint replacements seldom lasted more than a year or so.

John Charnley, a British orthopedic surgeon, is credited with perfecting in the 1960s the modern hip replacement technique. He organized a fund-raising campaign for the Wrightington Hospital in Lancastershire, where he had a surgical appointment, to set up a hip surgical center. He built a laboratory for testing materials and tools as well as components for hip replacements. He was a remarkably dedicated physician working tirelessly and often testing the biocompatibility of candidate materials by injecting them into his own body. (A biocompatible material is nontoxic and not rejected by the body.)

The hip joint is essentially a ball-and-socket structure. The ball is at the head of the femur (the large thigh bone) fitting into the concave socket in the pelvis (called acetabulum). In a healthy joint, cartilage tissue, coated with a layer of synovial fluid, separates the bones comprising the ball and socket of the joint. Pain and joint dysfunction arise when the cartilage is damaged or wears out resulting in bone-on-bone contact.

Charnley experimented and tested a range of materials and techniques and settled on the following procedure. In his full hip replacement technique, the head of the femur (which is usually damaged) is removed and a ball on a metallic rod is implanted into the femur and fixed with bone cement. (See Fig. 5.8.) A variety of metals have been used, among them titanium alloys and tantalum. Bone cement is not glue, rather it is a flexible Plexiglas-like substance that fills the space between the implant and the bone and forms a strong elastic bond between the implanted metal and the bone. (See Exercise 5.11.) The socket is machined out of high-molecular-weight polyethylene and implanted into the pelvis. The patient's own muscles and ligaments, which have been pushed aside during the surgical procedure, stabilize and control the joint.

A variety of metals has been used in joint transplant, among them stainless steel, titanium, and tantalum. Titanium (often alloyed with about 4% aluminum and vanadium) is now most commonly used in joint replacement procedures. Titanium has several advantages in this application. It is

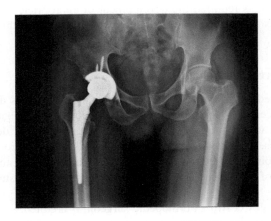

**FIGURE 5.8  Hip joint implant.** *(Source: http://ChooChin/Shutterstock.com.)*

biocompatible, strong, relatively light, resists corrosion, and has elastic properties (Young's modulus), most closely resembling that of bone. This last property is particularly important. As a result of applied forces, the bone as well as the metal implant bend. The smaller the difference in the amount of bending between the bone and the implanted metal, the less is the strain on the bone-metal bond and therefore less the likelihood of bond failure.

Postoperative infections were a serious problem in early joint replacement procedures. Here again, Charnley's contributions were key to reducing the problem. First, he introduced antibiotics into the bone cement. The medication was released as the bone cement set. This method helped but did not solve the problem. Charnley then focused on radical antiseptic procedures during the operation, including ventilated negative pressure surgical gowns for the staff, and most advanced air filtration available in the operating room. In its basic form, Charnley's full hip replacement technique remains in use to this day. Typically, depending on the nature of the patient's activities, a successful hip replacement lasts 20 years or more.

### 5.11.2 Knee replacement

The knee is the largest joint in the human body and is significantly more complex than the hip. After hip replacement techniques were perfected, it took another fifteen years or so before routinely successful knee replacements were possible.

Three bones are involved in a total knee replacement: The lower part of the thighbone (femur), the upper part of the tibia (shin bone), which is the larger and stronger of the two lower leg bones and bears the weight of the body, and the patella (knee cap), which protects the knee joint as a whole. These three bones are the ones usually worn and damaged. The second bone in the lower leg, the fibula, is the other bone in the shin. It is thinner than the tibia and its role is to stabilize the ankle and the knee joint configuration, rather than bear weight. Usually, it is not involved in knee replacement. Between the femur and the tibia are two C-shaped pieces of cartilage called menisci (singular, meniscus) coated with a thin layer of synovial fluid providing cushioning and smooth connection between the two bones. (See Exercise 5.12).

The basic knee replacement procedure is similar to hip replacement but more complex. Similar materials and antiseptic precautions are used. The knee joint is surgically opened and damaged cartilage bone surfaces are removed. Metal components are cemented and press fit into the ends of the prepared femur and tibia. A plastic high-density polyethylene spacer is placed between the metal covered bones to provide a smoothly gliding surface connection. If required, the patella is resurfaced with a plastic insert. As in hip replacement, the structure is held together with the person's own muscles and tendons that have been pushed out of the way during the surgical procedure. In the absence of complications, a knee replacement operation takes about two hours.

In spite of precautions, postoperative infections occur in about 1% of joint replacements. Such cases require medical interventions that may range from a simple course of antibiotics to redoing the joint replacement. The average risk of infection varies somewhat from hospital to hospital as well as from person to person. (See Exercise 5.13.)

### 5.11.3 Osseointegration

A process, now called *osseointegration*, was first reported in 1940 when it was observed that bone tended to fuse with titanium, a bioinert mechanically strong metal. While several researchers noted the possible medical application of the phenomenon, it was not till the 1960s that such applications were implemented. A key researcher in this field was Per-Ingvar Brånemark, a Swedish orthopedic surgeon, who first applied the process to dentistry. A titanium post is threaded into the jaw bone. After a period of typically 3 months, the bone integrates with the titanium rod, and the implant is completed by finishing the tooth with a crown. Such implants have a typical lifespan of 10 to 15 years. It is estimated that in the United States currently 2 to 3 million dental implants are inserted annually.

Although the basic nature of osseointegration is now known, the details of the process are not yet fully understood. The surface of the implant material (in most cases titanium) must be somewhat porous. Pores about 100 μm in size with nanoscale roughening promote strong osseointegration. The integration of bone to implant begins with proteins adhering to the implant surface. This is followed by undifferentiated cells aggregating at the interface. These cells become bone producing and form an integrated interface between bone and implant. At the forefront of current research in this field is the elucidation of the effect of surface preparation and coating on the strength of the implant-bone integration.

### 5.11.4 Attachment of artificial limbs

Conventional prosthetic limbs are attached to the remaining limb via a cup in contact with the stump of the amputated limb. In such an arrangement, the forces of the prostheses are applied to the soft tissue of the stump. Some people with amputations cannot tolerate such an arrangement. The cup in many cases causes abrasions resulting in recurrent infections. Further, in some amputations, the remaining stump is too short to support an artificial limb. In the 1990s, osseointegration techniques were applied to attaching artificial limbs to the remaining stump. Where conventional prosthetic attachment methods fail, osseointegration techniques have become important and transformative.

Osseointegration to support an artificial limb is usually a two-step process (although single-step techniques are being developed). First, the bone of the stump is exposed and a titanium implant is threaded into the bone with part of the post remaining outside the bone. The stump is then temporarily closed by the skin flap. Osseointegration occurs in 3 to 6 months. After the integration of the implant is deemed satisfactory, the soft tissue at the end of the stump is cut open and the prosthesis is attached to the protruding metal post. With this technique, the forces exerted by the prosthesis are borne by the bone rather than the tissue, thus eliminating the abrasive action of a conventional cup-connected prosthetic device. Users report that the prosthesis feels more natural and more easily controlled. There are also serious disadvantages to the technique—among them, the long duration of the insertion process and the omnipresent risk of infection in the region where the metal post protrudes through the skin. Still, for some people, the introduction of this new technique provides a viable alternative.

### 5.11.5 Osseointegration in joint replacement

As was discussed, the most common way to secure metal components to bone in conventional joint replacement procedures is using bone cement. More recently, some surgeons have substituted osseointegration for that purpose. Although bone cement has the advantage of drying quickly, within about 10 minutes, and can be used even with bones weakened by osteoporosis, some surgeons have switched to osseointegration because they feel that cementless attachment is likely to be longer lasting and will eliminate potential problems that might be caused by breakdown or flaking of aged cement.

## Exercises

5.1. Assume that a 50-kg runner trips and falls on his extended hand. If the bones of one arm absorb all the kinetic energy (neglecting the energy of the fall), what is the minimum speed of the runner that will cause a fracture of the arm bone? Assume that the length of arm is 1 m and that the area of the bone is 4 cm$^2$.

5.2. Repeat the calculations in Exercise 5.1 using impulsive force considerations. Assume that the duration of impact is $10^{-2}$ seconds and the area of impact is 4 cm$^2$. Repeat the calculation with area of impact $= 1$ cm$^2$.

5.3. From what height can a 1-kg falling object cause fracture of the skull? Assume that the object is hard, that the area of contact with the skull is 1 cm$^2$, and that the duration of impact is $10^{-3}$ seconds.

5.4. Calculate the duration of the collision between the passenger and the inflated bag of the collision protection device discussed in this chapter.

5.5. In a rear-end collision, the automobile that is hit is accelerated to a velocity v in $10^{-2}$/sec. What is the minimum velocity at which there is danger of neck fracture from whiplash? Use the data provided in the text, and assume that the area of the cervical vertebra is 1 cm$^2$ and the mass of the head is 5 kg.

**5.6.** Calculate the average decelerating impact force if a person falling with a terminal velocity of 62.5 m/sec is decelerated to zero velocity over a distance of 1 m. Assume that the person's mass is 70 kg and that she lands flat on her back so that the area of impact is 0.3 m$^2$. Is this force below the level for serious injury? (For body tissue, this is about $5 \times 10^6$ dyn/cm$^2$.)

**5.7.** A boxer punches a 50-kg bag. Just as his fist hits the bag, it travels at a speed of 7 m/sec. As a result of hitting the bag, his hand comes to a complete stop. Assuming that the moving part of his hand weighs 5 kg, calculate the rebound velocity and kinetic energy of the bag. Is kinetic energy conserved in this example? Why? (Use conservation of momentum.)

**5.8.** Describe the process of hardening of a photopolymer by ultraviolet light. Good relatively simple descriptions are found on the internet. In the process of learning about photohardening, you will encounter some chemistry terms that may be new to you. Learn their meaning before proceeding.

**5.9.** What are some of the anticipated future applications of 3-D printing?

**5.10.** What are some of the societal concerns about 3-D printing?

**5.11.** What is the difference between bone cement and glue? In the process of answering this question, explain the bonding action of glue and that of bone cement.

**5.12.** Referring to material in Chapter 2, design an experiment to measure the coefficient of friction between high-molecular-weight polyethylene and titanium.

**5.13.** (a) Some people are more at risk of postoperative joint replacement infections. What factors increase the personal risk of infection? (b) Does the percentage of postoperative infections vary from hospital to hospital? If so, why?

# Chapter 6

# Insect flight

In this chapter, we will analyze some aspects of insect flight. In particular, we will consider the hovering flight of insects, using many of the concepts introduced in the previous chapters in our calculations. The parameters required for the computations were in most cases obtained from the literature, but some had to be estimated because they were not readily available. The size, shape, and mass of insects vary widely. We will perform our calculations for an insect with a mass of 0.1 g, which is about the size of a bee.

In general, the flight of birds and insects is a complex phenomenon. A complete discussion of flight would take into account aerodynamics as well as the changing shape of the wings at the various stages of flight. Differences in wing movements between large and small insects have only recently been demonstrated. The following discussion is highly simplified but nevertheless illustrates some of the basic physics of insect flight.

## 6.1 Hovering flight

Many insects (and also some small birds) can beat their wings so rapidly that they are able to hover in air over a fixed spot. The wing movements in a hovering flight are complex. The wings are required to provide sideways stabilization as well as the lifting force necessary to overcome the force of gravity. The lifting force results from the downward stroke of the wings. As the wings push down on the surrounding air, the resulting reaction force of the air on the wings forces the insect up. The wings of most insects are designed so that during the upward stroke, the force on the wings is small. The lifting force acting on the wings during the wing movement is shown in Fig. 6.1. During the upward movement of the wings, the gravitational force causes the insect to drop. The downward wing movement then produces an upward force that restores the insect to its original position. The vertical position of the insect thus oscillates up and down at the frequency of the wingbeat.

The distance the insect falls between wingbeats depends on how rapidly its wings are beating. If the insect flaps its wings at a slow rate, the time interval during which the lifting force is zero is longer, and therefore the insect falls farther than if its wings were beating rapidly.

We can easily compute the wingbeat frequency necessary for the insect to maintain a given stability in its amplitude. To simplify the calculations, let us

Physics in Biology and Medicine. https://doi.org/10.1016/B978-0-443-21558-2.00006-7

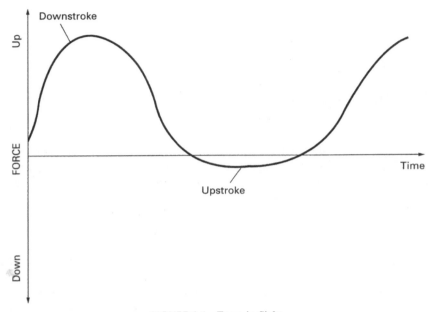

FIGURE 6.1    **Force in flight.**

assume that the lifting force is at a finite constant value while the wings are moving down and that it is zero while the wings are moving up. During the time interval $\Delta t$ of the upward wingbeat, the insect drops a distance h under the action of gravity. From Eq. 3.5, this distance is

$$h = \frac{g(\Delta t)^2}{2} \tag{6.1}$$

The upward stroke then restores the insect to its original position. Typically, it may be required that the vertical position of the insect change by no more than 0.1 mm (i.e., h = 0.1 mm). The maximum allowable time for free fall is then

$$\Delta t = \left(\frac{2h}{g}\right)^{1/2} = \sqrt{\frac{2 \times 10^{-2} \text{ cm}}{980 \text{ cm/sec}^2}} = 4.5 \times 10^{-3} \text{ seconds}$$

Since the up movements and the down movements of the wings are about equal in duration, the period T for a complete up-and-down wing movement is twice $\Delta t$; that is,

$$T = 2\Delta t = 9 \times 10^{-3} \text{ seconds} \tag{6.2}$$

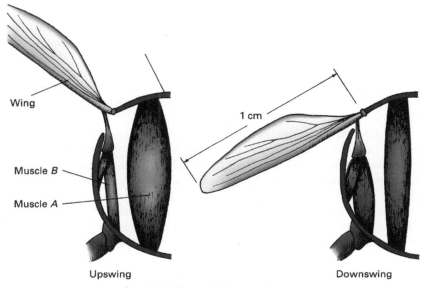

**FIGURE 6.2** **Wing muscles.**

The frequency of wingbeats f, that is, the number of wingbeats per second, is

$$f = \frac{1}{T} \tag{6.3}$$

In our example this frequency is 110 wingbeats per second. This is a typical insect wingbeat frequency, although some insects such as butterflies fly at a much lower frequency, about 10 wingbeats per second (they cannot hover), and other small insects produce as many as 1000 wingbeats per second. To restore the vertical position of the insect during the downward wing stroke, the average upward force, $F_{av}$ on the body of the insect must be equal to twice the weight of the insect (see Exercise 6.1). Note that since the upward force on the insect body is applied only for half the time, the average upward force on the insect is simply its weight.

## 6.2 Insect wing muscles

A number of different wing-muscle arrangements occur in insects. One arrangement, found in the dragonfly, is shown, in a highly simplified manner, in Fig. 6.2. The wing movement is controlled by many muscles, which are here represented by muscles $A$ and $B$. The upward movement of the wings is produced by the contraction of muscle $A$, which depresses the upper part of the thorax and causes the attached wings to move up. While muscle $A$ contracts, muscle $B$ is relaxed. Note that the force produced by muscle $A$ is applied to the wing by means of a Class 1 lever. The fulcrum here is the wing joint marked by the small circle in Fig. 6.2.

The downward wing movement is produced by the contraction of muscle $B$ while muscle $A$ is relaxed. Here the force is applied to the wings by means of a Class 3 lever. In our calculations, we will assume that the length of the wing is 1 cm.

The physical characteristics of insect flight muscles are not peculiar to insects. The amount of force per unit area of the muscle and the rate of muscle contraction are similar to the values measured for human muscles. Yet insect wing muscles are required to flap the wings at a very high rate. This is made possible by the lever arrangement of the wings. Measurements show that during a wing swing of about 70°, muscles $A$ and $B$ contract only about 2%. Assuming that the length of muscle $B$ is 3 mm, the change in length during the muscle contraction is 0.06 mm (this is 2% of 3 mm). It can be shown that under these conditions, muscle $B$ must be attached to the wing 0.052 mm from the fulcrum to achieve the required wing motion (see Exercise 6.2).

If the wingbeat frequency is 110 wingbeats per second, the period for one up-and-down motion of the wings is $9 \times 10^{-3}$ seconds. The downward wing movement produced by muscle $B$ takes half this length of time, or $4.5 \times 10^{-3}$ seconds. Thus, the rate of contraction for muscle $B$ is 0.06 mm divided by $4.5 \times 10^{-3}$ seconds, or 13 mm/sec. Such a rate of muscle contraction is commonly observed in many types of muscle tissue.

## 6.3 Power required for hovering

We will now compute the power required to maintain hovering. Let us consider again an insect with mass $= 0.1$ g. As is shown in Exercise 6.1, the average force, $F_{av}$, applied by the two wings during the downward stroke is $2W$. Because the pressure applied by the wings is uniformly distributed over the total wing area, we can assume that the force generated by each wing acts through a single point at the midsection of the wings. During the downward stroke, the center of the wings traverses a vertical distance d (see Fig. 6.3). The total work done by the insect during each downward stroke is the product of force and distance; that is,

$$\text{Work} = F_{av} \times d = 2Wd \tag{6.4}$$

If the wings swing through an angle of 70°, then in our case, for the insect with 1-cm-long wings, d is 0.57 cm. Therefore, the work done during each stroke by the two wings is

$$\text{Work} = 2 \times 0.1 \times 980 \times 0.57 = 112 \text{ erg}$$

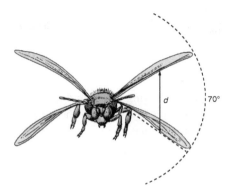

**FIGURE 6.3** **Insect wing motion.**

Let us now examine where this energy goes. In our example, the mass of the insect has to be raised 0.1 mm during each downstroke. The energy E required for this task is

$$E = mgh = 0.1 \times 980 \times 10^{-2} = 0.98 \text{ erg} \qquad (6.5)$$

This is a negligible fraction of the total energy expended. Clearly, most of the energy is expended on other processes. A more detailed analysis of the problem shows that the work done by the wings is converted primarily into kinetic energy of the wings and of the air that is accelerated by the downward stroke of the wings.

*Power* is the amount of work done in 1 second. Our insect makes 110 downward strokes per second; therefore, its power output P is

$$P = 112 \text{ erg} \times 110/\text{sec} = 1.23 \times 10^{4} \text{ erg}/\text{sec} = 1.23 \times 10^{-3} \text{ W} \qquad (6.6)$$

## 6.4 Kinetic energy of wings in flight

In our calculation of the power used in hovering, we have neglected the kinetic energy of the moving wings. The wings of insects, as light as they are, have a finite mass; therefore, as they move, they possess kinetic energy. Because the wings are in rotary motion, the maximum kinetic energy during each wing stroke is

$$KE = \frac{1}{2} I \omega_{max}^{2} \qquad (6.7)$$

Here, I is the moment of inertia of the wing and $\omega_{max}$ is the maximum angular velocity during the wing stroke. To obtain the moment of inertia for the wing, we will assume that the wing can be approximated by a thin rod pivoted at one end. The moment of inertia for the wing is then

$$I = \frac{m\ell^3}{3} \tag{6.8}$$

where $\ell$ is the length of the wing (1 cm in our case) and m is the mass of two wings, which may be typically $10^{-3}$ g. The maximum angular velocity $\omega_{max}$ can be calculated from the maximum linear velocity $v_{max}$ at the center of the wing

$$\omega_{max} = \frac{v_{max}}{\ell/2} \tag{6.9}$$

During each stroke, the center of the wings moves with an average linear velocity $v_{av}$ given by the distance d traversed by the center of the wing divided by the duration $\Delta t$ of the wing stroke. From our previous example, $d = 0.57$ cm and $\Delta t = 4.5 \times 10^{-3}$ seconds. Therefore,

$$v_{av} = \frac{d}{\Delta t} = \frac{0.57}{4.5 \times 10^{-3}} = 127 \text{ cm/sec} \tag{6.10}$$

The velocity of the wings is zero both at the beginning and at the end of the wing stroke. Therefore, the maximum linear velocity is higher than the average velocity. If we assume that the velocity varies sinusoidally along the wing path, the maximum velocity is twice as high as the average velocity. Therefore, the maximum angular velocity is

$$\omega_{max} = \frac{254}{\ell/2}$$

The kinetic energy is

$$KE = \frac{1}{2}I\omega_{max}^2 = \frac{1}{2}\left(10^{-3}\frac{\ell^2}{3}\right)\left(\frac{254}{\ell/2}\right)^2 = 43 \text{ erg}$$

Since there are two wing strokes (up and down) in each cycle of the wing movement, the kinetic energy is $2 \times 43 = 86$ erg. This is about as much energy as is consumed in hovering itself.

## 6.5 Elasticity of wings

As the wings are accelerated, they gain kinetic energy, which is of course provided by the muscles. When the wings are decelerated toward the end of the stroke, this energy must be dissipated. During the downstroke, the kinetic energy is dissipated by the muscles themselves and is converted into heat. (This heat is used to maintain the required body temperature of the insect.) Some insects are able to utilize the kinetic energy in the upward movement of

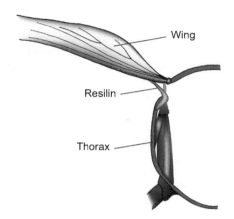

**FIGURE 6.4** **Resilin in the wing.**

the wings to aid in their flight. The wing joints of these insects contain a pad of elastic, rubber-like protein called *resilin* (Fig. 6.4). During the upstroke of the wing, the resilin is stretched. The kinetic energy of the wing is converted into potential energy in the stretched resilin, which stores the energy much like a spring. When the wing moves down, this energy is released and aids in the downstroke.

Using a few simplifying assumptions, we can calculate the amount of energy stored in the stretched resilin. Although the resilin is bent into a complex shape, we will assume in our calculation that it is a straight rod of area A and length $\ell$. Furthermore, we will assume that throughout the stretch, the resilin obeys Hooke's law. This is not strictly true as the resilin is stretched by a considerable amount and therefore both the area and Young's modulus change in the process of stretching.

The energy E stored in the stretched resilin is, from Eq. 5.9,

$$E = \frac{1}{2} \frac{YA\Delta\ell^2}{\ell} \tag{6.11}$$

Here, Y is the Young's modulus for resilin, which has been measured to be $1.8 \times 10^7 \ \text{dyn/cm}^2$.

Typically, in an insect the size of a bee the volume of the resilin may be equivalent to a cylinder $2 \times 10^{-2}$ cm long and $4 \times 10^{-4} \ \text{cm}^2$ in area. We will assume that the length of the resilin rod is increased by 50% when stretched. That is, $\Delta\ell$ is $10^{-2}$ cm. Therefore in our case the energy stored in the resilin of each wing is

$$E = \frac{1}{2} \frac{1.8 \times 10^7 \times 4 \times 10^{-4} \times 10^{-4}}{2 \times 10^{-2}} = 18 \ \text{erg}$$

The stored energy in the two wings is 36 erg, which is comparable to the kinetic energy in the upstroke of the wings. Experiments show that as much as 80% of the kinetic energy of the wing may be stored in the resilin. The utilization of resilin is not restricted to wings. The hind legs of the flea, for example, also contain resilin, which stores energy for jumping (see Exercise 6.3). A further application of energy storage in resilin is examined in Exercise 6.4.

## Exercises

**6.1.** Compute the force on the body of the insect that must be generated during the downward wing stroke to keep the insect hovering.

**6.2.** Referring to the discussion in the text, compute the point of attachment to the wing of muscle $B$ in Fig. 6.2. Assume that the muscle is perpendicular to the wing throughout the wing motion.

**6.3.** Assume that the shape of the resilin in each leg of the flea is equivalent to a cylinder $2 \times 10^{-2}$ cm long and $10^{-4}$ cm$^2$ in area. If the change in the length of the resilin is $\Delta\ell = 10^{-2}$ cm, calculate the energy stored in the resilin. The flea weighs $0.5 \times 10^{-3}$ g. How high can the flea jump utilizing only the stored energy?

**6.4.** Suppose that a 50-kg person was equipped with resilin pads in her joints. How large would these pads have to be in order for them to store enough energy for a $\frac{1}{2}$ m jump? Assume that the pad is cubic in shape and $\Delta\ell = \frac{1}{2}\ell$.

# Chapter 7

# Fluids

In the previous chapters, we have examined the behavior of solids under the action of forces. In the next three chapters, we will discuss the behavior of liquids and gases, both of which play an important role in the life sciences. The differences in the physical properties of solids, liquids, and gases are explained in terms of the forces that bind the molecules. In a solid, the molecules are rigidly bound; a solid therefore has a definite shape and volume. The molecules constituting a liquid are not bound together with sufficient force to maintain a definite shape, but the binding is sufficiently strong to maintain a definite volume. A liquid adapts its shape to the vessel in which it is contained. In a gas, the molecules are not bound to each other. Therefore a gas has neither a definite shape nor a definite volume—it completely fills the vessel in which it is contained. Both gases and liquids are free to flow and are called *fluids*. Fluids and solids are governed by the same laws of mechanics, but because of their ability to flow, fluids exhibit some phenomena not found in solid matter. In this chapter we will illustrate the properties of fluid pressure, buoyant force in liquids, and surface tension with examples from biology and zoology.

## 7.1 Force and pressure in a fluid

Solids and fluids transmit forces differently. When a force is applied to one section of a solid, this force is transmitted to the other parts of the solid with its direction unchanged. Because of a fluid's ability to flow, it transmits a force uniformly in all directions. Therefore, the pressure at any point in a fluid at rest is the same in all directions. The force exerted by a fluid at rest on any area is perpendicular to the area. A fluid in a container exerts a force on all parts of the container in contact with the fluid. A fluid also exerts a force on any object immersed in it.

The pressure in a fluid increases with depth because of the weight of the fluid above. In a fluid of constant density $\rho$, the difference in pressure, $P_2 - P_1$, between two points separated by a vertical distance h is

$$P_2 - P_1 = \rho g h \qquad (7.1)$$

Fluid pressure is often measured in millimeters of mercury, or *torr* (after Evangelista Torricelli [1608−1674], the first person to understand the nature of atmospheric pressure). One torr is the pressure exerted by a column of

Physics in Biology and Medicine. https://doi.org/10.1016/B978-0-443-21558-2.00007-9

mercury that is 1 mm high. *Pascal*, abbreviated as Pa, is another commonly used unit of pressure. The relationship between the torr and several of the other units used to measure pressure follows:

$$
\begin{aligned}
1 \text{ torr} &= 1 \text{ mm Hg} \\
&= 13.5 \text{ mm water} \\
&= 1.33 \times 10^3 \text{ dyn/cm}^2 \\
&= 1.32 \times 10^{-3} \text{ atm} \\
&= 1.93 \times 10^{-2} \text{ psi} \\
&= 1.33 \times 10^2 \text{ Pa } \left(\text{N/m}^2\right)
\end{aligned}
\tag{7.2}
$$

## 7.2 Pascal's principle

When a force $F_1$ is applied on a surface of a liquid that has an area $A_1$, the pressure in the liquid increases by an amount P (see Fig. 7.1), given by

$$
P = \frac{F_1}{A_1}
\tag{7.3}
$$

In an incompressible liquid, the increase in the pressure at any point is transmitted undiminished to all other points in the liquid. This is known as *Pascal's principle*. Because the pressure throughout the fluid is the same, the force $F_2$ acting on the area $A_2$ in Fig. 7.1 is

$$
F_2 = PA_2 = \frac{A_2}{A_1}F_1
\tag{7.4}
$$

The ratio $A_2/A_1$ is analogous to the mechanical advantage of a lever.

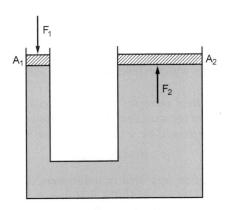

FIGURE 7.1   **An illustration of Pascal's principle.**

## 7.3 Hydrostatic skeleton

We showed in Chapter 1 that muscles produce movement by pulling on the bones of the skeleton. There are, however, soft-bodied animals (such as the sea anemone and the earthworm) that lack a firm skeleton. Many of these animals utilize Pascal's principle to produce body motion. The structure by means of which this is done is called the *hydrostatic skeleton*.

For the purpose of understanding the movements of an animal such as a worm, we can think of the animal as consisting of a closed elastic cylinder filled with a liquid; the cylinder is its hydrostatic skeleton. The worm produces its movements with the longitudinal and circular muscles running along the walls of the cylinder (see Fig. 7.2). Because the volume of the liquid in the cylinder is constant, contraction of the circular muscles makes the worm thinner and longer. Contraction of the longitudinal muscles causes the animal to become shorter and fatter. If the longitudinal muscles contract only on one side, the animal bends toward the contracting side. By anchoring alternate ends of its body to a surface and by producing sequential longitudinal and circular contractions, the animal moves itself forward or backward. Longitudinal contraction on one side changes the direction of motion.

Let us now calculate the hydrostatic forces inside a moving worm. Consider a worm that has a radius r. Assume that the circular muscles running around its circumference are uniformly distributed along the length of the worm and that the effective area of the muscle per unit length of the worm is $A_M$. As the circular muscles contract, they generate a force $f_M$, which, along each centimeter of the worm's length, is

$$f_M = SA_M \tag{7.5}$$

Here S is the force produced per unit area of the muscle. (Note that $f_M$ is in units of force per unit length.) This force produces a pressure inside the worm.

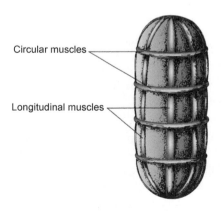

Circular muscles

Longitudinal muscles

FIGURE 7.2   The hydrostatic skeleton.

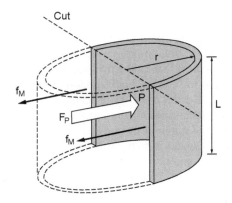

FIGURE 7.3  **Calculating pressure inside a worm.**

The magnitude of the pressure can be calculated with the aid of Fig. 7.3, which shows a section of the worm. The length of the section is L. If we were to cut this section in half lengthwise, as shown in Fig. 7.3, the force due to the pressure inside the cylinder would tend to push the two halves apart. This force is calculated as follows. The surface area A along the cut midsection is

$$A = L \times 2r \tag{7.6}$$

Because fluid pressure always acts perpendicular to a given surface area, the force $F_P$ that tends to split the cylinder is

$$F_P = P \times A = P \times L \times 2r \tag{7.7}$$

Here P is the fluid pressure produced inside the worm by contraction of the circular muscles.

In equilibrium, the force $F_P$ is balanced by the muscle forces acting along the two edges of the imaginary cut. Therefore,

$$F_P = 2f_M L$$

or

$$P \times L \times 2r = 2f_M L$$

and

$$P = \frac{f_M}{r} \tag{7.8}$$

To make the calculations specific, let us assume that the radius of the worm r is 0.4 cm, the area of the circular muscles per centimeter length of the worm is $A_M = 1.5 \times 10^{-3}$ cm$^2$, and S, the maximum force generated per unit area of the muscle, is $7 \times 10^6$ dyn/cm$^2$. (This is the value we used previously for human muscles.) Therefore, the pressure inside the worm under maximum contraction of the circular muscles is

$$P = \frac{f_M}{r} = \frac{SA_M}{r} = \frac{7 \times 10^6 \times 1.5 \times 10^{-3}}{0.4}$$

$$= 2.63 \times 10^4 \ dyn/cm^2 = 19.8 \ torr$$

This is a relatively high pressure. It can raise a column of water to a height of 26.7 cm. The force $F_f$ in the forward direction generated by this pressure, which stretches the worm, is

$$F_f = P \times \pi r^2 = 1.32 \times 10^4 \ dyn$$

The action of the longitudinal muscles can be similarly analyzed.

## 7.4 Archimedes' principle

Archimedes' principle states that a body partially or wholly submerged in a fluid is buoyed upward by a force that is equal in magnitude to the weight of the displaced fluid. The derivation of this principle is found in basic physics texts. We will now use Archimedes' principle to calculate the power required to remain afloat in water and to study the buoyancy of fish.

## 7.5 Power required to remain afloat

Whether an animal sinks or floats in water depends on its density. If its density is greater than that of water, the animal must perform work in order not to sink. We will calculate the power P required for an animal of volume V and density $\rho$ to float with a fraction f of its volume submerged. This problem is similar to the hovering flight we discussed in Chapter 6, but our approach to the problem will be different.

Because a fraction f of the animal is submerged, the animal is buoyed up by a force $F_B$ given by

$$F_B = gfV\rho_w \tag{7.9}$$

where $\rho_w$ is the density of water. The force $F_D$ is simply the weight of the displaced water.

The net downward force $F_D$ on the animal is the difference between its weight $gV\rho$ and the buoyant force; that is,

$$F_D = gV\rho - gVf\rho_w = gV(\rho - f\rho_w) \tag{7.10}$$

To keep itself floating, the animal must produce an upward force equal to $F_D$. This force can be produced by pushing the limbs downward against the water. This motion accelerates the water downward and results in the upward reaction force that supports the animal.

If the area of the moving limbs is $A$ and the final velocity of the accelerated water is v, the mass of water accelerated per unit time in the treading motion is given by (see Exercise 7.1)

$$m = Av\rho_w \qquad (7.11)$$

Because the water is initially stationary, the amount of momentum imparted to the water each second is mv. (Remember that here m is the mass accelerated *per second*.)

$$\text{Momentum given to the water per second} = mv$$

This is the rate of change of momentum of the water. The force producing this change in the momentum is applied to the water by the moving limbs. The upward reaction force $F_R$, which supports the weight of the swimmer, is equal in magnitude to $F_D$ and is given by

$$F_R = F_D = gV(\rho - f\rho_w) = mv \qquad (7.12)$$

Substituting Eq. 7.11 for m, we obtain

$$\rho_w Av^2 = gV(\rho - f\rho_w)$$

or

$$v = \sqrt{\frac{gV(\rho - f\rho_w)}{A\rho_w}} \qquad (7.13)$$

The work done by the treading limbs goes into the kinetic energy of the accelerated water. The kinetic energy given to the water each second is half the product of the mass accelerated each second and the squared final velocity of the water. This kinetic energy imparted to the water each second is the power generated by the limbs; that is,

$$\text{KE/sec} = \text{Power generated by the limbs}, P = \frac{1}{2}mv^2$$

Substituting equations for m and v, we obtain (see Exercise 7.1)

$$P = \frac{1}{2}\sqrt{\frac{\left[W\left(1 - \frac{f\rho_w}{\rho}\right)\right]^3}{A\rho_w}} \qquad (7.14)$$

Here W is the weight of the animal ($W = gV\rho$).

It is shown in Exercise 7.2 that a 50-kg woman expends about 7.8 W to keep her nose above water. Note that, in our calculation, we have neglected the kinetic energy of the moving limbs. In Eq. 7.14 it is assumed that the density of the animal is greater than the density of water. The reverse case is examined in Exercise 7.3.

## 7.6 Buoyancy of aquatic animals

The bodies of some fish and other aquatic animals contain porous bones or air-filled swim bladders that decrease their average density and allow them to float in water without an expenditure of energy. The body of the cuttlefish, for example, contains a porous bone that has a density of 0.62 g/cm$^3$. The rest of its body has a density of 1.067 g/cm$^3$. We can find the percentage of the body volume X occupied by the porous bone that makes the average density of the fish the same as the density of sea water $(1.026 \text{ g/cm}^3)$ by using the following equation (see Exercise 7.4):

$$1.026 = \frac{0.62\text{X} + (100 - \text{X})\,1.067}{100} \tag{7.15}$$

In this case X = 9.2%.

The cuttlefish lives in the sea at a depth of about 150 m. At this depth, the pressure is 15 atm (see Exercise 7.5). The spaces in the porous bone are filled with gas at a pressure of about 1 atm. Therefore, the porous bone must be able to withstand a pressure of 14 atm. Experiments have shown that the bone can in fact survive pressures up to 24 atm.

In fish that possess swim bladders, the decrease in density is provided by the gas in the bladder. Because the density of the gas is negligible compared to the density of tissue, the volume of the swim bladder required to reduce the density of the fish is smaller than that of the porous bone. For example, to achieve the density reduction calculated in the preceding example, the volume of the bladder is only about 4% of the total volume of the fish (see Exercise 7.6).

Aquatic animals possessing porous bones or swim bladders can alter their density. The cuttlefish alters its density by injecting or withdrawing fluid from its porous bone. Fish with swim bladders alter their density by changing the amount of gas in the bladder. Another application of buoyancy is examined in Exercise 7.7.

## 7.7 Surface tension

The molecules constituting a liquid exert attractive forces on each other. A molecule in the interior of the liquid is surrounded by an equal number of neighboring molecules in all directions. Therefore, the net resultant intermolecular force on an interior molecule is zero. The situation is different, however, near the surface of the liquid. Because there are no molecules above the surface, a molecule here is pulled predominantly in one direction, toward the interior of the surface. This causes the surface of a liquid to contract and behave somewhat like a stretched membrane. This contracting tendency results in a surface tension that resists an increase in the free surface of the liquid. It can be shown (see Morgan, 1969) that surface tension is a force acting tangential to the surface, normal to a line of unit length on the surface (Fig. 7.4). The surface tension T of water at 25°C is 72.8 dyn/cm. The total force $F_T$ produced by surface tension tangential to a liquid surface of boundary length L is

$$F_T = TL \tag{7.16}$$

When a liquid is contained in a vessel, the surface molecules near the wall are attracted to the wall. This attractive force is called *adhesion*. At the same time, however, these molecules are also subject to the attractive cohesive force exerted by the liquid, which pulls the molecules in the opposite direction. If the adhesive force is greater than the cohesive force, the liquid wets the container wall, and the liquid surface near the wall is curved upward. If the opposite is the case, the liquid surface is curved downward (see Fig. 7.5). The angle $\theta$ in Fig. 7.5 is the angle between the wall and the tangent to the liquid surface at the point of contact with the wall. For a given liquid and surface material, $\theta$ is a well-defined constant. For example, the contact angle between glass and water is 25°.

If the adhesion is greater than the cohesion, a liquid in a narrow tube will rise to a specific height h (see Fig. 7.6a), which can be calculated from the following considerations. The weight W of the column of the supported liquid is

$$W = \pi R^2 h \rho g \tag{7.17}$$

where R is the radius of the column and $\rho$ is the density of the liquid. The maximum force $F_m$ due to the surface tension along the periphery of the liquid is

$$F_m = 2\pi RT \tag{7.18}$$

The upward component of this force supports the weight of the column of liquid (see Fig. 7.6a); that is,

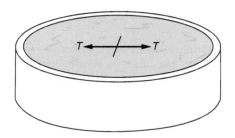

FIGURE 7.4   **Surface tension.**

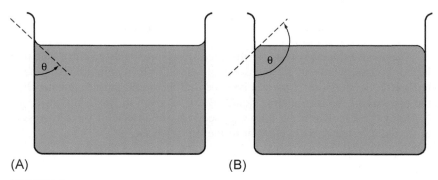

**FIGURE 7.5** Angle of contact when (A) liquid wets the wall and (B) liquid does not wet the wall.

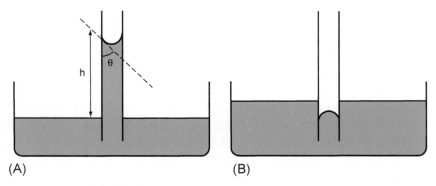

**FIGURE 7.6** (A) Capillary rise. (B) Capillary depression.

$$2\pi RT\cos\theta = \pi R^2 h\rho g \qquad (7.19)$$

Therefore, the height of the column is

$$h = \frac{2T\cos\theta}{R\rho g} \qquad (7.20)$$

If the adhesion is smaller than the cohesion, the angle $\theta$ is greater than 90°. In this case, the height of the fluid in the tube is depressed (Fig. 7.6b). Eq. 7.20 still applies, yielding a negative number for h. These effects are called *capillary action*.

Another consequence of surface tension is the tendency of liquid to assume a spherical shape. This tendency is most clearly observed in a liquid outside a container. Such an uncontained liquid forms into a sphere that can be noted in the shape of raindrops. The pressure inside the spherical liquid drop is higher than the pressure outside. The excess pressure $\Delta P$ in a liquid sphere of radius R is

$$\Delta P = \frac{2T}{R} \qquad (7.21)$$

This is also the expression for the excess pressure inside an air bubble in a liquid. In other words, to create gas bubble of radius R in a liquid with surface tension T, the pressure of the gas injected into the liquid must be greater than the pressure of the surrounding liquid by $\Delta P$ as given in Eq. 7.21.

As will be shown in the following sections, the effects of surface tension are evident in many areas relevant to the life sciences.

## 7.8 Soil water

Most soil is porous, with narrow spaces between the small particles. These spaces act as capillaries and in part govern the motion of water through the soil. When water enters soil, it penetrates the spaces between the small particles and adheres to them. If the water did not adhere to the particles, it would run rapidly through the soil until it reached solid rock. Plant life would then be severely restricted. Because of adhesion and the resulting capillary action, a significant fraction of the water that enters the soil is retained by it. For a plant to withdraw this water, the roots must apply a negative pressure, or suction, to the moist soil. The required negative pressure may be quite high. For example, if the effective capillary radius of the soil is $10^{-3}$ cm, the pressure required to withdraw the water is $1.46 \times 10^5$ dyn/cm$^2$, or 0.144 atm (see Exercise 7.8).

The pressure required to withdraw water from the soil is called the *soil moisture tension* (SMT). The SMT depends on the grain size of the soil, its moisture content, and the material composition of the soil. The SMT is an important parameter in determining the quality of the soil. The higher the SMT, the more difficult it is for the roots to withdraw the water necessary for plant growth.

The dependence of the SMT on the grain size can be understood from the following considerations. The spaces between the particles of soil increase with the size of the grains. Because capillary action is inversely proportional to the diameter of the capillary, finely grained soil will hold water more tightly than soil of similar material with larger grains (see Fig. 7.7).

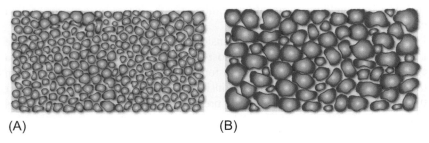

(A)                                    (B)

**FIGURE 7.7**   Fine-grained soil (A) holds water more tightly than coarse-grained soil (B).

When all the pores of the soil are filled with water, the surface moisture tension is at its lowest value. In other words, under these conditions, the required suction pressure produced by the plant roots to withdraw the water from the soil is the lowest. Saturated soil, however, is not the best medium for plant growth. The roots need some air, which is absent when the soil is fully saturated with water. As the amount of water in the soil decreases, the SMT increases. In loam, for example, with a moisture content of 20%, the SMT is about 0.19 atm. When the moisture content drops to 12%, the SMT increases to 0.76 atm.

The rise in SMT with decreasing moisture content can be explained in part by two effects. As the soil loses moisture, the remaining water tends to be bound into the narrower capillaries. Therefore the withdrawal of water becomes more difficult. In addition, as the moisture content decreases, sections of water become isolated and tend to form droplets. The size of these droplets may be very small. If, for example, the radius of a droplet decreases to $10^{-5}$ cm, the pressure required to draw the water out of the droplet is about 14.5 atm.

Capillary action also depends on the strength of adhesion, which in turn depends on the material composition of the capillary surface. For example, under similar conditions of grain size and moisture content, the SMT in clay may be ten times higher than in loam. There is a limit to the pressure that roots can produce in order to withdraw water from the soil. If the SMT increases above 15 atm, wheat, for example, cannot obtain enough water to grow. In hot, dry climates where vegetation requires more water, plants may wilt even at an SMT of 2 atm. The ability of a plant to survive depends not so much on the water content as it does on the SMT of the soil. A plant may thrive in loam and yet wilt in a clayey soil with twice the moisture content. Other aspects of SMT are treated in Exercises 7.9 and 7.10.

## 7.9 Insect locomotion on water

About 3% of all insects are to some extent aquatic. In one way or another, their lives are associated with water. Many of these insects are adapted to utilize the surface tension of water for locomotion. The surface tension of water makes it possible for some insects to stand on water and remain dry. Let us now estimate the maximum weight of an insect that can be supported by surface tension.

When the insect lands on water, the surface is depressed as shown in Fig. 7.8. The legs of such an insect, however, must not be wetted by water. A waxlike coating can provide the necessary water-repulsive property. The weight W of the insect is supported by the upward component of the surface tension; that is,

$$W = LT\sin\theta \tag{7.22}$$

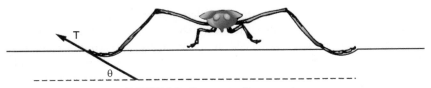

FIGURE 7.8   **Insect standing on water.**

where L is the combined circumference of all the insect legs in contact with the water.

To perform quantitative calculations, we must introduce some assumptions. We assume that the insect is in the shape of a cube with side dimensions $\ell$. The weight of the insect of density $\rho$ is then

$$W = \ell^3 \rho g \tag{7.23}$$

Let us further assume that the circumference of the legs in contact with water is approximately equal to the dimension of the cube; that is, from Eq. 7.23,

$$L = \ell = \left(\frac{W}{\rho g}\right)^{1/3} \tag{7.24}$$

The greatest supporting force provided by surface tension occurs at the angle $\theta = 90°$ (see Fig. 7.8). (At this point, the insect is on the verge of sinking.) The maximum weight $W_m$ that can be supported by surface tension is obtained from Eq. 7.22; that is,

$$W_m = LT = \left(\frac{W_m}{\rho g}\right)^{1/3} T$$

or

$$W_m^{2/3} = \frac{T}{(\rho g)^{1/3}} \tag{7.25}$$

If the density of the insect is $1 \ \text{g/cm}^3$, then with $T = 72.8 \ \text{dyn/cm}$, the maximum weight is

$$W_m^{2/3} = \frac{72.8}{(980)^{1/3}}$$

or

$$W_m = 19.7 \ \text{dyn}$$

The mass of the insect is therefore about $2 \times 10^{-2}$ g. The corresponding linear size of such an insect is about 3 mm.

As is shown in Exercise 7.11, a 70-kg person would have to stand on a platform about 10 km in perimeter to be supported solely by surface tension. (This is a disk about 3.2 km in diameter.)

More recent research has shown that the legs of some insects (water strider, for example) are covered with very fine hairs, in the size range of a micron or less, that increase the maximum weight supportable by surface tension by an order of magnitude (see Gao and Jiang, 1967).

## 7.10 Contraction of muscles

An examination of skeletal muscles shows that they consist of smaller muscle fibers, which in turn are composed of yet smaller units called *myofibrils*. Further, examination with an electron microscope reveals that the myofibril is composed of two types of threads, one made of *myosin*, which is about $160 \, \text{Å} \left(1 \, \text{Å} = 10^{-8} \, \text{cm}\right)$ in diameter, and the other made of *actin*, which has a diameter of about $50 \, \text{Å}$. Each myosin—actin unit is about 1 mm long. The threads are aligned in a regular pattern with spaces between threads so that the threads can slide past one another, as shown in Fig. 7.9.

Muscle contraction begins with an electrical nerve impulse that results in a release of $Ca^{2+}$ ions into the myosin—actin structure. The calcium ions in turn produce conformational changes that result in the sliding of the threads through each other, shortening the myosin—actin structure. The collective effect of this process is the contraction of the muscle.

Clearly, a force must act along the myosin—actin threads to produce such a contracting motion. The physical nature of this force is not fully understood. It has been suggested by Gamow and Ycas (1967) that this force may be due to surface tension, which is present not only in liquids but also in jellylike materials such as tissue cells. The motion of the threads is then similar to capillary movement of a liquid. Here the movement is due to the attraction

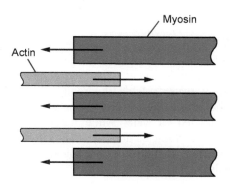

FIGURE 7.9   **Contraction of muscles.**

between the surfaces of the two types of thread. The surface attraction may be triggered by a release of the $Ca^{2+}$ ions. Let us now estimate the force per square centimeter of muscle tissue that could be generated by the surface tension proposed in this model.

If the average diameter of the threads is D, the number of threads N per square centimeter of muscle is approximately

$$N = \frac{1}{\frac{\pi}{4} \times D^2} \tag{7.26}$$

The maximum pulling force $F_f$ produced by the surface tension on each fiber is, from Eq. 7.16,

$$F_f = \pi DT \tag{7.27}$$

The total maximum force $F_m$ due to all the fibers in a $1 - cm^2$ area of muscle is

$$F_m = NF_f = \frac{4T}{D} \tag{7.28}$$

The average diameter D of the muscle fibers is about 100 Å $(10^{-6}\ cm)$. Therefore, the maximum contracting force that can be produced by surface tension per square centimeter of muscle area is

$$F_m = T \times 4 \times 10^6\ dyn/cm^2$$

A surface tension of 1.75 dyn/cm can account for the $7 \times 10^6\ dyn/cm^2$ measured force capability of muscles. Because this is well below surface tensions commonly encountered, we can conclude that surface tension could be the source of muscle contraction. This proposed mechanism, however, should not be taken too seriously. The actual processes in muscle contraction are much more complex and cannot be reduced to a simple surface tension model (see Rome, 1997).

## 7.11 Surfactants

Surfactants are molecules that lower surface tension of liquids. (The word is an abbreviation of surface active agent.) The most common surfactant molecules have one end that is water soluble (hydrophilic) and the other end water insoluble (hydrophobic) (see Fig. 7.10). As the word implies, the hydrophilic end is strongly attracted to water while the hydrophobic has very little attraction to water but is attracted and is readily soluble in oily liquids. Many different types of surfactant molecules are found in nature or as products of laboratory synthesis.

Hydrophobe                    Hydrophile

FIGURE 7.10   **Schematic of a surfactant molecule.**

When surfactant molecules are placed in water, they align on the surface with the hydrophobic end pushed out of the water, as shown in Fig. 7.11. Such an alignment disrupts the surface structure of water, reducing the surface tension. A small concentration of surfactant molecules can typically reduce surface tension of water from 73 dyn/cm to 30 dyn/cm. In oily liquids, surfactants are aligned with the hydrophilic end squeezed out of the liquid. In this case the surface tension of the oil is reduced.

The most familiar use of surfactants is soaps and detergents to wash away oily substances. Here the hydrophobic end of the surfactants dissolves into the oil surface while the hydrophilic end remains exposed to the surrounding water, as shown in Fig. 7.12. The aligned surfactant molecules reduce the surface tension of the oil. As a result, the oil breaks up into small droplets surrounded by the hydrophilic end of the surfactants. The small oil droplets are solubilized (that is, suspended or dissolved) in the water and can now be washed away.

Surfactants are widely used in experimental biochemistry. In certain types of experiments, for example, proteins that are hydrophobic such as membrane

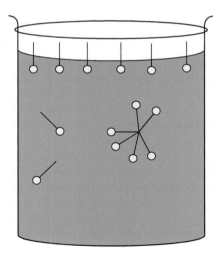

FIGURE 7.11   **Surface layer of surfactant molecules.**

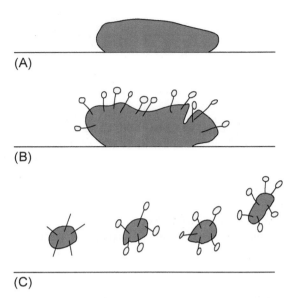

(A)

(B)

(C)

**FIGURE 7.12  Action of detergents.** (A) Oil drop on a wet spot. (B) The hydrophobic end of surfactant molecules enters the oil spot. (C) The oil spot breaks up into smaller sections surrounded by hydrophilic ends.

proteins and lipoproteins must be dissolved in water. Here surfactants are used to solubilize the proteins in a process similar to that illustrated in Fig. 7.12. The hydrophobic ends of the surfactant molecules dissolve into the surface of the protein. The aligned hydrophilic ends surround the protein, solubilizing it in the ambient water.

Some insects such as the *Microvelia* not only stand on water but also utilize surface tension for propulsion. They secrete a substance from their abdomen that reduces the surface tension behind them. As a result, they are propelled in a forward direction. Here the effect is similar to cutting a taut rubber membrane which then draws apart, each section moving away from the cut. This effect, known as Marangoni propulsion, can be demonstrated simply by coating one end of a toothpick with soap and placing it in water. The soap acting as the surfactant reduces the surface tension behind the coated end, resulting in the acceleration of the toothpick away from the dissolved soap.

Experiments have shown that the surfactant excreted by insects reduces the surface tension of water from 73 dyn/cm to about 50 dyn/cm. Measurements show that during Marangoni propulsion, *Microvelia* can attain peak speeds of 17 cm/sec. This value is in agreement with a simple calculation presented in Exercise 7.12.

The importance of surfactants in the process of breathing is described in Chapter 9.

## Exercises

**7.1.** Verify Eqs. 7.11 and 7.14.

**7.2.** With the nose above the water, about 95% of the body is submerged. Calculate the power expended by a 50-kg woman treading water in this position. Assume that the average density of the human body is about the same as water $\left(\rho = \rho_w = 1 \text{ g/cm}^3\right)$ and that the area $A$ of the limbs acting on the water is about 600 cm².

**7.3.** In Eq. 7.14, it is assumed that the density of the animal is greater than the density of the fluid in which it is submerged. If the situation is reversed, the immersed animal tends to rise to the surface, and it must expend energy to keep itself below the surface. How is Eq. 7.14 modified for this case?

**7.4.** Derive the relationship shown in Eq. 7.15.

**7.5.** Calculate the pressure 150 m below the surface of the sea. The density of sea water is 1.026 g/cm³.

**7.6.** Calculate volume of the swim bladder as a percent of the total volume of the fish in order to reduce the average density of the fish from 1.067 g/cm³ to 1.026 g/cm³.

**7.7.** The density of an animal is conveniently obtained by weighing it first in air and then immersed in a fluid. Let the weight in air and in the fluid be, respectively, $W_1$ and $W_2$. If the density of the fluid is $\rho_1$, the average density $\rho_2$ of the animal is

$$\rho_2 = \rho_1 \frac{W_1}{W_1 - W_2}$$

Derive this relationship.

**7.8.** Starting with Eq. 7.20, show that the pressure P required to withdraw the water from a capillary of radius R and contact angle $\theta$ is

$$P = \frac{2T\cos\theta}{R}$$

With the contact angle $\theta = 0°$, determine the pressure required to withdraw water from a capillary with a $10^{-3}$ cm radius. Assume that the surface tension $= 72.8$ dyn/cm.

**7.9.** If a section of coarse-grained soil is adjacent to a finer-grained soil of the same material, water will seep from the coarse-grained to the finer-grained soil. Explain the reason for this.

**7.10.** Design an instrument to measure the SMT. (You can find a description of one such device in Foth and Turk (1972).)

**7.11.** Calculate the perimeter of a platform required to support a 70-kg person solely by surface tension.

**7.12.** (a) Estimate the maximum acceleration of the insect that can be produced by reducing the surface tension as described in the text. Assume that the linear dimension of the insect is $3 \times 10^{-1}$ cm and its mass is $3 \times 10^{-2}$ g. Further, assume that the surface tension difference between the clean water and surfactant-altered water provides the force to accelerate the insect. Use surface tension values provided in the text. (b) Calculate the speed of the insect assuming that the surfactant release lasts 0.5 sec.

Chapter 8

# The motion of fluids

The study of fluids in motion is closely related to biology and medicine. In fact, one of the foremost workers in this field, L. M. Poiseuille (1799−1869), was a French physician whose study of moving fluids was motivated by his interest in the flow of blood through the body. In this chapter, we will review briefly the principles governing the flow of fluids and then examine the flow of blood in the circulatory system.

## 8.1 Bernoulli's equation

If frictional losses are neglected, the flow of an incompressible fluid is governed by *Bernoulli's equation*, which gives the relationship between velocity, pressure, and elevation in a line of flow. Bernoulli's equation states that at any point in the channel of a flowing fluid, the following relationship holds:

$$P + \rho gh + \frac{1}{2}\rho v^2 = \text{Constant} \tag{8.1}$$

Here P is the pressure in the fluid, h is the height, $\rho$ is the density, and v is the velocity at any point in the flow channel. The first term in the equation is the potential energy per unit volume of the fluid due to the pressure in the fluid. (Note that the unit for pressure, which is $dyn/cm^2$, is identical to $erg/cm^3$, which is energy per unit volume.) The second term is the gravitational potential energy per unit volume, and the third is the kinetic energy per unit volume.

Bernoulli's equation follows from the law of energy conservation. Because the three terms in the equation represent the total energy in the fluid, in the absence of friction their sum must remain constant no matter how the flow is altered.

We will illustrate the use of Bernoulli's equation with a simple example. Consider a fluid flowing through a pipe consisting of two segments with cross-sectional areas $A_1$ and $A_2$, respectively (see Fig. 8.1). The volume of fluid flowing per second past any point in the pipe is given by the product of the fluid velocity and the area of the pipe, $A \times v$. If the fluid is incompressible, in a unit time as much fluid must flow out of the pipe as flows into it. Therefore, the rates of flow in segments 1 and 2 are equal; that is,

Physics in Biology and Medicine. https://doi.org/10.1016/B978-0-443-21558-2.00008-0

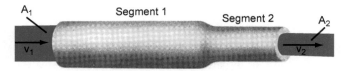

**FIGURE 8.1** **Flow of fluid through a pipe with two segments of different areas.**

$$A_1 v_1 = A_2 v_2 \quad \text{or} \quad v_2 = \frac{A_1}{A_2} v_1 \qquad (8.2)$$

In our case $A_1$ is larger than $A_2$, so we conclude that the velocity of the fluid in segment 2 is greater than in segment 1.

Bernoulli's equation states that the sum of the terms in Eq. 8.1 at any point in the flow is equal to the same constant. Therefore the relationship between the parameters P, $\rho$, h, and v at points 1 and 2 is

$$P_1 + \rho gh_1 + \frac{1}{2}\rho v_1^2 = P_2 + \rho gh_2 + \frac{1}{2}\rho v_2^2 \qquad (8.3)$$

where the subscripts designate the parameters at the two points in the flow. Because in our case the two segments are at the same height ($h_1 = h_2$), Eq. 8.2 can be written as

$$P_1 + \frac{1}{2}\rho v_1^2 = P_2 + \frac{1}{2}\rho v_2^2 \qquad (8.4)$$

Because $v_2 = (A_1/A_2)v_1$, the pressure in segment 2 is

$$P_2 = P_1 - \frac{1}{2}\rho v_1^2 \left[ \left( \frac{A_1}{A_2} \right)^2 - 1 \right] \qquad (8.5)$$

This relationship shows that while the flow velocity in segment 2 increases, the pressure in that segment decreases.

## 8.2 Viscosity and Poiseuille's law

Frictionless flow is an idealization. In a real fluid, the molecules attract each other; consequently, relative motion between the fluid molecules is opposed by a frictional force, which is called *viscous friction*. Viscous friction is proportional to the velocity of flow and to the coefficient of viscosity for the given fluid. As a result of viscous friction, the velocity of a fluid flowing through a pipe varies across the pipe. The velocity is highest at the center and decreases toward the walls; at the walls of the pipe, the fluid is stationary. Such fluid flow is called *laminar*. Figure 8.2 shows the velocity profile for laminar flow in a pipe. The lengths of the arrows are proportional to the velocity across the pipe diameter.

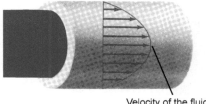

Velocity of the fluid

**FIGURE 8.2  Laminar flow.** The length of the arrows indicates the magnitude of the velocity of the fluid.

**TABLE 8.1** Viscosities of selected fluids.

| Fluid | Temperature (°C) | Viscosity (poise) |
|---|---|---|
| Water | 20 | 0.01 |
| Glycerin | 20 | 8.3 |
| Mercury | 20 | 0.0155 |
| Air | 20 | 0.00018 |
| Blood | 37 | 0.04 |

If viscosity is taken into account, it can be shown (see reference Richardson and Neergaard, 1972) that the rate of laminar flow Q through a cylindrical tube of radius R and length L is given by *Poiseuille's law,* which is

$$Q = \frac{\pi R^4 (P_1 - P_2)}{8\eta L} \text{ cm}^3/\text{sec} \tag{8.6}$$

where $P_1 - P_2$ is the difference between the fluid pressures at the two ends of the cylinder and $\eta$ is the coefficient of viscosity measured in units of dyn $(\text{sec}/\text{cm}^2)$, which is called a *poise.* The viscosities of some fluids are listed in Table 8.1. In general, viscosity is a function of temperature and increases as the fluid becomes colder.

There is a basic difference between frictionless and viscous fluid flow. A frictionless fluid will flow steadily without an external force applied to it. This fact is evident from Bernoulli's equation, which shows that if the height and velocity of the fluid remain constant, there is no pressure drop along the flow path. But Poiseuille's equation for viscous flow states that a pressure drop always accompanies viscous fluid flow. By rearranging Eq. 8.6, we can express the pressure drop as

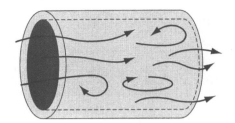

FIGURE 8.3    Turbulent fluid flow.

$$P_1 - P_2 = \frac{Q8\eta L}{\pi R^4} \qquad (8.7)$$

The expression $P_1 - P_2$ is the pressure drop that accompanies the flow rate Q along a length L of the pipe. The product of the pressure drop and the area of the pipe is the force required to overcome the frictional forces that tend to retard the flow in the pipe segment. Note that for a given flow rate, the pressure drop required to overcome frictional losses decreases as the fourth power of the pipe radius. Thus, even though all fluids are subject to friction, if the area of the flow is large, frictional losses and the accompanying pressure drop are small and can be neglected. In these cases, Bernoulli's equation may be used with little error.

## 8.3 Turbulent flow

If the velocity of a fluid is increased past a critical point, the smooth laminar flow shown in Fig. 8.2 is disrupted. The flow becomes turbulent with eddies and whirls disrupting the laminar flow (see Fig. 8.3). In a cylindrical pipe the critical flow velocity $v_c$, above which the flow is turbulent, is given by

$$v_c = \frac{\Re\eta}{\rho D} \qquad (8.8)$$

Here D is the diameter of the cylinder, $\rho$ is the density of the fluid, and $\eta$ is the viscosity. The symbol $\Re$ is the *Reynold's number*, which for most fluids has a value between 2000 and 3000. The frictional forces in turbulent flow are greater than in laminar flow. Therefore, as the flow turns turbulent, it becomes more difficult to force a fluid through a pipe.

## 8.4 Circulation of the blood

The circulation of blood through the body is often compared to a plumbing system, with the heart as the pump and the veins, arteries, and capillaries as the pipes through which the blood flows. This analogy is not entirely correct. Blood is not a simple fluid; it contains cells that complicate the flow, especially

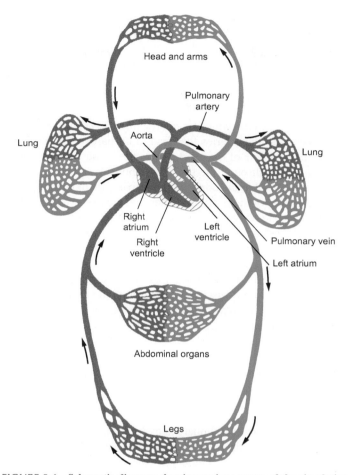

**FIGURE 8.4** Schematic diagram showing various routes of the circulation.

when the passages become narrow. Furthermore, the veins and arteries are not rigid pipes but are elastic and alter their shape in response to the forces applied by the fluid. Still, it is possible to analyze the circulatory system with reasonable accuracy using the concepts developed for simple fluids flowing in rigid pipes.

Figure 8.4 is a drawing of the human circulatory system. The blood in the circulatory system brings oxygen, nutrients, and various other vital substances to the cells and removes the metabolic waste products from the cells. The blood is pumped through the circulatory system by the heart, and it leaves the heart through vessels called *arteries* and returns to it through *veins*.

The mammalian heart consists of two independent pumps, each made of two chambers called the *atrium* and the *ventricle*. The entrances to and exits from these chambers are controlled by valves that are arranged to maintain the flow of blood in the proper direction. Blood from all parts of the body except the lungs enters the right atrium, which contracts and forces the blood into the right ventricle. The ventricle then contracts and drives the blood through the pulmonary artery into the lungs. In its passage through the lungs, the blood releases carbon dioxide and absorbs oxygen. The blood then flows into the left atrium via the pulmonary vein. The contraction of the left atrium forces the blood into the left ventricle, which on contraction drives the oxygen-rich blood through the aorta into the arteries that lead to all parts of the body except the lungs. Thus, the right side of the heart pumps the blood through the lungs, and the left side pumps it through the rest of the body.

The large artery, called the *aorta*, which carries the oxygenated blood away from the left chamber of the heart, branches into smaller arteries, which lead to the various parts of the body. These in turn branch into still smaller arteries, the smallest of which are called *arterioles*. As we will explain later, the arterioles play an important role in regulating the blood flow to specific regions in the body. The arterioles branch further into narrow capillaries that are often barely wide enough to allow the passage of single blood cells.

The capillaries are so profusely spread through the tissue that nearly all the cells in the body are close to a capillary. The exchange of gases, nutrients, and waste products between the blood and the surrounding tissue occurs by diffusion through the thin capillary walls (see Chapter 9). The capillaries join into tiny veins called *venules*, which in turn merge into larger and larger veins that lead the oxygen-depleted blood back to the right atrium of the heart.

## 8.5 Blood pressure

The contraction of the heart chambers is triggered by electrical pulses that are applied simultaneously both to the left and to the right halves of the heart. First the atria contract, forcing the blood into the ventricles; then the ventricles contract, forcing the blood out of the heart. Because of the pumping action of the heart, blood enters the arteries in spurts or pulses. The maximum pressure driving the blood at the peak of the pulse is called the *systolic pressure*. The lowest blood pressure between the pulses is called the *diastolic pressure*. In a young, healthy individual the systolic pressure is about 120 torr (mm Hg) and the diastolic pressure is about 80 torr, which combined are usually expressed as 120/80. Therefore the average pressure of the pulsating blood at heart level is 100 torr.

As the blood flows through the circulatory system, its initial energy, provided by the pumping action of the heart, is dissipated by two loss mechanisms: Losses associated with the *expansion and contraction* of the arterial walls and *viscous friction* associated with the blood flow. Due to these energy losses, the initial pressure fluctuations are smoothed out as the blood flows away from the heart, and the average pressure drops. By the time the blood reaches the capillaries, the flow is smooth and the blood pressure is only about 30 torr. The pressure drops still lower in the veins and is close to zero just before returning to the heart. In this final stage of the flow, the movement of blood through the veins is aided by the contraction of muscles that squeeze the blood toward the heart. One-way flow is assured by unidirectional valves in the veins.

The main arteries in the body have a relatively large radius. The radius of the aorta, for example, is about 1 cm; therefore, the pressure drop along the arteries is small. We can estimate this pressure drop using Poiseuille's law (Eq. 8.7). However, to solve the equation, we must know the rate of blood flow. The rate of blood flow Q through the body depends on the level of physical activity. At rest, the total flow rate is about 5 L/min. During intense activity, the flow rate may rise to about 25 L/min. Exercise 8.1 shows that at peak flow, the pressure drop per centimeter length of the aorta is only 42.5 $dyn/cm^2$ $(3.19 \times 10^{-2}$ torr$)$, which is negligible compared to the total blood pressure.

Of course, as the aorta branches, the size of the arteries decreases, resulting in an increased resistance to flow. Although the blood flow in the narrower arteries is also reduced, the pressure drop is no longer negligible (see Exercise 8.2). The average pressure at the entrance to the arterioles is about 90 torr. Still, this is only a 10% drop from the average pressure at the heart. The flow through the arterioles is accompanied by a much larger pressure drop, about 60 torr. As a result, the pressure at the capillaries is only about 30 torr.

Since the pressure drop in the main arteries is small, when the body is horizontal, the average arterial pressure is approximately constant throughout the body. The arterial blood pressure, which is on average 100 torr, can support a column of blood 129 cm high (see Eq. 7.1 and Exercise 8.3). This means that if a small tube were introduced into the artery, the blood in it would rise to a height of 129 cm (see Fig. 8.5).

If a person is standing erect, the blood pressure in the arteries is not uniform in the various parts of the body. The weight of the blood must be taken into account in calculating the pressure at various locations. For example, the average pressure in the artery located in the head, 50 cm above the heart (see Exercise 8.4a), is $P_{head} = P_{heart} - \rho gh = 61$ torr. In the feet, 130 cm below the heart, the arterial pressure is 200 torr (see Exercise 8.4b).

The cardiovascular system has various flow-control mechanisms that can compensate for the large arterial pressure changes that accompany shifts in the position of the body. Still, it may take a few seconds for the system to compensate. Thus, a person may feel momentarily dizzy as he/she jumps up from a prone position. This is due to the sudden decrease in the blood pressure of the brain arteries, which results in a temporary decrease in blood flow to the brain.

The same hydrostatic factors operate also in the veins, and here their effect may be more severe than in the arteries. The blood pressure in the veins is

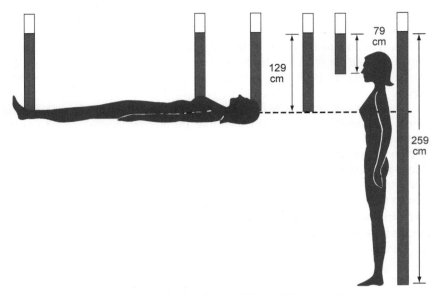

FIGURE 8.5    **Blood pressure in a reclining and in an erect person.**

lower than in the arteries. When a person stands motionless, the blood pressure is barely adequate to force the blood from the feet back to the heart. Thus when a person sits or stands without muscular movement, blood gathers in the veins of the legs. This increases the pressure in the capillaries and may cause temporary swelling of the legs. With some exceptions, the blood pressure of most animals is in the same range as that of humans. For example, the systolic blood pressure of a pig, cat, and dog is for all three about 120 torr. The giraffe with its head positioned high above the heart is an exception. Its blood pressure is significantly higher, typically 240/160.

## 8.6 Control of blood flow

The pumping action of the heart (i.e., blood pressure, flow volume, and rate of heartbeat) is regulated by a variety of hormones. Hormones are molecules, often proteins, that are produced by organs and tissues in different parts of the body. They are secreted into the blood stream and carry messages from one part of the body to another. Hormones affecting the heart are produced in response to stimuli such as need for more oxygen, changes in body temperature, and various types of emotional stress.

The flow of blood to specific parts of the body is controlled by the arterioles. These small vessels that receive blood from the arteries have an average diameter of about 0.1 mm. The walls of the arterioles contain smooth muscle fibers that contract when stimulated by nerve impulses and hormones. The contraction of the arterioles in one part of the body reduces the blood flow to that region and diverts it to another. Since the radius of the arterioles is small, constriction is an effective method for controlling blood flow. Poiseuille's equation shows that if the pressure drop remains constant, a 20% decrease in the radius reduces the blood flow by more than a factor of 2 (see Exercise 8.5).

A stress-induced heart condition called stress cardiomyopathy (broken heart syndrome) has only recently been clearly identified by Western medicine. The syndrome occurs most frequently after a sudden intense emotional trauma such as death in the family, an experience of violence, or extreme anger. The symptoms are similar to an acute heart attack, but the coronary arteries are found to be normal and the heart tissue is not damaged. It has been suggested that the condition is triggered by an excessive release of stress-related hormones called catecholamines.

## 8.7 Energetics of blood flow

For an individual at rest, the rate of blood flow is about 5 L/min. This implies that the average velocity of the blood through the aorta is 26.5 cm/sec (see Exercise 8.6). However, the blood in the aorta does not flow continuously. It moves in spurts. During the period of flow, the velocity of the blood is about three times as high as the overall average value calculated in Exercise 8.6. Therefore, the kinetic energy per cubic centimeter of flowing blood is

$$KE = \frac{1}{2}\rho v^2 = \frac{1}{2}(1.05) \times (79.5)^2 = 3320 \text{ erg/cm}^3$$

We mentioned earlier that energy density (energy per unit volume) and pressure are measured by the same unit (i.e., $1 \text{ erg/cm}^3 = 1 \text{ dyn/cm}^2$); therefore, they can be compared to each other. The kinetic energy of $3320 \text{ erg/cm}^3$ is equivalent to 2.50 torr pressure; this is small compared to the blood pressure in the aorta (which is on the average 100 torr). The kinetic energy in the smaller arteries is even less because, as the arteries branch, the overall area increases and, therefore, the flow velocity decreases. For example, when the total flow rate is 5 L/min, the blood velocity in the capillaries is only about 0.33 mm/sec.

The kinetic energy of the blood becomes more significant as the rate of blood flow increases. For example, if during physical activity, the flow rate increases to 25 L/min, the kinetic energy of the blood is $83,300 \text{ erg/cm}^3$, which is equivalent to a pressure of 62.5 torr. This energy is no longer negligible compared to the blood pressure measured at rest. In healthy arteries, the increased velocity of blood flow during physical activity does not present a problem. During intense activity, the blood pressure rises to compensate for the pressure drop.

## 8.8 Turbulence in the blood

Eq. 8.8 shows that if the velocity of a fluid exceeds a specific critical value, the flow becomes turbulent. Through most of the circulatory system, the blood flow is laminar. Only in the aorta does the flow occasionally become turbulent. Assuming a Reynold's number of 2000, the critical velocity for the onset of turbulence in the 2-cm-diameter aorta is, from Eq. 8.8,

$$V_c = \frac{\Re\eta}{\rho D} = \frac{2000 \times 0.04}{1.05 \times 2} = 38 \text{ cm/sec}$$

For the body at rest, the flow velocity in the aorta is below this value. But as the level of physical activity increases, the flow in the aorta may exceed the critical rate and become turbulent. In the other parts of the body, however, the flow remains laminar unless the passages are abnormally constricted.

Laminar flow is quiet, but turbulent flow produces noises due to vibrations of the various surrounding tissues, which indicate abnormalities in the circulatory system. These noises, called *bruit*, can be detected by a stethoscope and can help in the diagnosis of circulatory disorders.

## 8.9 Arteriosclerosis and blood flow

Arteriosclerosis is the most common of cardiovascular diseases. In the United States, an estimated 200,000 people die annually as a consequence of this disease. In arteriosclerosis, the arterial wall becomes thickened, and the artery is narrowed by deposits called *plaque*. This condition may seriously impair the functioning of the circulatory system. A 50% narrowing (*stenosis*) of the arterial area is considered moderate. About 60% to 70% is considered severe, and a narrowing above 80% is deemed critical. One problem caused by stenosis is made clear by Bernoulli's equation. The blood flow through the region of constriction is sped up. If, for example, the radius of the artery is narrowed by a factor of three, the cross-sectional area decreases by a factor of nine, which results in a nine-fold increase in velocity. In the constriction, the kinetic energy increases by $9^2$, or 81. The increased kinetic energy is at the expense of the blood pressure; that is, in order to maintain the flow rate at the higher velocity, the potential energy due to pressure is converted to kinetic energy. As a result, the blood pressure in the constricted region drops. For example, if in the unobstructed artery the flow velocity is 50 cm/sec, then in the constricted region, where the area is reduced by a factor of 9, the velocity is 450 cm/sec. Correspondingly, the pressure is decreased by about 80 torr (see Exercise 8.8). Because of the low pressure inside the artery, the external pressure may actually close off the artery and block the flow of blood. When such a blockage occurs in the coronary artery, which supplies blood to the heart muscle, the heart stops functioning.

Stenosis above 80% is considered critical because at this point the blood flow usually becomes turbulent with inherently larger energy dissipation than is associated with laminar flow. As a result, the pressure drop in the situation presented earlier is even larger than calculated using Bernoulli's equation. Further, turbulent flow can damage the circulatory system because parts of the flow are directed toward the artery wall rather than parallel to it, as in laminar flow. The blood impinging on the arterial wall may dislodge some of the plaque deposit, which downstream may clog a narrower part of the artery. If such clogging occurs in a cervical artery, blood flow to some part of the brain is interrupted, causing an *ischemic stroke.*

There is another problem associated with arterial plaque deposit. The artery has a specific elasticity; therefore, it exhibits certain springlike properties. Specifically, in analogy with a spring, the artery has a natural frequency at which it can be readily set into vibrational motion. (See Chapter 5, Section 5.2.) The natural frequency of a healthy artery is in the range of 1 to 2 kHz. Deposits of plaque cause an increase in the mass of the arterial wall and a decrease in its elasticity. As a result, the natural frequency of the artery is significantly decreased, often down to a few hundred hertz. Pulsating blood flow contains frequency components in the range of 450 Hz. The plaque-coated artery with its lowered natural frequency may now be set into resonant vibrational motion, which may dislodge plaque deposits or cause further damage to the arterial wall.

## 8.10 Power produced by the heart

The energy in the flowing blood is provided by the pumping action of the heart. We will now compute the power generated by the heart to keep the blood flowing in the circulatory system.

The power $P_H$ produced by the heart is the product of the flow rate Q and the energy E per unit volume of the blood; that is,

$$P_H = Q \left( \frac{cm^3}{sec} \right) \times E \left( \frac{erg}{cm^2} \right) = Q \times E \text{ erg/sec} \tag{8.9}$$

At rest, when the blood flow rate is 5 L/min, or 83.4 cm$^3$/sec, the kinetic energy of the blood flowing through the aorta is $3.33 \times 10^3$ erg/cm$^3$. (See previous section.) The energy corresponding to the systolic pressure of 120 torr is $160 \times 10^3$ erg/cm$^3$. The total energy is $1.63 \times 10^5$ erg/cm$^3$—the sum of the kinetic energy and the energy due to the fluid pressure. Therefore, the power P produced by the left ventricle of the heart is

$$P = 83.4 \times 1.63 \times 10^5 = 1.35 \times 10^7 \text{ erg/sec} = 1.35 \text{ W}$$

Exercise 8.9 shows that during intense physical activity, when the flow rate increases to 25 L/min, the peak power output of the left ventricle increases to 10.1 W.

The flow rate through the right ventricle, which pumps blood through the lungs, is the same as the flow through the left ventricle. Here, however, the blood pressure is only one-sixth the pressure in the aorta. Therefore, as shown in Exercise 8.10, the power output of the right ventricle is 0.25 W at rest and 4.5 W during intense physical activity. Thus, the total peak power output of the heart is between 1.9 and 14.6 W, depending on the intensity of the physical activity. While in fact the systolic blood pressure rises with increased blood flow, in these calculations we have assumed that it remains at 120 torr.

## 8.11 Measurement of blood pressure

The arterial blood pressure is an important indicator of the health of an individual. Both abnormally high and abnormally low blood pressures indicate some disorders in the body that require medical attention. High blood pressure, which may be caused by constrictions in the circulatory system, certainly implies that the heart is working harder than usual and that it may be endangered by the excess load. Blood pressure can be measured most directly by inserting a vertical glass tube into an artery and observing the height to which the blood rises (see Fig. 8.5). This was, in fact, the way blood pressure was first measured in 1733 by Reverend Stephen Hales, who connected a long vertical glass tube to an artery of a horse. Although sophisticated modifications of this technique are still used in special cases, this method is obviously not satisfactory for routine clinical examinations. Routine measurements of blood pressure are now most commonly performed by the cut-off method. Although this method is not as accurate as direct measurements, it is simple and in most cases adequate. In this technique, a cuff containing an inflatable balloon is placed tightly around the upper arm. The balloon is inflated with a bulb, and the pressure in the balloon is monitored by a pressure gauge. The initial pressure in the balloon is greater than the systolic pressure, and the flow of blood through the artery is therefore cut off. The observer then allows the pressure in the balloon to fall slowly by releasing some of the air. As the pressure drops, she listens with a stethoscope placed over the artery downstream from the cuff. No sound is heard until the pressure in the balloon decreases to the systolic pressure. Just below this point, the blood begins to flow through the artery; however, since the artery is still partially constricted, the flow is turbulent and is accompanied by a characteristic sound. The pressure recorded at the onset of the sound is the systolic blood pressure. As the pressure in the balloon drops further, the artery expands to its normal size, the flow becomes laminar, and the noise disappears. The pressure at which the sound begins to fade is taken as the diastolic pressure.

In clinical measurements, the variation of the blood pressure along the body must be considered. The cut-off blood pressure measurement is taken with the cuff placed on the arm approximately at heart level.

## 8.12 Microfluidics

Laboratory experiments in biology, biochemistry, or medicine are most often performed in fluids, usually aqueous solutions of components to be studied. Examples of such experiments may be studies of interactions between specific proteins and selected cells, or the growth of a microbial population as a function of a variety of nutrients. Studies of this type using conventional techniques require milliliter quantities of fluid ($\sim cm^3$ volumes) flowing typically through tubing several millimeters in diameter. Much of our current understanding of biochemical processes has been obtained using such conventional techniques. There are, however, limitations to these techniques. Foremost among them are given as follows. (1) Flow through millimeter-diameter channels is often turbulent. As a result, when fluids containing two or more species (reagents) are brought together for the study of their interactions, the mixing is accompanied by eddies and swirls associated with turbulent flow. (See Section 8.3.) Such turbulence can disrupt weak bonds between components being studied, for example, bonds between enzymes attached to a substrate suspended in the flowing fluid. Disruption of bonds makes studies of such systems difficult, if not impossible. Further, as a result of turbulence, the mixing region is not well defined, resulting in uncertainties in the measurement of interaction times and in local reagent concentrations. (2) The relatively large quantities of reagent required using conventional techniques are often difficult to obtain, limiting the types of studies that can be conducted. (3) In some cases, the relatively large quantities of reagents require an inconveniently long time for their interactions to reach completion.

In the 1980s, several techniques were brought together to create a new technology now called *microfluidics*. Microfluidics provides a means for accurately controlling, manipulating, and facilitating interactions of small volumes of fluids flowing in small channels, typically of dimensions 5 to 100 μm. The volume of the interacting fluids is on the order of μL ($10^{-6}$ L) or smaller. In some applications, fluid volumes can be as small as a pL ($10^{-12}$ L).

Microfluidics became realizable by merging two technologies developed mainly in the 1960s and 1970s. They were (a) development of microanalytical techniques that can detect the small quantities of reagents used in microfluidics and (b) development of photolithographic techniques to form small structures for solid-state electronics. These fabrication techniques are used to form small, often intricately connected channels, usually in plastics, for the flow of interacting fluids.

In Fig. 8.6, we show a simple two-port microfluidics component. Often the lithographic techniques employed make it convenient to form rectangular

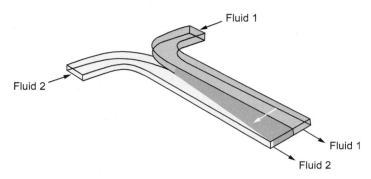

**FIGURE 8.6 A simple microfluidics component.** *(From http://faculty.washington.edu/yagerp/microfluidicstutorial/tutorialhome.htm.)*

channels, as shown in the figure. Typically, the width $w$ of the channel may be 100 μm and the corresponding depth d $=$ 10 μm. (1 μm $=$ $10^{-6}$ m $=$ 10 $\times$ $10^{-4}$ cm.) The dimensions vary depending on the application.

As stated earlier, an important feature of microfluidic structures is the laminar flow of the fluids. As a consequence, two fluids can flow side by side without turbulent mixing. The fluids do mix via diffusion, a process that is described in the following chapter. Such diffusive mixing of the components of fluids 1 and 2 does not perturb the laminar nature of the fluid flows and allows an accurate determination of reactant and product concentrations along the flow using a variety of optical techniques.

Whether the flow is laminar or turbulent is determined by Reynold's number R introduced in Section 8.3. The flow is laminar for R $<$ 2000 and turbulent for R $>$ 3000. Rearranging Eq. 8.8, we obtain an expression for R as.

$$R = v\rho L / \eta \qquad (8.9)$$

Here instead of tube diameter D, we have substituted a more general term, the characteristic linear dimension "L." As shown in Exercise 8.12a, for a rectangular cross-section channel such as shown in Fig. 8.6, L is one-half the depth "d" of the channel. The other parameters are as before: v is the average flow speed that in microfluidics experiments is typically 1 cm/s, ρ is the density of the fluid, and η is the viscosity of the fluid. Using the typical values for the parameters in Eq. 8.9, a simple calculation yields R $=$ 5 $\times$ $10^{-2}$, placing the flow firmly in the laminar domain (see Exercise 8.12b).

The potential of microfluidics is just starting to be realized. However, the possibilities are already clearly evident. Some inexpensive microfluidic disposable devices are now commercially available for detection and diagnosis of HIV, cholera toxins, and salmonella, among other pathogens and toxins. Many of these detection techniques utilize the interaction of an enzyme with the specific antibody associated with the pathogen to be identified. The interaction, if present, produces an easily identified color change. Micro-fluidics are also beginning to be utilized by pharmaceutical companies in drug discovery and development. Potentially useful drugs can be quickly screened for interaction with specific target proteins. Typically, testing for interactions that would take many hours using standard techniques can be completed in minutes using microfluidics.

## Exercises

**8.1.** Calculate the pressure drop per centimeter length of the aorta when the blood flow rate is 25 L/min. The radius of the aorta is about 1 cm, and the coefficient of viscosity of blood is $4 \times 10^{-2}$ poise.

**8.2.** Compute the drop in blood pressure along a 30-cm length of artery of radius 0.5 cm. Assume that the artery carries blood at a rate of 8 L/min.

**8.3.** How high a column of blood can an arterial pressure of 100 torr support? (The density of blood is $1.05 \text{ g/cm}^3$.)

**8.4.** (a) Calculate the arterial blood pressure in the head of an erect person. Assume that the head is 50 cm above the heart. (The density of blood is $1.05 \text{ g/cm}^3$.) (b) Compute the average arterial pressure in the legs of an erect person, 130 cm below the heart.

**8.5.** (a) Show that if the pressure drop remains constant, reduction of the radius of the arteriole from 0.1 to 0.08 mm decreases the blood flow by more than a factor of 2. (b) Calculate the decrease in the radius required to reduce the blood flow by 90%.

**8.6.** Compute the average velocity of the blood in the aorta of radius 1 cm if the flow rate is 5 L/min.

**8.7.** When the rate of blood flow in the aorta is 5 L/min, the velocity of the blood in the capillaries is about 0.33 mm/sec. If the average diameter of a capillary is 0.008 mm, calculate the number of capillaries in the circulatory system.

**8.8.** Compute the decrease in the blood pressure of the blood flowing through an artery, the radius of which is constricted by a factor of 3. Assume that the average flow velocity in the unconstricted region is 50 cm/sec.

**8.9.** Using information provided in the text, calculate the power generated by the left ventricle during intense physical activity when the flow rate is 25 L/min.

**8.10.** Using information provided in the text, calculate the power generated by the right ventricle during (a) restful state; blood flow 5 L/min, and (b) intense activity; blood flow 25 L/min.

**8.11.** During each heartbeat, the blood from the heart is ejected into the aorta and the pulmonary artery. Since the blood is accelerated during this part of the heartbeat, a force in the opposite direction is exerted on the rest of the body. If a person is placed on a sensitive scale (or another force-measuring device), this reaction force can be measured. An instrument based on this principle is called the *ballistocardiograph*. Discuss the type of information that might be obtained from measurements with a ballistocardiograph, and estimate the magnitude of the forces measured by this instrument.

8.12. (a) In general terms, the characteristic linear dimension is defined as the ratio of the volume of the fluid to the surface area of that volume. Show that for a rectangular cross-section channel such as shown in Fig. 8.6 where the depth of the channel, d, is much smaller than its width, *w*, the characteristic linear dimension L is one-half the depth "d" of the channel. (Hint: Start by computing the volume-to-surface ratio for a specific length of the channel.) In computing the area of the channel, the area of the end sections is not included in the total area because the flow in the channel is continuous. (b) Calculate Reynold's number $R$ using the following values for the relevant parameters: The other parameters are as before: $v$ is the average flow speed, typically $=$ 1 cm/s; $\rho = 1$ g/cm$^3$; and $\eta$ is the fluid viscosity $= 0.01$ g/cm/s for water. From the text, L $= 5 \times 10^{-4}$ cm.

8.13. Describe a lithographic technique used in fabricating microfluidic devices.

Chapter 9

# Heat and kinetic theory

## 9.1 Heat and hotness

The sensation of hotness is certainly familiar to all of us. We know from experience that when two bodies, one hot and the other cold, are placed in an enclosure, the hotter body will cool and the colder body will heat until the degree of hotness of the two bodies is the same. Clearly something has been transferred from one body to the other to equalize their hotness. That which has been transferred from the hot body to the cold body is called *heat*. Heat may be transformed into work, and therefore it is a form of energy. Heated water, for example, can turn into steam, which can push a piston. In fact, heat can be defined as energy being transferred from a hotter body to a colder body.

In this chapter, we will discuss various properties associated with heat. We will describe the motion of atoms and molecules due to thermal energy and then discuss diffusion in connection with the functioning of cells and the respiratory system.

## 9.2 Kinetic theory of matter

To understand the present-day concept of heat, we must briefly explain the structure of matter. Matter is made of atoms and molecules, which are in continuous chaotic motion. In a gas, the atoms (or molecules) are not bound together. They move in random directions and collide frequently with one another and with the walls of the container. In addition to moving linearly, gas molecules vibrate and rotate, again in random directions. In a solid, where the atoms are bound together, the random motion is more restricted. The atoms are free only to vibrate and do so, again randomly, about some average position to which they are locked. The situation with regard to liquids is between these two extremes. Here the molecules can vibrate, but they also have some freedom to move and to rotate.

Because of their motion, the moving particles in a material possess kinetic energy. This energy of motion inside materials is called *internal energy*, and the motion itself is called *thermal motion*. What we have so far qualitatively called the hotness of a body is a measure of the internal energy; that is, in hotter bodies, the random motion of atoms and molecules is faster than in

Physics in Biology and Medicine. https://doi.org/10.1016/B978-0-443-21558-2.00009-2

colder bodies. Therefore, the hotter an object, the greater its internal energy. The physical sensation of hotness is the effect of this random atomic and molecular motion on the sensory mechanism. *Temperature* is a quantitative measure of hotness. The internal energy of matter is proportional to its temperature.

Using these concepts, it is possible to derive the equations that describe the behavior of matter as a function of temperature. Gases are the simplest to analyze. The theory considers a gas made of small particles (atoms or molecules) which are in continuous random motion. Each particle travels in a straight line until it collides with another particle or with the walls of the container. After a collision, the direction and speed of the particle is changed randomly. In this way kinetic energy is exchanged among the particles.

The colliding particles exchange energy not only among themselves but also with the wall of the container (Fig. 9.1). For example, if initially the walls of the container are hotter than the gas, the particles colliding with the wall on average pick up energy from the vibrating molecules in the wall. As a result of the wall collisions, the gas is heated until it is as hot as the walls. After that, there is no net exchange of energy between the walls and the gas. This is an equilibrium situation in which, on average, as much energy is delivered to the wall by the gas particles as is picked up from it.

The speed and corresponding kinetic energy of the individual particles in a gas vary over a wide range. Still it is possible to compute an average kinetic energy for the particles by adding the kinetic energy of all the individual particles in the container and dividing by the total number of particles (for details, see for example Morgan, 1968). Many of the properties of a gas can be simply derived by assuming that each particle has this same average energy.

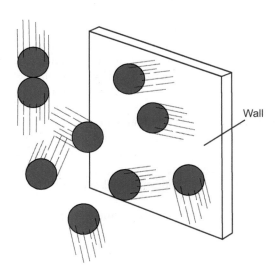

FIGURE 9.1   Collisions in a gas.

The internal energy in an ideal gas is in the form of kinetic energy,[1] and therefore the average kinetic energy $\left[\left(\frac{1}{2}mv^2\right)_{av}\right]$ is proportional to the temperature. The proportionality can be changed to an equality by multiplying the temperature T by a suitable constant which relates the temperature to the internal energy. The constant is designated by the symbol k, which is called the *Boltzmann constant*. For historical reasons, Boltzmann constant has been so defined that it has to be multiplied by a factor of $\frac{3}{2}$ to relate temperature to the average kinetic energy of a molecule; thus,

$$\left(\frac{1}{2}mv^2\right)_{av} = \frac{3}{2}kT \tag{9.1}$$

The temperature in this equation is measured on the absolute temperature scale in degrees Kelvin. The size of the degree division on the absolute scale is equal to the Celsius, or centigrade, degree, but the absolute scale is transposed so that $0°C = 273.15$ K. Since our calculations are carried only to three significant figures, we will use simply $0°C = 273$ K. The value of Boltzmann constant is

$$k = 1.38 \times 10^{-23} \text{ J/molecule K}$$

The velocity defined by Eq. 9.1 is called *thermal velocity*.

Each time a molecule collides with the wall, momentum is transferred to the wall. The change in momentum per unit time is a force. The pressure exerted by a gas on the walls of its container is due to the numerous collisions of the gas molecules with the container. The following relationship between pressure P, volume V, and temperature is derived in most basic physics texts:

$$PV = NkT \tag{9.2}$$

Here N is the total number of gas molecules in the container of volume *V*, and the temperature is again measured on the absolute scale.

In a closed container, the total number of particles N is fixed; therefore, if the temperature is kept unchanged, the product of pressure and volume is a constant. This is known as *Boyle's law*. (See Exercises 9.1 and 9.2.)

## 9.3 Definitions

### 9.3.1 Unit of heat

As discussed in Appendix A, heat is measured in *calories*. One calorie (cal) is the amount of heat required to raise the temperature of 1 g of water by 1 C°.[2]

---

1. The simple theory neglects the vibrational and rotational energy of the molecules.
2. For the symbol °C, read as degree Celsius. For the symbol C°, read as Celsius degree.

Actually, because this value depends somewhat on the initial temperature of the water, the calorie is defined as the heat required to raise the temperature of 1 g of water from 14.5°C to 15.5°C. One calorie is equal to 4.184 J. In the life sciences, heat is commonly measured in kilocalorie units, abbreviated Cal; 1 Cal is equal to 1000 cal.

## 9.3.2 Specific heat

Specific heat is the quantity of heat required to raise the temperature of 1 g of a substance by 1 degree. The specific heats of some substances are shown in Table 9.1.

The human body is composed of water, proteins, fat, and minerals. Its specific heat reflects this composition. With 75% water and 25% protein, the specific heat of the body would be

$$\text{Specific heat} = 0.75 \times 1 + 0.25 \times 0.4 = 0.85$$

The specific heat of the average human body is closer to 0.83 due to its fat and mineral content, which we have not included in the calculation.

## 9.3.3 Latent heats

In order to convert a solid to a liquid at the same temperature or to convert a liquid to a gas, heat energy must be added to the substance. This energy is called *latent heat*. The latent heat of fusion is the amount of energy required to change 1 g of solid matter to liquid. The latent heat of vaporization is the amount of heat required to change 1 g of liquid to gas.

**TABLE 9.1 Specific heat for some substances.**

| Substance | Specific heat (cal/g°C) |
| --- | --- |
| Water | 1 |
| Ice | 0.480 |
| Average for human body | 0.83 |
| Soil | 0.2 to 0.8, depending on water content |
| Aluminum | 0.214 |
| Protein | 0.4 |

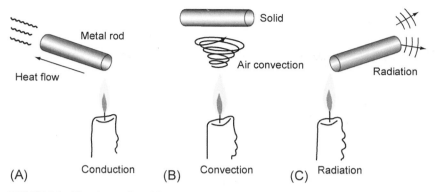

FIGURE 9.2   Heat is transferred from one region to another by: (A) conduction, (B) convection, and (C) radiation.

## 9.4 Transfer of heat

Heat is transferred from one region to another in three ways: by conduction, convection, and radiation (Fig. 9.2).

### 9.4.1 Conduction

If one end of a solid rod is placed in the proximity of a heat source such as a fire, after some time, the other end of the rod will become hot. In this case, heat has been transferred from the fire through the rod by conduction. The process of heat conduction involves the increase of internal energy in the material. The heat enters one end of the rod and increases the internal energy of the atoms near the heat source. In a solid material, the internal energy is in the vibration of the bound atoms and in the random motion of free electrons, which exist in some materials. The addition of heat increases both the random atomic vibrations and the speed of the electrons. The increased vibrational motion is transferred down the rod through collisions with neighboring atoms. However, because the atoms in the solid are tightly bound, their motion is restricted. Therefore, heat transfer via atomic vibrations is slow.

In some materials, the electrons in the atoms have enough energy to break loose from a specific nucleus and move freely through the material. The electrons move rapidly through the material so that, when they gain energy, they transfer it quickly to adjacent electrons and atoms. In this way, free electrons transfer the increase in the internal energy down the rod. Materials such as metals, which contain free electrons, are good conductors of heat; materials such as wood, which do not have free electrons, are insulators.

The amount of heat $H_c$ conducted per second through a block of material (see Fig. 9.3) is given by

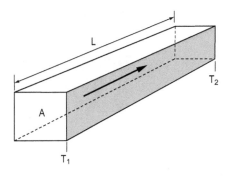

FIGURE 9.3    **Heat flow through a block of material.**

$$H_c = \frac{K_c A}{L} (T_1 - T_2) \qquad (9.3)$$

Here A is the area of the block perpendicular to the heat flux, L is its length, and $T_1 - T_2$ is the temperature difference between the two ends. The constant $K_c$ is the *coefficient of thermal conductivity* of the material, defined as the quantity of heat conducted per unit time through a unit area of a slab of unit length *L* (or thickness) when the temperature difference between its ends is 1 C° (or 1 K). In most physics publications, $K_c$ is given in units of J/s m K = W/m K. However, for problems involving living systems, it is often more convenient to express $K_c$ in units of $\frac{(Cal\ cm)}{(m^2\ h\ C°)}$. This is the amount of heat (in Cal units) that flows per 1 C° temperature difference per hour, through a slab of material 1 cm thick and 1 $m^2$ area. The thermal conductivity of a few materials is given in Table 9.2. (Note 1 C° = 1 K.)

## 9.4.2 Convection

In solids, heat transfer occurs by *conduction*; in fluids (gases and liquids), heat transfer proceeds primarily by *convection*. When a liquid or a gas is heated, the molecules near the heat source gain energy and tend to move away from the

TABLE 9.2 Thermal conductivity of some materials.

| Material | Thermal conductivity, $K_c$ (Cal cm/$m^2$ h°C) |
|---|---|
| Silver | $3.6 \times 10^4$ |
| Cork | 3.6 |
| Tissue (unperfused) | 18 |
| Felt and down | 0.36 |
| Aluminum | $1.76 \times 10^4$ |

heat source. Therefore, the fluid near the heat source becomes less dense. Fluid from the denser region flows into the rarefied region, causing convection currents. These currents carry energy away from the heat source. When the energetic molecules in the heated convection current come in contact with a solid material, they transfer some of their energy to the atoms of the solid, increasing the internal energy of the solid. In this way, heat is coupled into a solid. The amount of heat transferred by convection per unit time $H'_c$ is given by

$$H'_c = K'_c A(T_1 - T_2) \tag{9.4}$$

Here $A$ is the area exposed to convective currents perpendicular to the heat flux, $(T_1 - T_2)$ is the temperature difference between the surface and the convective fluid, and $K'_c$ is the *coefficient of convection*, which is usually a function of the velocity of the convective fluid.

### 9.4.3 Radiation

Vibrating electrically charged particles emit electromagnetic radiation, which propagates away from the source at the speed of light. Electromagnetic radiation is itself energy (called *electromagnetic energy*), which in the case of a moving charge is obtained from the kinetic energy of the charged particle.

Because of internal energy, particles in a material are in constant random motion. Both the positively charged nuclei and the negatively charged electrons vibrate and, therefore, emit electromagnetic radiation. In this way, internal energy is converted into radiation, called *thermal radiation*. Due to the loss of internal energy, the material cools. The amount of radiation emitted by vibrating charged particles is proportional to the speed of vibration. Hot objects, therefore, emit more radiation than cold ones. Because the electrons are much lighter than the nuclei, they move faster and emit more radiant energy than the nuclei.

When a body is relatively cool, the radiation from it is in the long-wavelength region to which the eye does not respond. As the temperature (i.e., the *internal energy*) of the body increases, the wavelength of the radiation decreases. At high temperatures, some of the electromagnetic radiation is in the visible region, and the body is observed to glow.

When electromagnetic radiation impinges on an object, the charged particles (electrons) in the object are set into motion and gain kinetic energy. Electromagnetic radiation is, therefore, transformed into internal energy. The amount of radiation absorbed by a material depends on its composition. Some materials, such as carbon black, absorb most of the incident radiation. These materials are easily heated by radiation. Other materials, such as quartz and certain glasses, transmit the radiation without absorbing much of it. Metallic surfaces also reflect radiation without much absorption. Such reflecting and transmitting materials cannot be heated efficiently by radiation. The rate of emission of radiant energy $H_r$ by a unit area of a body at temperature T is

$$H_r = e\sigma T^4 \tag{9.5}$$

Here $\sigma$ is the *Stefan–Boltzmann constant*, which is $5.67 \times 10^{-8}$ W/m$^2$ K$^4$ or $5.67 \times 10^{-5}$ erg/cm$^2$ $-°$K$^4$ $-$ sec. The temperature is measured on the absolute scale, and $e$ is the *emissivity* of the surface, which depends on the temperature and nature of the surface. The value of the emissivity varies from 0 to 1. Emission and absorption of radiation are related phenomena; surfaces that are highly absorptive are also efficient emitters of radiation and have an emissivity close to 1. Conversely, surfaces that do not absorb radiation are poor emitters with a low value of emissivity.

A body at temperature $T_1$ in an environment at temperature $T_2$ will both emit and absorb radiation. The rate of energy emitted per unit area is $e\sigma T_1^4$, and the rate of energy absorbed per unit is $e\sigma T_2^4$. The values for $e$ and $\sigma$ are the same for both emission and absorption.

If a body at a temperature $T_1$ is placed in an environment at a lower temperature $T_2$, the net loss of energy from the body is

$$H_r = e\sigma\left(T_1^4 - T_2^4\right) \tag{9.6}$$

If the temperature of the body is lower than the temperature of the environment, the body gains energy at the same rate.

### 9.4.4 Diffusion and random walk

If a drop of colored solution is introduced into a still liquid, we observe that the color spreads gradually throughout the volume of the liquid. The molecules of color spread from the region of high concentration (of the initially introduced drop) to regions of lower concentration. This process is called *diffusion*.

Diffusion is the main mechanism for the delivery of oxygen and nutrients into cells and for the elimination of waste products from cells. On a large scale, diffusive motion is relatively slow (it may take hours for the colored solution in our example to diffuse over a distance of a few centimeters), but on the small scale of tissue cells, diffusive motion is fast enough to provide for the life function of cells.

Diffusion is the direct consequence of the random thermal motion of molecules. Although a detailed treatment of diffusion is beyond our scope, some of the features of diffusive motion can be deduced from simple kinetic theory.

Consider a molecule in a liquid or a gas that is moving away from the starting point 0. The molecule has a thermal velocity v and travels on average a distance L before colliding with another molecule (see Fig. 9.4). As a result of the collision, the direction of motion of the molecule is changed randomly. The

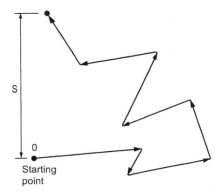

**FIGURE 9.4    Random walk.**

path may be only slightly deflected, or it may be altered substantially. On average, however, after a certain number of collisions, the molecule will be found at a distance S from the starting point. A statistical analysis of this type of motion shows that after N collisions, the distance of the molecule from the starting point is, on average,

$$S = L\sqrt{N} \tag{9.7}$$

The average distance (L) traveled between collisions is called the *mean free path*. This type of diffusive motion is called a *random walk*.

A frequently used illustration of the random walk examines the position of a drunkard walking away from a lamppost. He starts off in a particular direction, but with each step, he changes his direction of motion randomly. If the length of each step is 1 m, after taking 100 steps, he will be only 10 m away from the lamppost even if he has walked a total of 100 m. After 10,000 steps, having walked 10 km, he will be still only 100 m (on average) from his starting point.

Let us now calculate the length of time required for a molecule to diffuse a distance S from the starting point. From Eq. 9.7 the number of steps or collisions that take place while diffusing through a distance S is

$$N = \frac{S^2}{L^2} \tag{9.8}$$

The total distance traveled is the product of the number of steps and the length of each step; that is,

$$\text{Total distance} = NL = \frac{S^2}{L} \tag{9.9}$$

If the average velocity of the particle is $v$, the time $t$ required to diffuse a distance S is

$$t = \frac{\text{Total distance}}{v} = \frac{S^2}{Lv} \tag{9.10}$$

Although our treatment of diffusion has been simplified, Eq. 9.10 does lead to reasonable estimates of diffusion times. In a liquid such as water, molecules are close together. Therefore, the mean free path of a diffusing molecule is short, about $10^{-8}$ cm (this is approximately the distance between atoms in a liquid). The velocity of the molecule depends on the temperature and on its mass. At room temperature, the velocity of a light molecule may be about $10^4$ cm/sec. From Eq. 9.10, the time required for molecules to diffuse a distance of 1 cm is

$$t = \frac{S^2}{Lv} = \frac{(1)^2}{10^{-8} \times 10^4} = 10^4 \text{ seconds} = 2.8 \text{ hours}$$

However, the time required to diffuse a distance of $10^{-3}$ cm, which is the typical size of a tissue cell, is only $10^{-2}$ seconds (see Exercise 9.3a).

Gases are less densely packed than liquids; consequently, in gases the mean free path is longer and the diffusion time shorter. In a gas at 1 atm pressure, the mean free path is on the order of $10^{-5}$ cm—the exact value depends on the specific gas. The time required to diffuse a distance of 1 cm is about 10 seconds. Diffusion through a distance of $10^{-3}$ cm takes only $10^{-5}$ seconds (see Exercise 9.3b).

### 9.4.5 Random walk and signal-to-noise ratio

The random walk concept that we applied in the previous section to the process of diffusion has numerous other applications. It has been used to model the behavior of stock prices and to simulate growth of epidemics and the spreading of fake news. In fact, the random walk approach can be applied to any process that contains a random component.

In obtaining a solution to the diffusion problem we made several simplifying assumption. As a result, the answer we obtained is only approximate. Still this approximate approach provides a deeper insight into the nature of the process and yields a solution that is usually adequately close to the more difficult exact solution.

Here we will use the solution we obtained in the previous section for diffusion to analyze the nature of the signal-to-noise parameter. Most signals that we detect, for example, a voltage generated by a biological process, have some noise associated with the signal of interest. The quality of the signal, that is, how unambiguously we can interpret the signal, is usually measured by the signal-to-noise ratio. Here we will consider the signal and noise as two separate entities. In turning to the diffusion process we encountered in the previous section (9.4.4), we will consider the signal as analogous to a person walking in a straight line away from the starting point. In the language of the diffusion problem, the size of the signal ($s_s$), that is, the straight line distance traveled, is: $s_s = NL$. Here N is the number of steps during the measurement time t, and L is the length of the step (i.e., the mean free path). The size of the noise signal ($s_n$) from Eq. 9.7 is $s_n = L/N^{1/2}$. Therefore the signal-to-noise ratio is

$$\text{Signal-to-noise ratio} = s_s/s_n = NL/L/N^{1/2} = N^{1/2}$$

Now N has to be expressed in terms of the measurement time t, $t = NL/v$. In the diffusion problem NL is the total linear distance traveled and v is the speed of travel. Because in this model both the mean free path L and the speed v are constant through the measurement, the ratio v/L is a constant, here designated as k. Therefore,

$$N^{1/2} = kt^{1/2}, \text{ where } k = (v/L)^{1/2} \text{ and the signal-to-noise ratio}$$
$$= s_s/s_n = N^{1/2} = kt^{1/2}$$

Thus the signal-to-noise ratio is proportional to square root of the measurement time t. Were we to increase the measurement time, say from 1 second to 100 seconds, the signal-to-noise ratio would increase by a factor of 10.

## 9.5 Transport of molecules by diffusion

We will now calculate the number of molecules transported by diffusion from one region to another. Consider a cylinder containing a nonuniform distribution of diffusing molecules or other small particles (see Fig. 9.5). At position $x = 0$, the density of the diffusing molecules is $C_1$. At a small distance $\Delta x$ away from this point, the concentration is $C_2$. We can define a diffusion velocity $V_D$ as the average speed of diffusion from $x = 0$ to $x = \Delta x$. This velocity is simply the distance $\Delta x$ divided by the average time for diffusion t; that is,

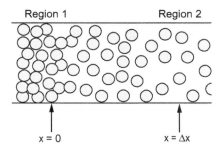

Region 1 Region 2

$x = 0$ $x = \Delta x$

**FIGURE 9.5** **Diffusion.**

$$V_D = \frac{\Delta x}{t}$$

Substituting $t = (\Delta x)^2/Lv$ from Eq. 9.10, we obtain

$$V_D = \frac{\Delta x}{(\Delta x)^2/Lv} = \frac{Lv}{\Delta x} \qquad (9.11)$$

(Remember that $v$ here is the thermal velocity.) The number of molecules J arriving per second per unit area, from region 1, where the density is $C_1$, to region 2 (see Exercise 9.4) is

$$J_1 = \frac{V_D C_1}{2} \qquad (9.12)$$

The factor of 2 in the denominator accounts for the fact that molecules are diffusing both toward and away from region 2. The term J is called the *flux*, and it is in units $(cm^{-2}s^{-1})$.

At the same time, molecules are also diffusing from region 2, where the density is $C_2$ toward region 1. This flux $J_2$ is

$$J_2 = \frac{V_D C_2}{2}$$

The net flux of molecules into region 2 is the difference between the arriving and the departing flux, which is

$$J = J_1 - J_2 = \frac{V_D (C_1 - C_2)}{2}$$

Substituting for $V_D = Lv/\Delta x$, we obtain

$$J = \frac{Lv (C_1 - C_2)}{2\Delta x} \qquad (9.13)$$

This derivation assumes that the velocities $v$ in the two regions are the same. Although this solution for the diffusion problem is not exact, it does illustrate the nature of the diffusion process. (For a more rigorous treatment, see, for example, Morgan, 1969). The net flux from one region to another depends on the difference in the density of the diffusing particles in the two regions. The flux increases with thermal velocity $v$ and decreases with the distance between the two regions.

$$J = \frac{D}{\Delta x} (C_1 - C_2) \qquad (9.14)$$

where D is called the *diffusion coefficient*. In our case, the diffusion coefficient is simply

$$D = \frac{Lv}{2} \tag{9.15}$$

In general, however, the diffusion coefficient is a more complex function because the mean free path L depends on the size of the molecule and the viscosity of the diffusing medium. In our previous illustration of diffusion through a fluid, where $L = 10^{-8}$ cm and $v = 10^4$ cm/sec, the diffusion coefficient calculated from Eq. 9.15 is $5 \times 10^{-5}$ cm$^2$/sec. By comparison, the measured diffusion coefficient of salt (NaCl) in water, for example, is $1.09 \times 10^{-5}$ cm$^2$/sec. Thus, our simple calculation gives a reasonable estimate for the diffusion coefficient. Larger molecules, of course, have a smaller diffusion coefficient. The diffusion coefficients for biologically important molecules are in the range from $10^{-7}$ to $10^{-6}$ cm$^2$/sec.

## 9.6 Diffusion through membranes

So far, we have discussed only free diffusion through a fluid, but the cells constituting living systems are surrounded by membranes that impede free diffusion. Oxygen, nutrients, and waste products must pass through these membranes to maintain the life functions. In the simplest model, the biological membrane can be regarded as porous, with the size and the density of the pores governing the diffusion through the membrane. If the diffusing molecule is smaller than the size of the pores, the only effect of the membrane is to reduce the effective diffusion area and thus decrease the diffusion rate. If the diffusing molecule is larger than the size of the pores, the flow of molecules through the membrane may be barred. (See Fig. 9.6.) (Some molecules may still get through the membrane, however, by dissolving into the membrane material.)

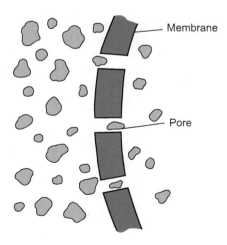

FIGURE 9.6 **Diffusion through a membrane.**

The net flux of molecules J flowing through a membrane is given in terms of the permeability of the membrane P:

$$J = P(C_1 - C_2) \tag{9.16}$$

This equation is similar to Eq. 9.14 except that the term D is replaced by the permeability P, which includes the diffusion coefficient as well as the effective thickness $\Delta x$ of the membrane. The permeability depends, of course, on the type of membrane as well as on the diffusing molecule. Permeability may be nearly zero (if the molecules cannot pass through the membrane) or as high as $10^{-4}$ cm/sec.

The dependence of permeability on the diffusing species allows the cell to maintain a composition different from that of the surrounding environment. Many membranes, for example, are permeable to water but do not pass molecules dissolved in water. As a result, water can enter the cell, but the components of the cell cannot pass out of the cell. Such a one-way passage of water is called *osmosis*.

In the type of diffusive motion we have discussed so far, the movement of the molecules is due to their thermal kinetic energy. Some materials, however, are transported through membranes with the aid of electric fields that are generated by charge differences across the membrane. This type of transport will be discussed in Chapter 13.

We have shown that over distances larger than a few millimeters, diffusion is a slow process. Therefore, large living organisms must use circulating systems to transport oxygen, nutrients, and waste products to and from the cells. The evolution of the respiratory system in animals is a direct consequence of the inadequacy of diffusive transportation over long distances.

## 9.7 The respiratory system

As will be shown in the following two chapters, animals require energy to function. This energy is provided by food, which is oxidized by the body. On average, 0.207 L of oxygen at 760 torr is required for every Cal of energy released by the oxidation of food in the body. At rest, an average 70-kg adult requires about 70 Cal of energy per hour, which implies a consumption of 14.5 L of $O_2$ per hour, which is about $10^{20}$ oxygen molecules per second (see Exercise 9.5).

The simplest way to obtain the required oxygen is by diffusion through the skin. This method, however, cannot supply the needs of large animals. It has been determined that in a person only about 2% of oxygen consumed at rest is obtained by diffusion through the skin. The rest of the oxygen is obtained through the lungs.

The lungs can be thought of as an elastic bag suspended in the chest cavity (see Fig. 9.7). When the diaphragm descends, the volume of the lungs

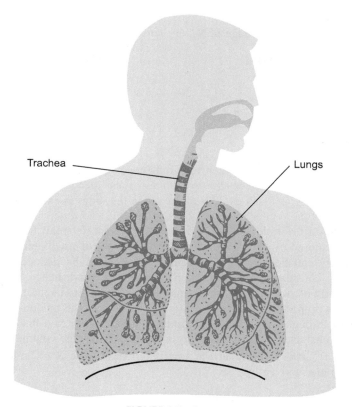

FIGURE 9.7   Lungs.

increases, causing a reduction in gas pressure inside the lungs. As a result, air enters the lungs through the trachea. The trachea branches into smaller and smaller tubes, which finally terminate at tiny cavities called *alveoli*. It is here that gas is exchanged by diffusion between the blood and the air in the lungs. The lungs of an adult contain about 300 million alveoli with diameters ranging between 0.1 and 0.3 mm. The total alveolar area of the lungs is about 100 m², which is about 50 times larger than the total surface area of the skin. The barrier between the alveolar air and the blood in the capillaries is very thin, only about $4 \times 10^{-5}$ cm. Therefore, the gas exchange of oxygen into the blood and $CO_2$ out of the blood is very fast.

The lungs are not fully emptied and filled with each breath. In fact, the full volume of the lungs is about 6 L, and at rest, only about $\frac{1}{2}$ L is exchanged during each breath. The composition of inspired and expired air is shown in Table 9.3.

Using the experimental data in Table 9.3, we can easily show that about 10.5 breaths per minute satisfy the oxygen requirements for a resting person

**TABLE 9.3** The percentage of $N_2$, $O_2$, and $CO_2$ in inspired and expired air for a resting person.

|  | $N_2$ | $O_2$ | $CO_2$ |
|---|---|---|---|
| Inspired air | 79.02 | 20.94 | 0.04 |
| Expired air | 79.2 | 16.3 | 4.5 |

(see Exercise 9.6). The oxygen requirement, of course, rises with increased physical activity, which results in both faster and deeper breathing. During deep breathing, as much as 70% of the air in the lungs is exchanged in each breath.

While diffusion through the skin can supply only a small fraction of the oxygen required by large animals, the oxygen needs of small animals may be completely satisfied through this channel. This can be deduced from the following considerations. The energy consumption and, hence, the oxygen requirement of an animal is approximately proportional to its mass.[3] The mass in turn is proportional to the volume of the animal. The amount of oxygen diffusing through the skin is proportional to the surface area of the skin. Now, if R is a characteristic linear dimension of the animal, the volume is proportional to $R^3$, and the skin surface area is proportional to $R^2$. The surface-to-volume ratio is given by

$$\frac{\text{Surface area}}{\text{Volume}} = \frac{R^2}{R^3} = \frac{1}{R} \tag{9.17}$$

Therefore, as the size of the animal R decreases, its surface-to-volume ratio increases; that is, for a unit volume, a small animal has a greater surface area than a large animal.

It is possible to obtain an estimate for the maximum size of the animal that can get its oxygen entirely by skin diffusion. A highly simplified calculation outlined in Exercise 9.7 shows that the maximum linear size of such an animal is about 0.5 cm. Therefore, only small animals, such as insects, can rely entirely on the diffusion transfer to provide them with oxygen. However, during hibernation, when the oxygen requirements of the animal are reduced to a very low value, larger animals such as frogs can obtain all the necessary oxygen through their skin. In fact, some species of frog hibernate through the winter at the bottom of lakes where the temperature is constantly at 4°C. The required oxygen enters the frog's body by diffusion from the surrounding water, which contains dissolved oxygen.

---

3. This is an approximation. A more detailed discussion is found in Rose (1967).

## 9.8 Surfactants and breathing

The discussion in the previous section neglected an important aspect of breathing: the size of the alveoli. As was stated, the diameters of the alveoli range from about 0.1 to 0.3 mm (radius R from 0.05 to 0.15 mm). The inner wall of the alveoli is coated with a thin layer of water that protects the tissue. The surface tension of this water layer tends to minimize the surface, thereby shrinking the alveolar cavity. When the diaphragm descends, the incoming air has to enter the alveoli and expand them to their full size. Because the alveoli are embedded in a moist medium, expanding the alveoli is analogous to creating a bubble inside a liquid. As was discussed in Section 7.7, to create a gas bubble of radius R in a liquid with surface tension T, the pressure of the gas injected into the liquid must be greater than the pressure of the surrounding liquid by $\Delta P$ as given in Eq. 7.21. As is shown in Exercise 9.8, $\Delta P$ required to expand an alveolus to its full volume is 0.0287 atm. This is the minimum pressure required to open a 0:05-mm-radius alveolus that has its walls coated with plain water. In a typical well-functioning lung the pressure difference between exhale and inhale is only about 0.0013 atm, which is not adequate to inflate a water-coated alveolus without surfactant.

Breathing is made possible by surfactants that cover the alveolar water layer and greatly reduce its surface tension. These surfactant molecules are a complex mixture of lipids and proteins produced by special cells in the alveoli and they can reduce surface tension by as much as a factor of 70 (to about 1 dyn/cm), but more typically by a factor of 50. With the surfactant, $\Delta P$ required to expand an alveolus to its full volume is 0.0287/50 = 0.000574 atm. With the surfactant coating, the alveoli $\Delta P$ is adequate to expand them.

The lungs of premature infants often fail to produce adequate amounts of surfactants required for breathing. This life-threatening condition called infant respiratory distress syndrome can now be treated with artificial lung surfactants developed in the 1980s. When introduced into the lungs of the infant, these surfactants often stabilize breathing till the alveoli begin to produce surfactants on their own.

Cold-blooded animals such as frogs, snakes, and lizards do not need lung surfactants for breathing. Such animals do not use energy to heat their bodies. As a result, they require about a factor of 10 less oxygen than warm-blooded animals of comparable size. Therefore, cold-blooded animals can function with correspondingly smaller lung surface area. The alveolar radii of these animals are 10 times larger than those of warm-blooded animals (see Exercise 9.9). An alveolus of larger radius requires correspondingly lower pressure to overcome surface tension eliminating the need for lung surfactants.

## 9.9 Diffusion and contact lenses

Most parts of the human body receive the required oxygen from the circulating blood. However, the *cornea*, which is the transparent surface layer of the eye, does not contain blood vessels (this allows it to be transparent). The cells in the cornea receive oxygen by diffusion from the surface layer of tear fluid, which contains dissolved oxygen. This fact allows us to understand why most contact lenses should not be worn during sleep. The contact lens is fitted so that blinking rocks the lens slightly. This rocking motion brings fresh oxygen-rich tear fluid under the lens. Of course, when people sleep, they do not blink; therefore, the corneas under their contact lenses are deprived of oxygen. This may result in a loss of corneal transparency.

## Exercises

**9.1.** Fish using air bladders to control their buoyancy are less stable than those using porous bones. Explain this phenomenon using the gas equation (Eq. 9.2). (*Hint*: What happens to the air bladder as the fish sinks to a greater depth?)

**9.2.** A scuba diver breathes air from a tank that has a pressure regulator that automatically adjusts the pressure of the inhaled air to the ambient pressure. If a diver 40 m below the surface of a deep lake fills his lungs to the full capacity of 6 L and then rises quickly to the surface, to what volume will his lungs expand? Is such a rapid ascent advisable?

**9.3.** (a) Calculate the time required for molecules to diffuse in a liquid at a distance of $10^{-3}$ cm. Assume that the average velocity of the molecules is $10^4$ cm/sec and that the mean free path is $10^{-8}$ cm. (b) Repeat the calculation for diffusion in a gas at 1 atm pressure, where the mean free path is $10^{-5}$ cm.

**9.4.** Consider a beam of particles traveling at a velocity $V_D$. If the area of the beam is A and the density of particle in the beam is C, show that the number of particles that pass by a given point each second is $V_D \times C \times A$.

**9.5.** A consumption of 14.5 L of oxygen per hour is equivalent to how many molecules per second? (The number of molecules per cubic centimeter at $0°C$ and 760 torr is $2.69 \times 10^{19}$.)

**9.6.** Using the data in the text and in Table 9.3, calculate the number of breaths per minute required to satisfy the oxygen needs of a resting person.

**9.7.** (a) We stated in the text that the oxygen consumption at rest for a 70-kg person is 14.5 L/h and that 2% of this requirement is provided by the diffusion of oxygen through the skin. Assuming that the skin surface area of the person is 1.7 $m^2$, calculate the diffusion rate for oxygen through the skin in L/h $cm^2$. (b) What is the maximum linear size of an animal whose oxygen requirements at rest can be provided by diffusion through the skin?

Use the following assumptions:

    i. The density of animal tissue is 1 $g/cm^3$.

    ii. Per unit volume, all animals require the same amount of oxygen.

    iii. The animal is spherical in shape.

**9.8.** Calculate the excess pressure $\Delta P$ required to expand a 0.05 mm radius alveolus to its full volume.

**9.9.** Show that if the oxygen requirement of an animal is reduced by a factor of 10, then within the same lung volume, alveolar radius can be increased by a factor of 10.

# Chapter 10

# Thermodynamics

Thermodynamics is the study of the relationship between heat, work, and the associated flow of energy. After many decades of experience with heat phenomena, scientists have formulated two fundamental laws as the foundation of thermodynamics. The first law of thermodynamics states that energy, which includes heat, is conserved; that is, one form of energy can be converted into another, but energy can neither be created nor destroyed. This implies that the total amount of energy in the universe is a constant.[1] The second law, more complex than the first, can be stated in a number of ways which, although they appear different, can be shown to be equivalent. Perhaps the simplest statement of the second law of thermodynamics is that *spontaneous change in nature occurs from a state of order to a state of disorder.*

## 10.1 First Law of Thermodynamics

One of the first to state the law of energy conservation was the German physician Robert Mayer (1814–1878). In 1840 Mayer was the physician on the schooner *Java*, which sailed for the East Indies. While aboard the ship, he was reading a treatise by the French scientist Laurent Lavoisier in which Lavoisier suggested that the heat produced by animals is due to the slow combustion of food in their bodies. Lavoisier further noted that less food is burned by the body in a hot environment than in a cold one.

When the ship reached the tropics, many of its crew became sick with fever. Applying the usual remedy for fever, Mayer bled his patients. He noticed that the venous blood, which is normally dark red, was nearly as red as arterial blood. He considered this a verification of Lavoisier's suggestion. Because in the tropics less fuel is burned in the body, the oxygen content of the venal blood is high, giving it a brighter color. Mayer then went beyond Lavoisier's theory and suggested that in the body, there is an exact balance of energy (which he called *force*). The energy released by the food is balanced by the lost body heat and the work done by the body. Mayer wrote in an article published in 1842, "Once in existence, force [energy] cannot be annihilated—it can only change its form."

---

1. It has been shown by the theory of relativity that the conservation law must include matter which is convertible to energy.

Physics in Biology and Medicine. https://doi.org/10.1016/B978-0-443-21558-2.00010-9

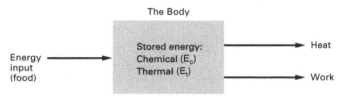

FIGURE 10.1 **The energetics of the body.**

Considerably more evidence had to be presented before conservation of energy was accepted as a law, but it is interesting that such a fundamental physical law was first suggested from the observation of human physiology.

Conservation of energy is implicit in all our calculations of energy balance in living systems. Consider, for example, the energetics for the functioning of an animal (see Fig. 10.1). The body of an animal contains internal thermal energy $E_t$, which is the product of the mass and specific heat, and chemical energy $E_c$ stored in the tissue of the body. In terms of energy, the activities of an animal consist of simply eating, working, and rejecting excess heat by means of various cooling mechanisms (radiation, convection, etc.). Without going into detailed calculations, the first law allows us to draw some conclusions about the energetics of the animal. For example, if the internal temperature and the weight of the animal are to remain constant (i.e., $E_c$ and $E_t$ constant), over a given period of time, the energy intake must be exactly equal to the sum of the work done and the heat lost by the body. An imbalance between intake and output energy implies a change in the sum $E_c + E_t$. The first law of thermodynamics is implicit in all the numerical calculations presented in Chapter 11.

## 10.2 Second Law of Thermodynamics

There are many imaginable phenomena that are not forbidden by the first law of thermodynamics but still do not occur. For example, when an object falls from a table to the ground, its potential energy is first converted into kinetic energy; then, as the object comes to rest on the ground, the kinetic energy is converted into heat. The first law of thermodynamics does not forbid the reverse process, whereby the heat from the floor enters the object and gets converted into kinetic energy, causing the object to jump back on the table. Yet this event does not occur. Experience has shown that certain types of events are irreversible. Broken objects do not mend by themselves. Spilled water does not collect itself back into a container. The irreversibility of these types of events is intimately connected with the probabilistic behavior of systems comprised of a large ensemble of subunits.

As an example, consider three coins arranged heads up on a tray. We will consider this an ordered arrangement. Suppose that we now shake the tray so that each coin has an equal chance of landing on the tray with either head or tail up. The possible arrangements of coins that we may obtain are shown in Table 10.1. Note that there are eight possible outcomes of tossing the three coins. Of these, only one yields the original ordered arrangement of three heads (H, H, H). Because the probabilities of obtaining any one of the coin arrangements in Table 10.1 are the same, the probability of obtaining the three-head arrangement after shaking the tray once is 1/8, or 0.125; that is, on average, we must toss the coins eight times before we can expect to see the three-head arrangement again.

As the number of coins in the experiment is increased, the probability of returning to the ordered arrangement of all heads decreases. With ten coins on the tray, the probability of obtaining all heads after shaking the tray is 0.001. With 1000 coins, the probability of obtaining all heads is so small as to be negligible. We could shake the tray for many years without seeing the ordered arrangement again. In summary, the following is to be noted from this illustration: The number of possible coin arrangements is large, and only one of them is the ordered arrangement; therefore, although any one of the coin arrangements—including the ordered one—is equally likely, the probability of returning to an ordered arrangement is small. As the number of coins in the ensemble increases, the probability of returning to an ordered arrangement decreases. In other words, if we disturb an ordered arrangement, it is likely to become disordered. This type of behavior is characteristic of all events that involve a collective behavior of many components.

The second law of thermodynamics is a statement about the type of probabilistic behavior illustrated by our coin experiment. One statement of the second law is: *The direction of spontaneous change in a system is from an arrangement of lesser probability to an arrangement of greater probability*;

**TABLE 10.1 The ordering of three coins.**

| Coin 1 | Coin 2 | Coin 3 |
| --- | --- | --- |
| H | H | H |
| H | H | T |
| H | T | H |
| T | H | H |
| H | T | T |
| T | H | T |
| T | T | H |
| T | T | T |

that is, from order to disorder. This statement may seem to be so obvious as to be trivial, but once the universal applicability of the second law is recognized, its implications are seen to be enormous. We can deduce from the second law the limitations on information transmission, the meaning of time sequence, and even the fate of the universe. These subjects, however, are beyond the scope of our discussion.

One important implication of the second law is the limitation on the conversion of heat and internal energy to work. This restriction can be understood by examining the difference between heat and other forms of energy.

## 10.3 Difference between heat and other forms of energy

We defined heat as energy being transferred from a hotter to a colder body. Yet when we examined the details of this energy transfer, we saw that it could be attributed to transfer of a specific type of energy such as kinetic, vibrational, electromagnetic, or any combination of these (see Chapter 9). For this reason, it may not seem obvious why the concept of heat is necessary. It is, in fact, possible to develop a theory of thermodynamics without using the concept of heat explicitly, but we would then have to deal with each type of energy transfer separately, and this would be difficult and cumbersome. In many cases, energy is being transferred to or from a body via different methods, and keeping track of each of these is often not possible and usually not necessary. No matter how energy enters the body, its effect is the same. It raises the internal energy of the body. The concept of heat energy is, therefore, very useful.

The main feature that distinguishes heat from other forms of energy is the random nature of its manifestations. For example, when heat flows via conduction from one part of the material to another part, the flow occurs through the sequential increase in the internal energy along the material. This internal energy is in the form of random chaotic motion of atoms. Similarly, when heat is transferred by radiation, the propagating waves travel in random directions. The radiation is emitted over a wide wavelength (color) range, and the phases of the wave along the wave front are random. By comparison, other forms of energy are more ordered. Chemical energy, for example, is present by virtue of specific arrangements of atoms in a molecule. Potential energy is due to the well-defined position, or configuration, of an object.

While one form of energy can be converted to another, heat energy, because of its random nature, cannot be completely converted to other forms of energy. We will use the behavior of a gas to illustrate our discussion. First, let us examine how heat is converted to work in a heat engine (e.g., the steam engine). Consider a gas in a cylinder with a piston (see Fig. 10.2). Heat flows into the gas; this increases the kinetic energy of the gas molecules and, therefore, raises the internal energy of the gas. The molecules moving in the direction of the piston collide with the piston and exert a force on it. Under the

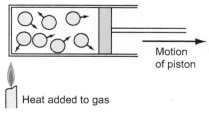

Motion
of piston

Heat added to gas

FIGURE 10.2   **The motion of a piston.**

influence of this force, the piston moves. In this way, heat is converted into work via internal energy.

The heat added to the gas causes the molecules in the cylinder to move in random directions, but only the molecules that move in the direction of the piston can exert a force on it. Therefore, the kinetic energy of only the molecules that move toward the piston can be converted into work. For the added heat to be completely converted into work, all the gas molecules would have to move in the direction of the piston motion. In a large ensemble of molecules, this is very unlikely.

The odds against the complete conversion of 1 cal of heat into work can be expressed in terms of a group of monkeys who are hitting typewriter keys at random and who by chance type out the complete works of Shakespeare without error. The probability that 1 cal of heat would be completely converted to work is about the same as the probability that the monkeys would type Shakespeare's works 15 quadrillion times in succession. (This example is taken from Angrist (1968).)

The distinction between work and heat is this: In work, the energy is in an ordered motion; in heat, the energy is in random motion. Although some of the random thermal motion can be ordered again, the ordering of all the motion is very improbable. Because the probability of completely converting heat to work is vanishingly small, the second law of thermodynamics states categorically that it is impossible.

Heat can be partially converted to work as it flows from a region of higher temperature $T_1$ to a region of lower temperature $T_2$ (see Fig. 10.3). A quantitative treatment of thermodynamics shows (see for e.g., Casey, 1962) that the maximum ratio of work to the input heat is

$$\frac{\text{Work}}{\text{Heat input}} = 1 - \frac{T_2}{T_1} \tag{10.1}$$

Here the temperature is measured on the absolute scale.

From this equation, it is evident that heat can be completely converted into work only if the heat is rejected into a reservoir at absolute zero temperature. Although objects can be cooled to within a very small fraction of absolute zero, absolute zero cannot be attained. Therefore, heat cannot be completely

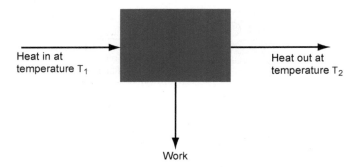

Heat in at temperature $T_1$

Heat out at temperature $T_2$

Work

**FIGURE 10.3   Conversion of heat to work.**

converted into work. Exercises 10.1 and 10.2 provide further exploration of the Second Law of Thermodynamics.

## 10.4 Thermodynamics of living systems

It is obvious that animals need food to live, but the reason for this is less obvious. The idea that animals need energy because they consume energy is, strictly speaking, incorrect. We know from the first law of thermodynamics that energy is conserved. The body does not consume energy; it changes it from one form to another. In fact, the first law could lead us to the erroneous conclusion that animals should be able to function without a source of external energy. The body takes in energy that is in the chemical bonds of the food molecules and converts it to heat. If the weight and the temperature of the body remain constant and if the body performs no external work, the energy input to the body equals exactly the heat energy leaving the body. We may suppose that if the heat outflow could be stopped—by good insulation, for example—the body could survive without food. As we know, this supposition is wrong. The need for energy is made apparent by examining the functioning of the body in light of the second law of thermodynamics.

The body is a highly ordered system. A single protein molecule in the body may consist of a million atoms bound together in an ordered sequence. Cells are even more complex. Their specialized functions within the body depend on a specific structure and location. We know from the second law of thermodynamics that such a highly ordered system, left to itself, tends to become disordered, and once it is disordered, it ceases to function. Work must be done on the system continuously to prevent it from falling apart. For example, the blood circulating in veins and arteries is subject to friction, which changes kinetic energy to heat and slows the flow of blood. If a force were not applied to the blood, it would stop flowing in a few seconds. The concentration of minerals inside a cell differs from that in the surrounding environment. This represents an ordered arrangement. The natural tendency is toward an equalization with the environment. Work must be done to prevent the contents of the cell from leaking out. Finally, cells that die must be replaced, and if the animal is growing, new tissue must be manufactured. For such replacement and growth, new proteins and other cell constituents must be put together from smaller, relatively more random subcomponents. Thus, the process of life consists of building and maintaining ordered structures. In the face of the natural tendency toward disorder, this activity requires work. The situation is somewhat analogous to a pillar made of small, slippery, uneven blocks that tend to slide out of the structure. The pillar remains standing only if blocks are continuously pushed back.

The work necessary to maintain the ordered structures in the body is obtained from the chemical energy in food. Except for the energy utilized in external work done by the muscles, all the energy provided by food is ultimately converted into heat by friction and other dissipative processes in the body. Once the temperature of the body is at the desired level, all the heat generated by the body must leave through the various cooling mechanisms of the body (see Chapter 11). The heat must be dissipated because, unlike heat engines (such as the turbine or the steam engine), the body does not have the ability to obtain work from heat energy. The body can obtain work only from chemical energy. Even if the body did have mechanisms for using heat to perform work, the amount of work it could obtain in this way would be small. Once again, the second law sets the limit. The temperature differences in the body are small—not more than about 7°C between the interior and the exterior. With the interior temperature $T_1$ at 310 K (37°C) and the exterior temperature $T_1$ at 303 K, the efficiency of heat conversion to work would be (from Eq. 10.1) at most only about 2%.

Of all the various forms of energy, the body can utilize only the chemical binding energy of the molecules which constitute food. The body does not have a mechanism to convert the other forms of energy into work. A person could bask in the sun indefinitely, receiving large quantities of radiant energy, and yet die of starvation. Plants, on the other hand, are able to utilize radiant energy. Just as animals use chemical energy, plants utilize solar radiation to provide the energy for the ordering processes necessary for life.

The organic materials produced in the life cycle of plants provide food energy for herbivorous animals, which in turn are food for the carnivorous animals that eat them. The sun is, thus, the ultimate source of energy for life on Earth.

Since living systems create order out of relative disorder (e.g., by synthesizing large complex molecules out of randomly arranged subunits), it may appear at first glance that they violate the second law of thermodynamics, but this is not the case. To ascertain that the second law is valid, we must examine the whole process of life, which includes not only the living unit but also the energy that it consumes and the by-products that it rejects. To begin with, the food that is consumed by an animal contains a considerable degree of order. The atoms in the food molecules are not randomly arranged but are ordered in specific patterns. When the chemical energy in the molecular bindings of the food is released, the ordered structures are broken down. The eliminated waste products are considerably more disordered than the food taken in. The ordered chemical energy is converted by the body into disordered heat energy.

The amount of disorder in a system can be expressed quantitatively by means of a concept called *entropy*. Calculations show that, in all cases, the increase in the entropy (disorder) in the surroundings produced by the living system is always greater than the decrease in entropy (i.e., ordering) obtained in the living system itself. The total process of life, therefore, obeys the second law. Thus, living systems are perturbations in the flow toward disorder. They keep themselves ordered for a while at the expense of the environment. This is a difficult task requiring the use of the most complex mechanisms found in nature. When these mechanisms fail, as they eventually must, the order falls apart, and the organism dies. Exercises 10.3 provides further exploration of the Second Law of Thermodynamics, life and information.

## 10.5 Information and the second law

We have stressed earlier that work must be done to create and maintain the highly ordered local state of life. We now turn to the question, what else is needed for such local ordering to occur? Perhaps we can get an insight into this issue from a simple everyday experience. In the course of time, our apartment becomes disordered. Books, which had been placed neatly, in alphabetical order, on a shelf in the living room, are now strewn on the table and some are even under the bed. Dishes that were clean and neatly stacked in the cupboard are now dirty with half-eaten food and are on the living room table. We decide to clean up, and in 15 minutes or so the apartment is back in order. The books are neatly shelved, and the dishes are clean and stacked in the kitchen. The apartment is clean.

Two factors were necessary for this process to occur. First, as was already stated, energy was required to do the work of gathering and stacking the books and cleaning and ordering the dishes. Second, and just as important, information was required to direct the work in the appropriate direction. We had to know where to place the books and how to clean the dishes and stack them just so. The concept of information is of central importance here.

In the 1940s, Claude Shannon developed a quantitative formulation for the amount of information available in a given system. Shannon's formula for information content is shown to be equivalent to the formula for entropy—the measure of disorder—except, with a negative sign. This mathematical insight formally shows that if energy and information are available, the entropy in a given locality can be decreased by the amount of information available to engage in the process of ordering. In other words, as in our example of the messy living room, order can be created in a disordered system by work that is directed by appropriate information. The second law, of course, remains valid:

*the overall entropy of the universe increases.* The work required to perform the ordering, one way or another, causes a greater disorder in the surroundings than the order that was created in the system itself. It is the availability of information and energy that allows living systems to replicate, grow, and maintain their structures.

The chain of life begins with plants that possess information in their genetic material on how to utilize the energy from the sun to make highly ordered complex structures from the simple molecules available to them: principally water, carbon dioxide, and an assortment of minerals. The process is, in essence, similar in human beings and other animals. All the information required for the function of the organism is contained in the intricate structure of DNA. Human DNA consists of about a billion molecular units in a well-determined sequence. Utilizing the energy obtained from the food that is consumed by the organism, the information in the DNA guides the assembly of the various proteins and enzymes required for the functioning of the organism. Exercises 10.4 and 10.5 provide further exploration of the Second Law of Thermodynamics, fractals and chaos.

## 10.6 Fractals, chaos, and the second law of thermodynamics

In this last section of the chapter on thermodynamics, we discuss two topics that are relatively new to the physics curriculum: fractals and chaos theory. These concepts relate to the second law of thermodynamics, help in the formulation of many configurations found in nature, and are likely to become useful in medical diagnosis. Our treatment of these subjects is only qualitative with the aim of acquainting students with these interesting and useful ideas.

### 10.6.1 Fractals

Conventional geometry has been highly successful in describing, manipulating, and solving problems involving regularly shaped objects such as lines, a variety of curves, surfaces, and 3-dimensional shapes such as spheres and cones. With the invention of calculus in the mid-1600s by Newton and Leibniz, it became possible to analyze and manipulate curves, surfaces, and volumes of any shape or complexity provided they were adequately smooth and could be expressed in terms of independent parameters such as time or space or other relevant parameters. However, most objects in nature are irregular to the extent that their treatment using conventional mathematical techniques is not possible. Many such irregular objects exhibit self-similar geometric patterns, that is, shapes that look similar as the scale of observation changes. Such self-similar patterns are called *fractals*. Trees and coastlines are two frequently given examples illustrating self-similarity or fractal patterns.

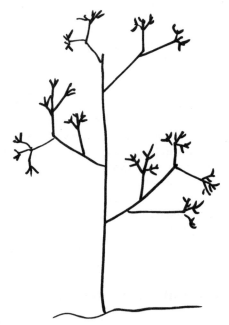

**FIGURE 10.4** **Schematic illustration of the fractal tree.**

As the tree grows, branches sprout from the trunk. Branching continues, with branches becoming smaller as one follows the growth up the tree. The process is illustrated schematically in Fig. 10.4, which shows a schematically typical branching of a tree. If one examines a small branch in isolation, without regard to the size of the branch, the pattern exhibited by the small branch is similar to the pattern of all the other branches and of the tree as a whole.

A coastline provides another often-used example of a fractal pattern. A series of photographs taken of a section of coastline from a satellite, an airplane, or a height right above ground likewise exhibit similar patterns, so much so that without some additional identifying scale, one cannot tell the height from which the photograph was taken.

Fractal configurations are everywhere in nature. Clouds, mountain ranges, rocks, crystals, and snowflakes, all display fractal configurations. In our own bodies, the blood circulation system with the arteries and veins branching into channels of decreasing diameter forms a fractal pattern, as does the structure of the lungs and the configuration of the nervous system.

Reportedly, Leibniz introduced the concept of self-similarity, and over the following two centuries, this concept was explored by several mathematicians. However, interest in this area of mathematics was minimal and progress was slow. The availability of modern computers in the 1960s greatly accelerated studies in this field. Naturally occurring self-similar patterns of the type we discussed are only approximately self-similar and the repetition of self-similarity is limited. For example, as we examine decreasing-sized tree branches for self-similarity, we are halted when inevitably we reach a point where branching is no longer occurring.

Using a variety of algorithms, computers can generate exact self-similar structures of arbitrary complexity. Benoit Mandelbrot was a pioneer in the rebirth of this field. In the 1960s and 1970s he introduced the terms fractal and fractal dimension. The concept of fractal dimension made it possible to provide a more quantitative description of self-similarity.

In conventional representation, lines are 1-dimensional, surfaces 2-dimensional, and volumes 3-dimensional. On the other hand, fractals have noninteger values of dimension. A curve may have values of fractal dimension between 1 and nearly 2, a surface between 2 and nearly 3. Fractal dimensions measure the complexity of the fractal properties, that is, the complexity of self-similarity of the pattern. A curve with a small fractal dimension, say 1.1, displays a low level of complexity and appears almost as a line. On the other hand, a curve with a fractal dimension of 1.9 displays a large complexity and appears almost as a surface. Similarly, a surface of fractal dimension 2.1 appears very much like a conventional surface, while a surface of fractal dimension, say, 2.9 is closer to a volume in space.

Several methods have been formulated for calculating fractal dimensions. A description of these techniques is beyond the scope of our discussion. Here we simply present some calculated fractal dimensions published in the literature. Coastline of Great Britain: 1.25; Cross-section of broccoli: 2.7; Surface of human brain: 2.79; Cross-section of cauliflower: 2.8; Lung surface: 2.97.

The question arises, why are so many structures found in nature fractal? There are no definitive, provable answers to this question. Several modeling approaches have been formulated to explain the fractal aspects of both biological and geological systems in terms of optimization of specific flow properties, most often efficiency of transport. Some of these formulations are quite complex. (See review by Miguel (2014).) One attribute of fractals leads to a simple possible hypothesis for biological systems. That is, fractal structures utilize a given space very effectively, providing such structures with an evolutionary advantage. The fractal structure of trees, for example, provides a highly effective utilization of sunlight required for the process of photosynthesis. The leaves sprouting out of the fractal structured branches intercept nearly all the sunlight shining on the tree. This is evident by the fully shaded area under a leaved tree. The fractal structure of lungs yields about 100 m$^2$ of area (about half a tennis court) packed in a volume of about 2000 cm$^3$ (2 L). By virtue of its fractal structure, the arteries that occupy a very small fraction of the body (about 3%) supply blood to every cell in the body.

## 10.6.2 Chaos

In the context of science, a *chaotic system* is defined as one that is highly sensitive to initial conditions. Commonly, the word *chaos* is used differently. The dictionary defines chaos as "a state of utter confusion or disorder; a total lack of organization or order." The consequences of the high sensitivity to initial conditions lead to some apparent attributes that are associated with the common usage of the term chaos. However, chaos in the context of science is not random. The process can be approached more systematically. The process can be modeled by mathematical equations and to some extent approached quantitatively. This type of chaos is often called deterministic chaos. The individual steps in a deterministic chaotic process are governed by well-known equations of classical physics. However, the sequentially combined effect of many interactions results in an outcome that is unpredictable.

Here an important distinction is to be made: The difference between linear and nonlinear processes. A linear process is one where the system response is linearly proportional to the input. That is, if the input is doubled, the output is likewise doubled. For example, if a soccer ball is kicked twice as hard, it will travel twice as far. In a nonlinear system, the output does not bear such a simple relationship to the input. If the input is increased by a factor of two, the output may increase by more or by less than a factor of two. A chaotic system involves nonlinear responses.

Interactions of two bodies, for example, a planet orbiting around a sun, are not excessively sensitive to initial conditions so that Newtonian mechanics can predict the motion of the two bodies with very high accuracy. However, when more than two bodies are involved in the interaction, sensitivity to initial conditions increases; as a result, the outcome of even a relatively simple event, governed by basic Newtonian mechanics, may become unpredictable. Here again, the availability of high-speed computers clarifies the issue. The following example is given by Angrist (1968). Consider a collection of billiard balls consisting of 15 balls and a cue ball, as shown in Fig. 10.5. The player starts the game with one shot executed as precisely as possible. With perfect control over the stroke and no frictional forces acting on the balls, they hit each other and bounce back and forth, balls hitting the sides of the table and each other. Hypothetically, in the absence of friction, the balls bounce back and forth forever. The question is posed: How far into the future could the player predict the trajectory of the cue ball? Computer simulations show that if the player did not take into account a force even as small as the attraction of one electron at the edge of our galaxy, the prediction would fail after about 1 minute or on the order of ten collisions per billiard ball. In other words, even with such a tiny uncertainty in the initial conditions, the movement and therefore the position of the cue ball, as well as the other billiard balls, becomes unpredictable after 1 minute. The reason for the unpredictability is that the slightest uncertainty in the point of impact of curved balls becomes amplified exponentially and very quickly the ability to predict the movement

**FIGURE 10.5   Billiard balls arranged in an ordered pattern.**

of the balls is lost. To restate this observation somewhat differently, two identical configurations of billiard balls at starting points that, within our ability to control, are identical will end up very quickly in unpredictably different configurations.

An important feature of chaotic systems is their fractal character in time. To reveal the self-similar fractal structure, the time behavior of the system has to be analyzed with specific mathematical techniques. As an example, consider a stream flowing past a boulder. At slow flow rates, the flow is laminar and its pattern is completely predictable. The motion of a cork carried by a laminar flow can be precisely described.

As the flow speed increases, the flow becomes turbulent, giving rise to waves and swirls as shown in Fig. 10.6. Turbulent flow is highly sensitive to initial conditions such as the angle at which the water encounters the boulder, the exact speed and volume of water that encounters the boulder, etc. There will be peaks and valleys in the flow, but we cannot tell where and when they will occur. Nor can we predict the height of the peaks and valleys. A cork floating in a turbulent stream bobs up and down and moves in unpredictable ways. Turbulent flow exhibits deterministic chaos.

Now suppose we focus on a specific position in the stream. We measure the height of the peaks in the flow at that point at a specific chosen interval, say, every 5 seconds. We plot the measured height as a function of time for periods of, say, 5 minutes, 10 minutes, 1 hour, and 1 day. Analysis of the plotted data,

FIGURE 10.6 **Turbulent flow.** *(Source: VarnakovR/Shutterstock.com.)*

which can often be complex, obtained on different time scales is then performed. The results display self-similarity, in other words, a fractal pattern. The fractal dimension obtained from such an analysis provides a quantitative measure of the chaotic nature of the process.

### 10.6.3 Entropy and chaos

In Sections 10.4 and 10.5, we defined *entropy* as a measure of disorder. Referring to the billiard ball example in Fig. 10.5, the initial configuration of billiard balls in the figure is well ordered and therefore it is a system of relatively low entropy. Now energy is introduced into the system via the cue ball striking the configuration. We know intuitively that the balls will disperse and the grouping will become more disordered. Because of the chaotic-deterministic nature of the collisions, that is, the high sensitivity to the initial conditions, we lose information about the position and velocity of the balls very quickly. As a result, the disorder, that is, entropy of the system consisting of the billiard balls, increases. As is pointed out by Michel Baranger in an article cited in this chapter, deterministic chaos is the dynamic process that brings about unpredictability of the outcome of a perturbation. The unpredictability is in effect the disordering. Increase in entropy is the measure of the disordering. Exercises 10.6 and 10.7 provide further exploration of fractals and chaotic processes.

### 10.6.4 Chaos and medicine

A variety of functions within our bodies display properties associated with deterministic chaos that appears to be indicators of our physical health. Here we will briefly describe two such functions that have been studied in some detail: heartbeat and gait. Electrical signals that monitor brain functions indicate that various aspects of these signals associated with physical, cognitive, and behavioral patterns also display fractal properties, although these relationships have not been as widely studied.

**Heartbeat.** Consider a person with an average pulse rate of, say, 60 beats per minute (bpm). In this case, the average interval between heartbeats is $1/60 = 1$ second. This is an average interval. In fact, the interval between heartbeats for a healthy person is not precisely 1 second. It varies from beat to beat, typically by $+/-$ 10%. That is, the time interval between beats at one point in time may be 0.9 seconds and the next beat may be 1.1 seconds later. When appropriately analyzed, the variations in the heartbeat as a function of time display fractal self-similarity, characteristic of a deterministic chaotic system. There is clear evidence that the variability in the heart rate is related to health. Clearly excessive variability indicates heart fibrillations that can be lethal. On the other hand, a heartbeat that lacks variability, that is, the interval between beats is nearly constant, often indicates congestive heart failure or some other serious heart disease. A heartbeat in a normal chaotic range indicates that the heart can respond quickly to changes in the demands on the blood circulatory system. By contrast, a heart that is locked to a fixed rate is nonresponsive to the changing demands encountered in life.

**Human gait.** The gait of a healthy person at first glance appears to be highly regular. However, as is the case with heartbeat, a more detailed examination of the various aspects of the gait of a healthy person reveals small variations from step to step in the various aspects of the gait; the length of the stride, the time between the steps, the relative positions of the knee, and hip joints as well as the pressures applied by the different parts of the foot. The variations are not random. They display properties characteristic of deterministic chaos including fractal self-similarity patterns. A large number of studies suggest that the chaotic variability in gait observed for healthy individuals allows for a quick and appropriate adaptation to unexpected stresses encountered in the process of walking. The gait variability of people with neurodegenerative disorders, such as Parkinson's and Huntington's disease, is significantly reduced. Such reduction in gait variability is associated with notable gait instability.

It is now evident that fractal (or chaotic) variability, in an appropriate range is a characteristic of healthy functioning of key processes in humans, in other animals, as well as in the microbial world. Deviations from an appropriate measure of chaotic behavior may signal disfunction. During the past 5 years (2018 to 2023) most of the studies of deterministic chaos and fractal variability systems associated with medicine and biology have been focused on obtaining a deeper understanding of several diseases and processes, among them Parkinson's disease (Dahs et al., 2020), cardiac arrhythmias (Gupta et al., 2021), multiple sclerosis (Akaishi et al., 2018), Psychology, (Schuldberg et al., 2022), and Neurology (Boudoue, 2020). The goal of these studies is to obtain models that could, where applicable, diagnose the disfunction, predict the course of the process, and in cases of disease, perhaps also to find a method of intervention.

At this point (2023), the observation that health is associated with fractal variability has not been utilized in medical practice. It is likely that the relationship between chaos and physical health will be more widely studied utilizing artificial intelligence programs that are highly capable recognizing significant patterns. This should greatly facilitate in the near future the use of chaos data in medical diagnosis. Exercise 10.8 provides further exploration of possible use of chaos theory in medicine.

## Exercises

**10.1.** Explain how the second law of thermodynamics limits conversion of heat to work.

**10.2.** From your own experience, give an example of the second law of thermodynamics.

**10.3.** Describe the connections between information, the second law of thermodynamics, and living systems.

**10.4.** Describe how to recognize a fractal pattern.

**10.5.** Describe two fractal patterns found in nature other than the ones discussed in the text.

**10.6.** How is a deterministic chaotic process different from chaos as defined in common usage?

**10.7.** Describe an experiment where a deterministic chaotic process was studied in an animal (other than human beings). The web is a good source of information.

**10.8.** Suggest ways that deterministic chaotic processes may be utilized in medical practice.

Chapter 11

# Heat and life

The degree of hotness, or temperature, is one of the most important environmental factors in the functioning of living organisms. The rates of the metabolic processes necessary for life, such as cell divisions and enzyme reactions, depend on temperature. Generally, the rates increase with temperature. A 10°C change in temperature may alter the rate by a factor of two.

Because liquid water is an essential component of living organisms as we know them, the metabolic processes function only within a relatively narrow range of temperatures, from about 2°C to 120°C. Only the simplest of living organisms can function near the extremes of this range.[1] Large-scale living systems are restricted to a much narrower range of temperatures.

The functioning of most living systems, plants, and animals is severely limited by seasonal variations in temperature. The life processes in reptiles, for example, slow down in cold weather to a point where they essentially cease to function. On hot sunny days, these animals must find shaded shelter to keep their body temperatures down.

For a given animal, there is usually an optimum rate for the various metabolic processes. Warm-blooded animals (mammals and birds) have evolved methods for maintaining their internal body temperatures at near-constant levels. As a result, warm-blooded animals are able to function at an optimum level over a wide range of external temperatures. Although this temperature regulation requires additional expenditures of energy, the adaptability achieved is well worth this expenditure.

In this chapter, we will examine energy consumption, heat flow, and temperature control in animals. Although most of our examples will be specific to people, the principles are generally applicable to all animals.

## 11.1 Energy requirements of people

All living systems need energy to function. In animals, this energy is used to circulate blood, obtain oxygen, repair cells, and so on. As a result, even at complete rest in a comfortable environment, the body requires energy to sustain its life functions. For example, a man weighing 70 kg lying quietly awake consumes

---

1. In deep oceans, the pressure is high and so is the boiling point of water. Here certain *thermophilic* bacteria can survive near thermal vents at significantly higher temperatures.

Physics in Biology and Medicine. https://doi.org/10.1016/B978-0-443-21558-2.00011-0
**159**

**TABLE 11.1** Metabolic rates for selected activities.

| Activity | Metabolic rate (Cal/m² h) |
|---|---|
| Sleeping | 35 |
| Lying awake | 40 |
| Sitting upright | 50 |
| Standing | 60 |
| Walking (3 mph) | 140 |
| Moderate physical work | 150 |
| Bicycling | 250 |
| Running | 600 |
| Shivering | 250 |

about 70 Cal/h (1 Cal $= 4.18$ J; 1000 Cal $= 1$ Cal; 1 Cal/h $= 1.16$ W). Of course, the energy expenditure increases with activity.

The amount of energy consumed by a person depends on the person's weight and build. It has been found, however, that the amount of energy consumed by a person during a given activity divided by the surface area of the person's body is approximately the same for most people. Therefore, the energy consumed for various activities is usually quoted in Cal/m² h. This rate is known as the *metabolic rate*. The metabolic rates for some human activities are shown in Table 11.1. To obtain the total energy consumption per hour, we multiply the metabolic rate by the surface area of the person. The following empirical formula yields a good estimate for the surface area.

$$\text{Area } (\text{m}^2) = 0.202 \times M^{0.425} \times H^{0.725} \tag{11.1}$$

Here M is the mass of the person in kilograms, and H is the height of the person in meters.

The surface area of a 70-kg man of height 1.55 m is about 1.70 m². His metabolic rate at rest is therefore $(40 \text{ Cal}/\text{m}^2 \text{ h}) \times 1.70 \text{ m}^2 = 68 \text{ Cal}/\text{h}$, or about 70 Cal/h, as stated in our earlier example. This metabolic rate at rest is called the *basal metabolic rate*.

## 11.1.1 Basal metabolic rate and body size

Larger animals have more cells requiring more energy to maintain them. Therefore, we expect the metabolic rate to increase with the size of the animal. Can this expectation be expressed mathematically? In 1883 the biologist Max

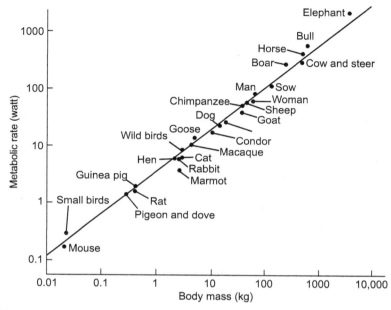

**FIGURE 11.1**   **Metabolic rates for mammals and birds, plotted as a function of body mass on a logarithmic scale.** *Adapted from Schmidt-Nielsen (1984).*

Rubner suggested that the basal metabolism, that is, the energy consumed by the animal at rest, ends up as heat and is therefore likely to be limited by the amount of heat the animal can eliminate. Using a highly simplified model, he proposed that the metabolic rate is proportional to $M^{2/3}$, where M is the mass of the animal. This expression is based on the assumption that the animal is spherical in shape. Because the mass is proportional to the volume, the radius R of the assumed sphere is proportional to $M^{1/3}$. The area that in this model sets the basal metabolism is proportional to $R^2$ or $M^{2/3}$.

This simple model was a good starting point but it did not adequately match subsequent experimental measurements. In 1932 Max Kleiber showed that for a wide range of species, the measured metabolic rate is proportional to $M^{3/4}$. This relationship is obtained from a plot such as shown in Fig. 11.1. Here, the metabolic rate is plotted versus the body mass for animals ranging in size from mouse (0.05 kg) to elephant (5000 kg), a factor of $10^5$ in mass. On a log-log scale, the plot is best fit by a straight line with a slope of 0.75, yielding Kleiber's "law" that is, metabolic rate is proportional to $M^{3/4}$. (See problem 11.1.) This relationship has been confirmed by many studies since the 1930s.

Unlike the more obvious but less applicable $M^{2/3}$ relationship derived by Rubner, there is no easily derivable principle leading to Kleiber's scaling law.

Several models of varying complexities have been proposed that yield the 3/4 exponent but none of the models are sufficiently compelling to have been universally accepted. Kleiber's scaling law remains a subject of ongoing research. The 2005 issue of the *Journal of Experimental Biology*, volume 208, is devoted to this field of inquiry.

It is universally observed that large animals live longer than small ones. For example, the lifespan of a mouse is estimated to be 1 to 3 years, while that of an elephant is about 70 years. The power law sheds some light on this observation, albeit only semiquantitatively. The issue of lifespan is best examined in terms of the *specific metabolic rate*, that is, the energy burned per unit mass. This parameter is obtained by dividing the basal metabolic rate by the mass of the animal. That is, the specific metabolic rate is proportional to $M^{3/4}/M = M^{-1/4}$. At this point, the assumption is made that the total energy consumption per unit mass of a creature during its lifetime is a constant. That is, (specific metabolic rate) $\times$ (lifespan) = constant. With this assumption, *lifetime* is inversely proportional to the specific metabolic rate (i.e., $= M^{1/4}$). With the mass of an elephant being a factor of $10^5$ greater than that of a mouse, its lifespan is expected to be longer by $\left(10^5\right)^{1/4} = 18$. This estimate yields a lifespan for an elephant of 18 to 54 years, which is not accurate, but in the right range.

## 11.2 Energy from food

The chemical energy used by animals is obtained from the oxidation of food molecules. The glucose sugar molecule, for example, is oxidized as follows:

$$C_6H_{12}O_6 + 6O_2 \rightarrow 6CO_2 + 6H_2O + \text{energy} \qquad (11.2)$$

For every gram of glucose ingested by the body, 3.81 Cal of energy is released for metabolic use.

The caloric value per unit weight is different for various foods. Measurements show that, on average, carbohydrates (sugars and starches) and proteins provide about 4 Cal/g; lipids (fats) produce 9 Cal/g; and the oxidation of alcohol produces 7 Cal/g.[2]

The oxidation of food, which releases energy, does not occur spontaneously at normal environmental temperatures. For oxidation to proceed at body

---

2. The high caloric content of alcohol presents a problem for people who drink heavily. The body utilizes fully the energy released by the oxidation of alcohol. Therefore, people who obtain a significant fraction of metabolic energy from this source reduce their intake of conventional foods. Unlike other foods, however, alcohol does not contain vitamins, minerals, and other substances necessary for proper functioning. As a result, chronic alcoholics often suffer from diseases brought about by nutritional deficiencies.

**TABLE 11.2** One day's metabolic energy expenditure.

| Activity | Energy expenditure (Cal/m$^2$) |
|---|---|
| 8 h sleeping (35 Cal/m$^2$ h) | 280 |
| 8 h moderate physical labor (150 Cal/m$^2$ h) | 1200 |
| 4 h reading, writing, TV watching (60 Cal/m$^2$ h) | 240 |
| 1 h heavy exercise (300 Cal/m$^2$ h) | 300 |
| 3 h dressing, eating (100 Cal/m$^2$ h) | 300 |
| Total expenditure | 2320 |

temperature, a catalyst must promote the reaction. In living systems, complex molecules, called enzymes, provide this function.

In the process of obtaining energy from food, oxygen is always consumed. It has been found that, independent of the type of food being utilized, 4.83 Cal of energy is produced for every liter of oxygen consumed. Knowing this relationship, one can measure with relatively simple techniques the metabolic rate for various activities (see Exercise 11.3).

The daily food requirements of a person depend on his or her activities. A sample schedule and the associated metabolic energy expenditure per square meter are shown in Table 11.2. Assuming, as before, that the surface area of the person whose activities are shown in Table 11.2 is 1.7 m$^2$, his/her total energy expenditure is 3940 Cal/day. If the person spent half the day sleeping and half the day resting in bed, the daily energy expenditure would be only 1530 Cal.

For most people, the energy expenditure is balanced by the food intake. For example, the daily energy needs of the person whose activities are shown in Table 11.2 (surface area 1.7 m$^2$) are met by the consumption of 400 g of carbohydrates, 200 g of protein, and 171 g of fat.

The composition and energy content of some common foods are shown in Table 11.3. Note that the sum of the weights of the protein, carbohydrates, and fat is smaller than the total weight of the food. The difference is due mostly to the water content of the food. The energy values quoted in the table reflect the fact that the caloric content of different proteins, carbohydrates, and fats deviates somewhat from the average values stated in the text.

If an excess of certain substances, such as water and salt, is ingested, the body is able to eliminate it. The body has no mechanism, however, for eliminating an excess in caloric intake. Over a period of time, the excess energy is used by the body to manufacture additional tissue. If the consumption of excess food occurs simultaneously with heavy exercise, the energy may be utilized to increase the weight of the muscles. Most often,

**TABLE 11.3 Composition and energy content of some common foods.**

| Food | Total weight (g) | Protein weight (g) | Carbohydrate weight (g) | Fat weight (g) | Total energy (Cal) |
|---|---|---|---|---|---|
| Whole milk, 1 quart | 976 | 32 | 48 | 40 | 660 |
| Egg, 1 | 50 | 6 | 0 | 12 | 75 |
| Hamburger, 1 | 85 | 21 | 0 | 17 | 245 |
| Carrots, 1 cup | 150 | 1 | 10 | 0 | 45 |
| Potato (1 med., baked) | 100 | 2 | 22 | 0 | 100 |
| Apple | 130 | 0 | 18 | 0 | 70 |
| Bread, rye, 1 slice | 23 | 2 | 12 | 0 | 55 |
| Doughnut | 33 | 2 | 17 | 7 | 135 |

however, the excess energy is stored in fatty tissue that is manufactured by the body. Conversely, if the energy intake is lower than the demand, the body consumes its own tissue to make up the deficit. While the supply lasts, the body first utilizes its stored fat. For every 9 Cal of energy deficit, about 1 g of fat is used. Under severe starvation, once the fat is used up, the body begins to consume its own protein. Each gram of consumed protein yields about 4 Cal. Consumption of body protein results in the deterioration of body functions, of course. A relatively simple calculation (see Exercise 11.6) shows that an average healthy person can survive without food but with adequate water for up to about 50 days. Overweight people can do better, of course. The "Guinness Book of World Records" states that Angus Barbieri of Scotland fasted from June 1965 to July 1966, consuming only tea, coffee, and water. During this period, his weight declined from 472 lb to 178 lb.

For a woman, the energy requirements increase somewhat during pregnancy due to the growth and metabolism of the fetus. As the following calculation indicates, the energy needed for the growth of the fetus is actually rather small. Let us assume that the weight gain of the fetus during the 270 days of gestation is uniform.[3] If at birth the fetus weighs 3 kg, each day it gains 11 g. Because 75% of tissue consists of water and inorganic minerals, only 2.75 g of the daily mass increase is due to organic materials, mainly protein. Therefore, the extra calories per day required for the growth of the fetus is

$$\text{Calories required} = \frac{2.75 \text{ g protein}}{\text{day}} \times \frac{4 \text{ Cal}}{\text{g protein}} = 11 \text{ Cal/day}$$

To this number, we must add the basal metabolic consumption of the fetus. At birth, the surface area of the fetus is about $0.13 \text{ m}^2$ (from Eq. 11.1); therefore, at most, the basal metabolic consumption of the fetus per day is about $0.13 \times 40 \times 24 = 125$ Cal. Thus, the total increase in the energy requirement of a pregnant woman is only about $(125 + 11) \text{ Cal/day} = 136 \text{ Cal/day}$. Actually, it may not even be necessary for a pregnant woman to increase her food intake, as the energy requirements of the fetus may be balanced by decreased physical activity during pregnancy. Various other aspects of metabolic energy balance are examined in Exercises 11.4 to 11.7.

## 11.3 Regulation of body temperature

People and other warm-blooded animals must maintain their body temperatures at a nearly constant level. For example, the normal internal body temperature of a person is about 37°C. A deviation of 1°C or 2°C in either

---

3. This is a simplification because the weight gain is not uniform. It is greatest toward the end of gestation.

direction may signal some abnormality. If the temperature-regulating mechanisms fail and the body temperature rises to 44°C or 45°C, the protein structures are irreversibly damaged. A fall in body temperature below about 28°C results in heart stoppage.

The body temperature is sensed by specialized nerve centers in the brain and by receptors on the surface of the body. The various cooling or heating mechanisms of the body are then activated in accordance with the temperature. The efficiency of muscles in performing external work is at best 20%. Therefore, at least 80% of the energy consumed in the performance of a physical activity is converted into heat inside the body. In addition, the energy consumed to maintain the basic metabolic processes is ultimately all converted to heat. If this heat were not eliminated, the body temperature would quickly rise to a dangerous level. For example, during moderate physical activity, a 70-kg man may consume 260 Cal/h. Of this amount, at least 208 Cal is converted to heat. If this heat remained within the body, the body temperature would rise by 3 C°/h, which implies that 2 hours of such an activity would cause complete collapse. Fortunately, the body possesses a number of highly efficient methods for controlling the heat flow out of the body, thereby maintaining a stable internal temperature.

Most of the heat generated by the body is produced deep in the body, far from the surfaces. In order to be eliminated, this heat must first be conducted to the skin. For heat to flow from one region to another, there must be a temperature difference between the two regions. Therefore, the temperature of the skin must be lower than the internal body temperature. In a warm environment, the temperature of the human skin is about 35°C. In a cold environment, the temperature of some parts of the skin may drop to 27°C.

The tissue of the body, without blood flowing through it, is a poor conductor. Its thermal conductivity is comparable to that of cork (see Table 9.2). ($K_c$ for tissue without blood is 18 Cal cm/(m$^2$ h C°).) Simple thermal conductivity through tissue is inadequate for elimination of the excess heat generated by the body. The following calculation illustrates this point. Assume that the thickness of the tissue between the interior and the exterior of the body is 3 cm and that the average area through which conduction can occur is 1.5 m$^2$. With a temperature difference T between the inner body and the skin of 2°C, the heat flow H per hour is, from Eq. 9.3,

$$H = \frac{K_c A \Delta T}{L} = \frac{18 \times 1.5 \times 2}{3} = 18 \text{ Cal/h} \tag{11.3}$$

In order to increase the conductive heat flow to a moderate level of, say, 150 Cal/h, the temperature difference between the interior body and the skin would have to increase to about 17 C°.

Fortunately, the body possesses another method for transferring heat. Most of the heat is transported from the inside of the body by blood in the circulatory system. Heat enters the blood from an interior cell by conduction. In this case, heat transfer by conduction is relatively fast because the distances between the capillaries and the heat-producing cells are small. The circulatory system carries the heated blood near the surface skin. The heat is then transferred to the outside surface by conduction. In addition to transporting heat from the interior of the body, the circulatory system controls the insulation thickness of the body. When the heat flow out of the body is excessive, the capillaries near the surface become constricted and the blood flow to the surface is greatly reduced. Because tissue without blood is a poor heat conductor, this procedure provides a heat-insulating layer around the inner body core.

## 11.4 Control of skin temperature

As was stated, for heat to flow out of the body, the temperature of the skin must be lower than the internal body temperature. Therefore, heat must be removed from the skin at a sufficient rate to ensure that this condition is maintained. Because the heat conductivity of air is very low (202 Cal cm/(m$^2$ h C°)), if the air around the skin is confined—for example, by clothing—the amount of heat removed by conduction is small. The surface of the skin is cooled primarily by convection, radiation, and evaporation. However, if the skin is in contact with a good thermal conductor such as a metal, a considerable amount of heat can be removed by conduction (see Exercise 11.8).

## 11.5 Convection

When the skin is exposed to open air or some other fluid, heat is removed from it by convection currents. The rate of heat removal is proportional to the exposed surface area and to the temperature difference between the skin and the surrounding air. The rate of heat transfer by convection $H_c'$ (see Eq. 9.4) is given by

$$H_c' = K_c' A_c (T_s - T_a) \tag{11.4}$$

where $A_c$ is the skin area exposed to the open air; $T_s$ and $T_a$ are the skin and air temperatures, respectively; and $K_c'$ is the convection coefficient, which has a value that depends primarily on the prevailing wind velocity. The value of $K_c'$ as a function of air velocity is shown in Fig. 11.2. As the plot shows, the convection coefficient initially increases sharply with wind velocity, and then the increase becomes less steep (see Exercise 11.9).

The exposed area $A_c$ is generally smaller than the total surface area of the body. For a naked person standing with legs together and arms close to the body, about 80% of the surface area is exposed to convective air currents. (The exposed area can be reduced by curling up the body.)

Note that heat flows from the skin to the environment only if the air is colder than the skin. If the opposite is the case, the skin is actually heated by the convective air flow.

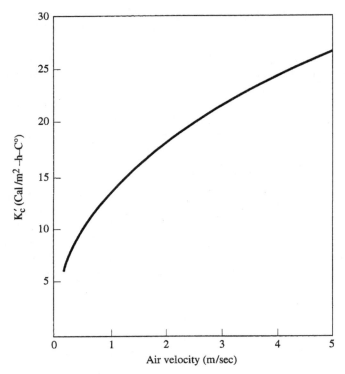

**FIGURE 11.2** **Convection coefficient as a function of air velocity.**

Let us now calculate the amount of heat removed from the skin by convection. Consider a naked person whose total surface area is 1.7 m². Standing straight, the exposed area is about 1.36 m². If the air temperature is 25°C and the average skin temperature is 33°C, the amount of heat removed is

$$H_c' = 1.36 K_c' \times 8 = 10.9 K_c' \ Cal/h$$

Under nearly windless conditions, $K_c'$ is about 6 Cal/(m² h C°)) (see Fig. 11.2), and the convective heat loss is 65.4 Cal/h. During moderate work, the energy consumption for a person of this size is about 170 Cal/h. Clearly, convection in a windless environment does not provide adequate cooling. The wind velocity has to increase to about 1.5 m/sec to provide cooling at a rate of 170 Cal/h.

## 11.6 Radiation

Eq. 9.6 shows that the energy exchange by radiation $H_r$ involves the fourth power of temperature; that is,

$$H_r = e\sigma(T_1^4 - T_2^4)$$

However, because in the environment encountered by living systems, the temperature on the absolute scale seldom varies by more than 15%, it is possible to use, without much error, a linear expression for the radiative energy exchange (see Exercise 11.10a and b); that is,

$$H_r = K_r A_r e(T_s - T_r) \tag{11.5}$$

where $T_s$ and $T_r$ are the skin surface temperature and the temperature of the nearby radiating surface, respectively; $A_r$ is the area of the body participating in the radiation; e is the emissivity of the surface; and $K_r$ is the radiation coefficient. Over a fairly wide range of temperatures, $K_r$ is, on average, about 6.0 Cal/(m$^2$ h C$°$) (see Exercise 11.10c).

The environmental radiating surface and skin temperatures are such that the wavelength of the thermal radiation is predominantly in the infrared region of the spectrum. The emissivity of the skin in this wavelength range is nearly unity, independent of the skin pigmentation. For a person with $A_r = 1.5$ m$^2$, $T_r = 25°C$, and $T_s = 32°C$. The radiative heat loss is 63 Cal/h.

If the radiating surface is warmer than the skin surface, the skin is heated by radiation. A person begins to feel discomfort due to radiation if the temperature difference between the exposed skin and the radiating environment exceeds about 6 C$°$. In the extreme case, when the skin is illuminated by the sun or some other very hot object like a fire, the skin is heated intensely. Because the temperature of the source is now much higher than the temperature of the skin, the simplified expression in Eq. 11.5 no longer applies.

## 11.7 Radiative heating by the sun

The intensity of solar energy at the top of the atmosphere is about 1150 Cal/m$^2$ h. Not all this energy reaches the surface of the Earth. Some of it is reflected by airborne particles and water vapor. A thick cloud cover may reflect as much as 75% of solar radiation. The inclination of the Earth's axis of rotation further reduces the intensity of solar radiation at the surface. However, in dry equatorial deserts, nearly all the solar radiation may reach the surface.

Because the rays of the sun come from one direction only, at most half the body surface is exposed to solar radiation. In addition, the area perpendicular to the solar flux is reduced by the cosine of the angle of incidence (see Fig. 11.3). As the sun approaches the horizon, the effective area for the interception of radiation increases, but at the same time, the radiation intensity decreases because the radiation passes through a thicker layer of air. Still, the amount of solar energy heating the skin can be very large. Assuming that the full intensity of solar radiation reaches the surface, the amount of heat $H_r$ that the human body receives from solar radiation is

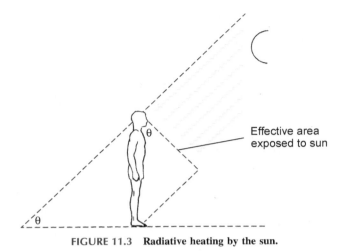

FIGURE 11.3    **Radiative heating by the sun.**

$$H_r = 1150/2 \times e \times A \cos\theta \; Cal/h \qquad (11.6)$$

Here, A is the skin area of the person, $\theta$ is the angle of incidence of sunlight, and e is the emissivity of the skin. The emissivity of the skin in the wavelength region of solar radiation depends on the pigmentation. Dark skin absorbs about 80% of the radiation, and light skin absorbs about 60%. From Eq. 11.6, a light-skinned person with a skin area of 1.7 m$^2$, subject to intense solar radiation incident at a 60° angle, receives heat at the rate of 294 Cal/ h. Radiative heating is decreased by about 40% if the person wears light-colored clothing. Radiative heating is also reduced by changing the orientation of the body with respect to the sun. Camels resting in the shadeless desert face the sun, which minimizes the skin area exposed to solar radiation.

## 11.8 Evaporation

In a warm climate, convection and radiation cannot adequately cool a person engaged in even moderate physical activity. A large fraction of cooling is provided by the evaporation of sweat from the skin surfaces. At normal skin temperatures, the latent heat of vaporization for water is 0.580 Cal/g. Therefore, about 580 Cal of heat is removed for each liter of sweat that evaporates from the skin. The body contains two types of sweat glands, the eccrine and the apocrine. The eccrine glands are distributed over the whole surface of the body, and they respond primarily to the nerve impulses generated by the thermoregulatory system of the body. As the heat load on the body rises, the sweat secreted by these glands increases proportionately. There is an exception to this. The *eccrine glands* in the palms of the hand and the soles of the feet are stimulated by elevated levels of adrenaline in the blood, which may result from emotional stress.

The *apocrine* sweat glands, found mostly in the pubic regions, are not associated with temperature control. They are stimulated by adrenaline in the blood stream, and they secrete a sweat rich in organic matter. The decomposition of these substances produces body odor.

The ability of the human body to secrete sweat is remarkable. For brief periods of time, a person can produce sweat at a rate of up to 4 L/h. Such a high rate of sweating, however, cannot be maintained. For longer periods, up to 6 hours, a sweating rate of 1 L/h is common in the performance of heavy work in a hot environment.

During prolonged heavy sweating, adequate amounts of water must be drunk; otherwise, the body becomes dehydrated. A person's functioning is severely limited when dehydration results in a 10% loss of body weight. Some desert animals can endure greater dehydration than humans; a camel, for example, may lose water amounting to 30% of its body weight without serious consequences.

Only sweat that evaporates is useful in cooling the skin. Sweat that rolls off or is wiped off does not provide significant cooling. Nevertheless, excess sweat does ensure full wetting of the skin. The amount of sweat that evaporates from the skin depends on ambient temperature, humidity, and air velocity. Evaporative cooling is most efficient in a hot, dry, windy environment.

There is another avenue for evaporative heat loss: Breathing. The air leaving the lungs is saturated by water vapor from the moist lining of the respiratory system. At a normal human breathing rate, the amount of heat removed by this avenue is small, less than 9 Cal/h (see Exercise 11.11); however, for furred animals that do not sweat, this method of heat removal is very important. These animals can increase heat loss by taking short, shallow breaths (panting) that do not bring excessive oxygen into the lungs but do pick up moisture from the upper respiratory tract.

By evaporative cooling, a person can cope with the heat generated by moderate activity even in a very hot, sunny environment. To illustrate this, we will calculate the rate of sweating required for a person walking nude in the sun at a rate of 3 mph, with the ambient temperature at 47°C (116.6°F).

With a skin area of 1.7 m$^2$, the energy consumed in the act of walking is about 240 Cal/h. Almost all this energy is converted to heat and delivered to the skin. In addition, the skin is heated by convection and by radiation from the environment and the sun. The heat delivered to the skin by convection is

$$H'_c = K'_c A_c (T_s - T_a)$$

For a 1 m/sec wind, $K'_c$ is 13 Cal/(m$^2$ h C°). The exposed area $A_c$ is about 1.5 m$^2$. If the skin temperature is 36°C,

$$H'_c = 13 \times 1.50 \times (47 - 36) = 215 \text{ Cal/h}$$

As calculated previously, the radiative heating by the sun is about 294 Cal/h. The radiative heating by the environment is

$$H'_c = K_r A_r e(T_r - T_s) = 6 \times 1.5 \times (47 - 36) = 99 \text{ Cal/h}$$

In this example, the only mechanism available for cooling the body is the evaporation of sweat. The total amount of heat that must be removed is $(240 + 215 + 294 + 99)$ Cal/h $= 848$ Cal/h. The evaporation of about 1.5 L/h of sweat will provide the necessary cooling. Of course, if the person is protected by light clothing, the heat load is significantly reduced. The human body is indeed very well equipped to withstand heat. In controlled experiments, people have survived a temperature of 125°C for a period of time that was adequate to cook a steak.

## 11.9 Resistance to cold

In a thermally comfortable environment, the body functions at a minimum expenditure of energy. As the environment cools, a point is reached where the basal metabolic rate increases to maintain the body temperature at a proper level. The temperature at which this occurs is called the *critical temperature*. This temperature is a measure of the ability of an animal to withstand cold.

Human beings are basically tropical animals. Unprotected, they are much better able to cope with heat than with cold. The critical temperature for humans is about 30°C. By contrast, the critical temperature for the heavily furred arctic fox is −40°C.

The discomfort caused by cold is due primarily to the increased rate of heat outflow from the skin. This rate depends not only on the temperature but also on the wind velocity and humidity. For example, at 20°C, air moving with a velocity of 30 cm/sec removes more heat than still air at 15°C. In this case, a mild wind at 30 cm/sec is equivalent to a temperature drop of more than 5°C.

The body defends itself against cold by decreasing the heat outflow and by increasing the production of heat. When the temperature of the body begins to drop, the capillaries leading to the skin become constricted, reducing the blood flow to the skin. This results in a thicker thermal insulation of the body. In a naked person, this mechanism is fully utilized when the ambient temperature drops to about 19°C. At this point, the natural insulation cannot be increased anymore.

Additional heat required to maintain the body temperature is obtained by increasing the metabolism. One involuntary response that achieves this is shivering. As shown in Table 11.1, shivering raises the metabolism to about 250 Cal/$m^2$ h. If these defenses fail and the temperature of the skin and underlying tissue fall below about 5°C, frostbite and eventually more serious freezing occur.

The most effective protection against cold is provided by thick fur, feathers, or appropriate clothing. At −40°C, without insulation, the heat loss is primarily convective and radiative. By convection alone in moderately moving air, the rate of heat removal per square meter of skin surface is about 660 Cal/ m² h (see Exercise 11.12). With a thick layer of fur or similar insulation, the skin is shielded from convection and the heat is transferred to the environment by conduction only. The thermal conductivity of insulating materials such as fur or down is $K_c = 0.36 \, \text{Cal cm}/(\text{m}^2 \, \text{h} \, \text{C}°)$; therefore, the heat transfer from the skin at 30°C to the ambient environment at −40°C through 1 cm of insulation is, from Eq. 9.3, 25.2 Cal/m² h. This is below the basal metabolic rate for most animals. Although body heat is lost also through radiation and evaporation, our calculation indicates that well-insulated animals, including a clothed person, can survive in cold environments.

As stated earlier, at moderate temperatures, the amount of heat removed by breathing at a normal rate is small. At very cold temperatures, however, the heat removed by this channel is appreciable. Although the heat removed by the evaporation of moisture from the lungs remains approximately constant, the amount of heat required to warm the inspired air to body temperature increases as the ambient air temperature drops. For a person at an ambient temperature of −40°C, the amount of heat removed from the body in the process of breathing is about 14.4 Cal/h (see Exercise 11.13). For a well-insulated animal, this heat loss ultimately limits its ability to withstand cold.

## 11.10 Heat and soil

Much of life depends directly or indirectly on biological activities near the surface of the soil. In addition to plants, there are worms and insects whose lives are soil bound (1 acre of soil may contain 500 kg of earthworms). Soil is also rich in tiny organisms such as bacteria, mites, and fungi whose metabolic activities are indispensable for the fertility of the soil. To all this life, the temperature of the soil is of vital importance.

The surface soil is heated primarily by solar radiation. Although some heat is conducted to the surface from the molten core of the Earth, the amount from this source is negligible compared to solar heating. The Earth is cooled by convection, radiation, and the evaporation of soil moisture. On average, over a period of a year, the heating and cooling are balanced; therefore, over this period of time, the average temperature of the soil does not change appreciably. However, over shorter periods of time, from night to day, from winter to summer, the temperature of the top soil changes considerably; these fluctuations govern the life cycles in the soil.

The variations in soil temperature are determined by the intensity of solar radiation; the composition and moisture content of the soil; the vegetation cover; and atmospheric conditions such as clouds, wind, and airborne particles

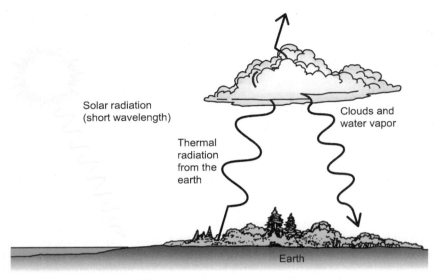

FIGURE 11.4   The greenhouse effect.

(see Exercises 11.14 and 11.15). Certain patterns, however, are general. During the day, while the sun is shining, more heat is delivered to the soil than is removed by the various cooling mechanisms. The temperature of the soil surface therefore rises during the day. In dry soil, the surface temperature may increase by 3 C°/h or 4 C°/h. The surface heating is especially intense in dry, unshaded deserts. Some insects living in these areas have evolved long legs to keep them removed from the hot surface.

The heat that enters the surface is conducted deeper into the soil. It takes some time, however, for the heat to propagate through the soil. Measurements show that a temperature change at the surface propagates into the soil at a rate of about 2 cm/h. At night, the heat loss predominates and the soil surface cools. The heat that was stored in the soil during the day now propagates to the surface and leaves the soil. Because of the finite time required for the heat to propagate through the soil, the temperature a few centimeters below the surface may still be rising while the surface is already cooling off. Some animals take advantage of this lag in temperature between the surface and the interior of the soil. They burrow into the ground to avoid the larger temperature fluctuations at the surface.

At the usual temperatures, the thermal radiation emitted by the soil is in the infrared region of the spectrum, which is strongly reflected by water vapor and clouds. As a result, on cloudy days, the thermal radiation emitted by the soil is reflected back, and the net outflow of heat from the soil is reduced—this is called the *greenhouse effect* (see Fig. 11.4). A similar effect is produced by the "greenhouse gases" in the atmosphere principally carbon dioxide ($CO_2$), methane $(CH_4)$, and ozone $(O_3)$. These gases absorb infrared radiation and emit it back to the Earth's surface, increasing the temperature of the planet.

## 11.11 Energy requirements of large carnivores

Several studies have been conducted recently to determine the *energy consumption* and hunting patterns of large *carnivores,* defined here as having an average body mass greater than about 20 kg (44 lb). These studies have been motivated in part by the globally declining populations of large carnivores such as for example lions, pumas, and wolves. Aside from the importance of preventing species extinction, studies by ecologists and biologists have made it evident that such large carnivores have an important role in maintaining ecosystem balance. (See Exercise 11.17.) According to Ripple and coworkers (see reference in Exercise 11.17), of the 31 large carnivores they studied worldwide, 75% are in continuing population decline, while 61% are in danger of extinction.

The population decline is due to several factors, among them decrease in the traditional habitat of the species due to increased farming and urbanization, increased hunting activity, and depletion of species that provide food for the carnivores.

To survive, an animal must obtain more energy from the food it consumes than it expends acquiring the food. To prevent collapse of carnivore populations, it is important to understand how they obtain the food needed for their survival. Carnivores expend considerably more energy satisfying their food requirements than do herbivores. They have to track their prey and then subdue and kill it. Often the hunted animal fights back, requiring additional energy expenditure by the hunter. This is particularly true for large carnivores. Observations as well as model calculations show that carnivores weighing more than about 20 kg (48 lb) hunt mostly large prey with a mass of about 45% or more of their own mass. Hunting animals smaller than that yields less energy for the hunter than is consumed in the hunt.

Carnivores employ a variety of hunting techniques, depending on the species. Some wolves, for example, hunt in packs. Other carnivores are predominantly lone hunters. Some, like pumas (also called mountain lions or cougars) as well as lions, sit most of the time and wait for the prey to enter their vicinity before they pounce. Cheetahs, on the other hand, seek out their prey and chase it at high speeds. None of the species restrict themselves to a single technique. Depending on conditions, specifically when prey is scarce, pumas are also in constant movement seeking prey. Recent studies using new monitoring technologies have measured the energetics of various hunting techniques. As expected, the seek-and-chase hunting method is far more energy intensive than the wait-and-pounce approach. However, when prey is scarce, the carnivore has no choice. It must resort to seeking out its food. A simple calculation (see Exercise 11.18) shows that a lion needs to consume the equivalent of about one antelope per week.

In the process of hunting, carnivores employ strategies that require split-second decision making. They evaluate the amount of effort they will have to exert to catch and subdue the prey. Then they decide whether the chase is worthwhile in light of the amount of food the kill will yield. Carnivores such as lions and cheetahs are often seen to abandon the chase when the prey is judged as too small, fast, or strong to make the pursuit worthwhile.

## 11.12 Physical exercise, calories, and weight loss

### 11.12.1 Background

Obesity and overweight are due to excessive fat accumulation in the body. The excessive body fat is associated with a host of health problems, among them high blood pressure, type 2 diabetes, high low-density cholesterol and tri-glycerides, low high-density cholesterol, coronary heart disease, stroke, cancer, osteoarthritis of the joints, and many other pathologies. Collectively the cluster of diseases associated with excessive body fat is called metabolic syndrome.

Overweight and obesity have become a serious health and economic problem for most industrial countries. In the United States (according to CDC numbers), in 2022, over 40% of the adult population was obese and another 35% was overweight Only about 25% of the adult population is in the normal weight category. In the early 1960s only about 13% of the population was obese. Over the same period of about 60 years, the percentage of overweight people remained constant at about 33% to 35%. Without awareness of the issue and intervention, in adulthood, most people tend to gain between one and two pounds a year. Hence the general trend is that overweight people tend to become obese and normal-weight people move toward overweight.

The terms obesity and overweight are measured quantitatively in terms of body mass index (BMI), defined as a person's mass m (weight) in kilograms divided by the square of the person's height (h) in meters. That is, $BMI = m/h^2$. The following World Health Organization (WHO) classifications are most often used:

| BMI | <18.5 | 18.5−24.9 | 25−29.9 | >30 |
|---|---|---|---|---|
| **Classification** | Underweight | Normal Weight | Overweight | Obese |

For clinical purposes, the "Obese" category is further subdivided into

| BMI | 30−34.9 | 35−39.9 | >40 |
|---|---|---|---|
| **Subdivision** | Class 1 | Class 2 | Class 3 |

Class 3 Obesity is sometimes categorized as "severe obesity."

These categorizations are approximate and are subject to modification. They are sometimes adjusted to take into account the specific physique of the individual. Muscle has a higher density than fat: 1.1 $g/cm^3$ versus 0.9 $g/cm^3$.

Thus, for example, a heavily muscled individual and an average person of the same height and weight may have the same calculated BMI, say, 29. In accordance with the formula, both are in the overweight range. However, with muscle being denser than fat, the muscled person will have a smaller fraction of fat in his body mass. As a result, the muscled person might be reclassified as having normal weight, whereas the other may remain in the overweight category.

While there are significant differences in BMI within population subgroups, the BMI of the US population as a whole is increasing precariously. Among industrialized countries, the United States has the highest percentage of overweight and obese people, but other developed countries are not far behind.

The severe problems, medical, financial, and social, caused by overweight and obesity have motivated considerable research over the past 20 years, focused on understanding the nature of weight loss, specifically the role of exercise, physical activity, and caloric intake on weight loss. To understand the effect of these factors on weight loss, we have to know the energy expenditure within the body and the input calories into the body. The caloric value of most foods is known from measurements using a variety of calorimetric methods. Therefore, the caloric value of the input into the body is easily calculated. Sample caloric food values are listed in Table 11.3.

Energies in food sciences are expressed in kilocalories symbolized by "Cal" (1 Cal = 4184 J). According to the US Department of Health and Human Services, the average adult woman of normal weight expends about 1600 to 2400 Cal per day. The average adult male of normal weight expends about 2000 to 3000 Cal per day. Both for women and men, these numbers depend on the age, size, and activities of the person.

Weight is usually stated in kilograms (although kg is actually a unit of mass). The conversion factor from kg to pounds (lb) (two-figure accuracy) is 1 kg = 2.2 lb. We will often use kg and lb, both commonly used units of weight, to promote familiarity with both of them.

In metabolic studies prior to the 1980s, the total energy expenditure, that is, all energy used (including by exercise) within the body of a person, was usually obtained by measuring the ratio of exhaled $O_2$ and $CO_2$, a by-product of all energy utilized by the body. In this type of measurement the studies had to be conducted in a laboratory where the gases were collected in a hood for analysis. In the 1980s a newer technique, the doubly labeled water (DLW) method, was adopted for metabolic studies. This technique made it possible to perform metabolic studies freed of laboratory confinement.

## 11.12.2 Energy expenditure measured by the DLW method

Currently, the DLW method is the most convenient and accurate technique for measuring the total energy expenditure of the body over a controlled period of time. The technique was developed in the 1950s by Nathan Lifson and colleagues at the University of Michigan, but initially, DLW was so expensive that it was not economically feasible till the 1980s to use the technique for the study of human metabolism.

In the experiments using the DLW method, the subject drinks a small bottle of water in which both the hydrogen atom (H) and the oxygen atom ($^{16}O$) have been replaced by their rare stable isotopes deuterium $^2H$ and $^{18}O$ (hence the name "doubly labeled water"). The isotope-labeled water is processed in the body identically with ordinary water. However, the isotope-labeled oxygen and hydrogen can be separately detected. Hydrogen passes through the body unchanged, but some fraction of oxygen is converted to carbon dioxide ($CO_2$) and breathed out. Measurement of the fraction of missing labeled oxygen in the urine yields the amount of breathed-out $CO_2$ during the period between drinking the labeled water and urine collection. Because the energy-producing reactions in the body all produce $CO_2$, the total energy expenditure of the body, including that due to exercise, can be computed for the period between the drinking of the water and the collection of the urine sample. Depending on the experiment, typical DLW measurement periods vary between 3 and 21 days. Most current metabolic energy expenditure studies are done using the DLW technique. The accuracy of such caloric expenditure measurements was shown to be about 5%.

### 11.12.3 Simple thermodynamic model for weight loss or gain

The initially used model for computing weight loss was guided by the first law of thermodynamics, that is, straightforward conservation of energy. Regarding the body as a simple thermodynamic system, the input is the energy provided by the food, drink, and nutrients taken in by the person. The categories making up total energy expenditures of a typical adult are shown in Fig. 11.5.

As is shown in the pie chart, the total energy expenditure is the sum of all the energy expenditures by the body, namely, basal metabolic rate, thermic effect of feeding, and thermic effect of physical activity. (a) Basal metabolic rate (sometimes called the resting metabolic rate), as discussed earlier (see Section 11.1), is the energy required to maintain a person at rest including the energy required for blood circulation; for breathing; for the functioning of the immune system; and for the various functions performed by the brain and other organs required for the functioning of the body. (b) Thermic effect of feeding is the energy expended in breaking down the food into components that the body can utilize for its functioning. (c) Thermic effect of physical activity is the energy expenditure due to physical activity including exercise.

If the total energy expenditure is greater than the input energy, the system is in calorie deficit, and the energy (calories) must come from some source within the body. In the simple thermodynamic model, the calorie deficit is

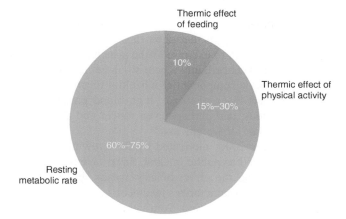

**FIGURE 11.5** **Total energy expenditure for an adult person.** Thermic effect of feeding is the energy required to process the food so that the body can utilize it for its various processes. The ranges shown are typical variations found in an average population. For a given person, the sum of the percentages in the various categories is, of course, 100%. *Reproduced with permission from Steinach and Gunga (2020).*

assumed to come from burning fat stored in the body (body fat). Of course, weight loss results if the calorie deficit is entirely or in part made up by the utilization of stored body fat. If the energy input is greater than the total energy expenditure, the system is in calorie excess. The excess calories are converted and stored as body fat. Steady-state weight of the person is attained when the food energy input into the body is on average equal to the total energy expenditure. This simple model can be expressed by the relationship:

(Calories from food into the body) − (total body energy expenditure)
= Change in weight

If this quantity is positive, the person gains weight. If it is negative, the person loses weight. If it is 0, the person's weight is in steady state.

Ways of losing weight (that is, producing a calorie deficit) are by exercise and by diet restriction, or by a combination of the two. Since the parameters on the left side of the equation can be measured or calculated, the calorie deficit for any activity including any exercise regimen can be obtained. The weight lost (or gained) is obtained using the well-established measurement that one pound of body fat contains approximately 3500 Calories. (1 kg of body fat contains 7700 Calories.) Such a conversion of calorie deficit to weight loss assumes that the calorie deficit is all made up by burning body fat.

As we will see, because of feedback mechanisms active in our bodies, the actual process of weight loss is far more complicated than the simple thermodynamic approach indicates.

## 11.12.4 Exercise and weight loss

A recent Gallup poll has shown that about 55% of US adults want to lose weight and people often look to exercise as the principal means to achieve the desired weight loss. There is certainly a general agreement and solid evidence that exercise is a necessary component of a healthy lifestyle. Regular exercise reduces the risk of heart disease, stroke, high blood pressure, osteoporosis, depression, inflammations, and perhaps even cancer and diabetes. But is exercise an effective method of weight loss?

Since about the year 2000, motivated by the growing problem of overweight and obesity, as well as public interest, there have been dozens of well-designed studies exploring the effect of exercise on weight loss. A review article by Swift et al. (2014) summarizes results of several such studies, among them results of weight loss resulting from aerobic exercise at public health recommendation levels. The public health exercise recommendation is 150 minutes of moderate exercise (such as walking briskly) per week, equivalent to walking 5 days per week, 30 minutes daily.

In one such study, a group of 464 postmenopausal women were divided into three groups, each group exercising respectively at 50%, 100%, and 150% of the public health exercise guidelines for 6 months. The average loss of body weight was measured to be as follows: for women exercising at 50% of public health recommendation levels, 0.4 kg = 0.88 lb; for women exercising at 100% of public health recommendation levels, 2.2 kg = 4.8 lb; and for women exercising at 150% of public health recommendation levels, 0.6 kg = 1.32 lb.

Based on the sample data in Section 11.1, we calculate that a person with a surface area of 1.7 $m^2$ walking at 3 mph expends 140 Cal/$m^2$ h $\times$ 1.7 $m^2$ = 240 Cal/h (two-figure accuracy). During the 30 min daily walk exercise, he or she expends 120 Cal/day or 600 Cal/wk. In 6 months the person expends, due to exercise, 120 Cal/day $\times$ 180 days = 21,600 Calories. Since one pound of body fat has been measured to contain approximately 3500 Calories, the simple thermodynamic model predicts that the exercising person will shed 21,600/3500 = 6.2 lb of body fat. Since in the model the body fat weight loss is linearly proportional to exercise exertion, the weight loss at 50% and 150% of the public health exercise guidelines are simply obtained by multiplying 6.2 lb by 0.5 and 1.5, respectively.

The measured average weight loss, together with the calculated weight loss based on the thermodynamic model, is shown below.

|  | 50% exertion 300 Cal/wk | 100% exertion 600 Cal/wk | 150% exertion 900 Cal/wk |
|---|---|---|---|
| Measured weight loss; 6 month period of exercise | 0.88 lb | 4.8 lb | 1.3 lb |
| Calculated weight loss; 6 month period of exercise | 3.1 lb | 6.2 lb | 9.3 lb |

The data in the table display three features evident in most weight-loss-through-exercise studies.

1. The measured weight loss is significantly smaller than predicted by the simple thermodynamic model.
2. The results of these weight loss studies are not in accordance with expectations. In many studies more exercise does not lead to greater weight loss.
3. Not discussed in this presentation is the substantial variability in weight loss data obtained by different groups doing similar studies under similar conditions. Further, significant variability in weight loss by different individuals is measured within the same group.

Observations such as shown in the table and in a number of other studies led several researchers to suggest that the body possesses a mechanism to conserve body fat. The development of such an *energy conservation* mechanism would have been an evolutionary process designed to preserve body fat for times when the species lacked food. This mechanism extracts energy from some other processes, such as the inflammatory response or immune system response providing energy that compensates in part for the energy expended in exercise and other physical activities, thus partially conserving body fat. Clearly the simple thermodynamic model that applies to a fixed heat engine is not applicable. It does not take into account that the human metabolism adjusts energy consumption in accordance with requirements. That is, as our physical activity increases, the body extracts at least part of the required energy from some other internal source rather than by burning body fat.

The process described can be regarded as homeostasis designed to keep body fat at as constant a level as possible. The mechanism for the proposed energy compensation is not known. Several possible energy compensation processes are discussed in the literature with no firm conclusion so far (Cox, 2017). The mechanism is probably a combination of both physiological and behavioral components. It should be noted that this energy compensation is highly variable among people and only part of the energy expenditure by physical activity including exercise is compensated by the process.

Studies reported by K.D. Flack in a 2018 publication (Flack et al., 2018) indicate that the limit for energy compensation in response to aerobic exercise is, on average, about 1000 Cal/wk. Flack et al. suggest that to achieve meaningful weight loss (of body fat) over a 12-week period, "exercise energy expenditure should exceed 1500 Cal/wk and likely be closer to 3000 Cal/week" to overcome energy compensation. A brief calculation shows that at 240 Cal/h expended in walking and with a 5-day-per-week exercise program, 3000 Cal/wk and 1500 Cal/wk exercise energy expenditure would require 2.5 and 1.25 hours of walking per day, respectively. Several studies listed by C.E. Cox show that exercise on that scale enables achieving weight loss in the range of 6 kg (13 lb) over a period of several months. However, researchers in this field suggest that such an amount of exercise may not be sustainable for people needing to achieve clinically significant weight loss.

Exercise may be effective for those who need to lose a few pounds, but for those who need to lose a clinically significant amount of weight, exercise alone is in most cases not likely to be an effective method of weight loss. It must be supplemented by dietary restriction. However, it needs to be emphasized and restated that while exercise may not be an effective means of weight loss, it is an essential component of a healthy lifestyle, as well as in the maintenance of a healthy weight.

## 11.12.5 Effect on weight loss of dietary restriction with and without exercise

A large number of studies have been conducted, all showing that dietary restriction is a far more effective way to lose weight than exercise. Here we will discuss weight loss resulting from dietary restrictions (a) without exercise and (b) with exercise.

A diet called the 1200-Calorie diet is frequently recommended as a low-calorie method of weight loss. In this regimen the per-day calorie intake is reduced over a few days to between 1200 and 1500 Cal/day for adult women and 1500 to 1800 Cal/day for adult men. This calorie regimen is considerably below the recommended values of 1600 to 2400 Cal/day for women and 2000 to 3000 Cal/day for men. Such a low-calorie diet results initially in a steep weight loss, peaking at about 6 months into the program. As the program continues, some of the weight is regained, likely due to a compensatory reduction in the body's metabolism that attempts to keep the body fat constant. Still after 1 year, the average weight loss for people following the 1200-Calorie diet is about 6 kg (~13 lb). At the end of 4 years, the weight loss seems to be held at about 6 lb, ~3 kg. However, without exercise, the weight loss is due to loss of both body fat and muscle tissue. Literature suggests that in a diet-alone weight loss program about 25% of weight loss is due to muscle mass reduction. Exercise is necessary to conserve muscle mass. Where exercise was added to dietary restriction, the exercise program was in most cases in the moderate to vigorous range of 5 days/wk for about 1 hour.

A useful systematic review of weight loss data under a variety of conditions has been published by M.J. Franz et al. (2007). In one set of results reported by Franz et al. over 18,000 individuals were involved. Some of the studies were followed out to 4 years. Common features are evident in both modalities (a) and (b), that is, without and with exercise. Initially weight loss increases (i.e., weight decreases) steeply for both modalities. Maximum weight loss occurs at about 6 months into the program. Here for modality (a), the weight loss was about 5 kg and for modality (b) (i.e., with exercise) 8 kg. Thereafter, the weight loss diminishes and the weight begins a slow rise. At 1 year, for modality (a), the weight loss is still about 5 kg, and for modality (b), about 7.5 kg. Four years into the program, the weight loss values with modalities (a) and (b) converge to nearly the same value of about 3 kg for (a) and 4 kg for (b). More severe diets bring about even greater initial weight loss followed by corresponding greater weight regain.

Dozens of studies have been conducted in both categories (a) and (b). Under similar conditions, they yield a variety of results significantly different from one another, but in essence, they convey a consistent message. Dietary restriction is a far more effective way to lose weight than exercise alone. Severe dietary restriction has to be done with care, acknowledging the inherent pitfalls of a method that drastically alters calorie intake essential to the functioning of the body. When exercise is added to the program, weight loss is initially significantly greater than with diet restriction alone. As the program continues, weight loss decreases and the value of weight loss with and without exercise converges. Data show that participants must continue both with the diet and exercise part of the program indefinitely; otherwise, the weight loss is not retained.

### 11.12.6 Extreme weight loss

In 2004 a TV program was initiated on NBC called *The Biggest Loser*. The program consisted of a 30-week competition for which 14 to 16 obese participants, both men and women, were selected from a large applicant pool. Typically men initially weighed 300 lb or over; women weighed over 200 lb. Most participants were in their 40s and 50s.

For the 30-week period, the applicants were subjected to a severe program of weight reduction. Typically the participant's calorie intake was reduced by about 65% and the participants were subject to 6 or 7 hours of intense exercise per day. The 30-week television program was interspersed with a variety of side events to maintain the interest of the audience. At the end of the 30-week series, the person who lost the most weight was declared the winner. The prize included publicity and a considerable sum of money.

Over the 30-week period, the participants lost a lot of weight, typically between 100 and 200 lb and some much more. The winner of season 8 shed 239 lb, going from 430 lb to 191 lb. At the end of *The Biggest Loser* program, participants started to gain weight. Within 2 years, on average, participants regained more than half the weight they lost during the program. People who work in this field suggest that eventually, 80% to 95% of program participants will regain all the weight they lost.

K.D. Hall (2021) studied the season 8 participants of *The Biggest Loser* program 6 years after the end of that season. He found that 14 out of 16 participants who completed the follow-up study after the 6-year period had regained about two-thirds of their initial weight loss. It is not surprising that during a period of calorie restriction, the body would decrease its energy consumption to conserve body fat. This could be most readily done by decreasing the resting metabolic rate (basal metabolic rate). Hall's surprise finding was that 6 years after the calorie restriction was over, the resting metabolic rate had not recovered. It was still down about 500 Cal/day, and in some cases still decreasing. This is of course a bad situation for anyone who wants to maintain weight loss. The 500 Cal/day conserved by the lowered metabolic rate is, as intended by the body, converted to body fat. At the same time, the levels of the hormone leptin were very low. Leptin signals to the body that enough food has been consumed. Low levels of the hormone produce intense feeling of hunger. The small minority who managed to maintain their hard-won weight loss were those who continued the intense exercise routine to consume the energy conserved by the metabolism.

### 11.12.7 What is the cause of the worldwide rise in obesity?

That obesity has been increasing in most regions of the world is evident. The reason for this increase is an important public health question. Three possibilities present themselves as an explanation for the growing obesity problem: (a) increased caloric intake combined with the evolution-driven functioning of our metabolism to conserve body fat, (b) widespread decrease in physical activity, and (c) a combination of the two.

In this connection, the field studies conducted by Herman Pontzer (2017) and his colleagues are relevant and interesting. Using the DLW technique, they measured the total energy expenditure of a group of Hadza people, members of a highly physically active Tanzanian tribe of hunter-gatherers, and compared their total energy expenditure to that of relatively sedentary Western populations. To their surprise, the total physical energy expenditure of the two groups was about the same, even though their physical activity levels were vastly different. Similar results had been obtained by other researchers. A possible mechanism explaining this phenomenon is discussed in Section 11.2.4 in connection with results of exercise and weight loss. That is, the hypothesis that the body possesses a mechanism, developed through evolution, to conserve body fat. This mechanism extracts energy from other processes, such as a reduction of the resting metabolic rate, inflammatory response, or immune system response, providing energy that compensates in part for the energy expended in physical activity exercise, thus partially conserving body fat. As a result, the total physical energy expenditure of the two groups was about the same, even though their physical activity levels were different. These studies led Pontzer to conclude that not decreased physical activity but rather the increased calorie-rich food intake is responsible for the growing obesity problem.

Not everyone agrees with this conclusion. Church and Martin (2018) suggest that the high calorie-rich food intake as the cause of widespread obesity has to be understood in connection with physical activity. They state that "there is a strong physiological drive to match caloric intake to caloric expenditure." However, they present evidence that the regulatory mechanism that matches caloric intake to caloric expenditure is active only at high to moderate physical activity. At low levels of physical activity, this balancing mechanism is disrupted and the caloric intake is not matched by caloric expenditure. Consequently, the calorie intake of people whose physical activity is low grows unchecked and they become obese. This is the case in developed countries, where the wide availability of labor-saving devices and the frequent use of mechanized transportation have greatly reduced the population's physical activities.

### 11.12.8 Conclusions

Studies show the importance of avoiding undesired weight gain. Once we gain weight, losing it is very difficult. Attempts to lose weight trigger our body's mechanisms to maintain the weight gain and any attempt to lose the weight puts us in conflict with processes developed by eons of evolution to maintain the weight. Should we succeed in achieving some weight loss, the body mobilizes to regain the weight.

Diet experts suggest employing a gradual and patient approach to weight loss that does not trigger metabolic compensations. Particularly severe dietary restrictions should be avoided. Such diets are most often not adequate to maintain the nutritional requirements of an average person. Severe diet restriction leads not only to body fat decrease but also decrease in muscle mass, nutrient deficiency, fatigue, and a host of undesirable metabolic changes, which then trigger the rebound. Severe diet restriction is bound to fail in achieving and maintaining long-term weight loss. A more patient approach to weight reduction with a reduced but well-balanced diet and reasonable exercise is more likely to lead to the achievement of long-term weight reduction. Successful weight reduction is a lifetime commitment and needs to be approached as such. There are several widely available diet programs with this approach.

We mention in passing that several prescription drugs are now available to aid weight loss. Mostly they act by suppressing the homeostatic action of the body to retain and conserve body fat. The safety and efficacy of these drugs has been improving over the years. However, a more detailed, reliable evaluation of these drugs is not yet available and is outside the scope of this presentation.

### Exercises

**11.1.** Show that Kleiber's law plotted on a log-log scale yields a straight line with a slope of 0.75.

**11.2.** Using results of a literature search, discuss the current state of the research and modeling of Kleiber's law.

**11.3.** Design an experiment that would measure the metabolic rate of walking at 5 km/h up a 20° slope.

**11.4.** How long can a man survive in an airtight room that has a volume of $27 \text{ m}^3$. Assume that his surface area is $1.70 \text{ m}^2$. Use data provided in the text.

**11.5.** A submarine carries an oxygen tank that holds oxygen at a pressure of 100 atm. What must the volume of the tank be to provide adequate oxygen for 50 people for 10 days? Assume that daily energy expenditure is as given in Table 11.2 and the average surface area of each person is $1.70 \text{ m}^2$.

**11.6.** Calculate the length of time that a person can survive without food but with adequate water. Obtain a solution under the following assumptions: (a) The initial weight and surface area of the person are 70 kg and $1.70 \text{ m}^2$, respectively. (b) The survival limit is reached when the person loses one-half his or her body weight. (c) Initially the body contains 5 kg of fatty tissue. (d) During the fast, the person sleeps 8 h/day and rests quietly the remainder of the time. (e) As the person loses weight, his or her surface area decreases (see Eq. 11.1). However, here, we assume that the surface area remains unchanged.

**11.7.** Suppose that a person of weight 60 kg and height 1.4 m reduces her sleep by 1 h/day and spends this extra time reading while sitting upright. If her food intake remains unchanged, how much weight will she lose in 1 year?

**11.8.** Assume that a person is sitting naked on an aluminum chair with $400 \text{ cm}^2$ area of the skin in contact with aluminum. If the skin temperature is $38°C$ and the aluminum is kept at $25°C$, compute the amount of heat transfer per hour from the skin. Assume that the body contacting the aluminum is insulated by a layer of unperfused fat tissue 0.5 cm thick $\left(K_c = 18 \text{ Cal cm}/(\text{m}^2 \text{ h C}°)\right)$ and that the heat conductivity of aluminum is very large. Is this heat transfer significant in terms of the metabolic heat consumption?

**11.9.** Explain qualitatively the functional dependence of $K'_c$ on the air velocity (see Fig. 11.2).

**11.10.** Show that $\left(T_s^4 - T_r^4\right) = \left(T_s^3 + T_s^2 T_r + T_s T_r^2 + T_r^3\right)(T_s - T_r)$. (b) Compute percentage change in the term $\left(T_s^3 + T_s^2 T_r + T_s T_r^2 + T_r^3\right)$ as the radiative temperature of the environment changes from $0°C$ to $40°C$. (Note that the temperatures in the computations must be expressed on the absolute scale. However, if the expression contains only the difference between two temperatures, either the absolute or the centigrade scale may be used.) (c) Calculate the value of $K_r$ in Eq. 11.5 under conditions discussed in the text, where $T_r = 25°C$ (298 K), $T_s = 32°C$ (305 K), and $H_r = 63 \text{ Cal}/h$.

**11.11.** A person takes about 20 breaths per minute with 0.5 L of air in each breath. How much heat is removed per hour by the moisture in the exhaled breath if the incoming air is dry and the exhaled breath is fully saturated? Assume that the water vapor pressure in the saturated exhaled air is 24 torr. Use data in Section 11.8.

**11.12.** Compute the heat loss per square meter of skin surface at $-40°C$ in moderate wind (about 0.5 m/sec, $K'_c = 10\,Cal/(m^2\,h\,C°)$). Assume that the skin temperature is 26°C.

**11.13.** Calculate the amount of heat required per hour to raise the temperature of inspired air from $-40°C$ to the body temperature of 37°C. Assume that the breathing rate is 600 L of air per hour. (This is the breathing rate specified in Exercise 11.11.) The amount of heat required to raise the temperature of 1 mole of air (22.4 L) by 1 C° at 1 atm is 29.2 J $(6.98 \times 10^{-3}\,Cal)$.

**11.14.** Explain why the daily temperature fluctuations in the soil are smaller (a) in wet soil than in dry soil, (b) in soil with a grass growth than in bare soil, and (c) when the air humidity is high.

**11.15.** Explain why the temperature drops rapidly at night in a desert.

**11.16.** The therapeutic effects of heat have been known since ancient times. Local heating, for example, relieves muscle pain and arthritic conditions. Discuss some effects of heat on tissue that may explain its therapeutic value.

**11.17.** Discuss the ecological effects of large carnivores using lions and wolves as specific examples. Information useful to this exercise can be found in J.R. William et al. (2014).

**11.18.** As stated in Chapter 11, the basal metabolic rate for a 70-kg man is about 70 Cal/h. Using Kleiber's law (Section 11.1.1), estimate the basal metabolic rate of a 190-kg male lion. The actual energy consumption of an active lion is likely to be twice as high. (b) The antelope is a favorite prey of lions. Assume the mass of an average antelope to be 40 kg. It is estimated that about 50% of the animal mass is consumable as food. (The rest is bone and hide that is not consumed by the carnivore.) The average caloric content of the edible portion is about 3 Cal/g. About how many antelopes must a lion kill and consume in a week to provide for its caloric needs?

**11.19.** What unexpected findings did you encounter in Section 11.12 of this chapter? The astronomer Carl Sagan stated, "Extraordinary claims require extraordinary evidence." Did the unexpected findings you encountered meet this criterion? Discuss in detail at least one issue that you selected.

# Chapter 12

# Waves and sound

Most of the information about our physical surroundings comes to us through our senses of hearing and sight. In both cases we obtain information about objects without being in physical contact with them. The information is transmitted to us in the first case by sound and in the second case by light. Although sound and light are very different phenomena, they are both waves. A wave can be defined as a disturbance that carries energy from one place to another without a transfer of mass. The energy carried by the waves stimulates our sensory mechanisms.

In this chapter, we will first explain briefly the nature of sound and then review some general properties of wave motion applicable to both sound and light. Using this background, we will examine the process of hearing and some other biological aspects of sound. Light will be discussed in Chapter 15.

## 12.1 Properties of sound

Sound is a mechanical wave produced by vibrating bodies. For example, when an object such as a tuning fork or the human vocal cords is set into vibrational motion, the surrounding air molecules are disturbed and are forced to follow the motion of the vibrating body. The vibrating molecules in turn transfer their motion to adjacent molecules, causing the vibrational disturbance to propagate away from the source. When the air vibrations reach the ear, they cause the eardrum to vibrate; this produces nerve impulses that are interpreted by the brain.

All matter transmits sound to some extent, but a material medium is needed between the source and the receiver to propagate sound. This is demonstrated by the well-known experiment of the bell in the jar. When the bell is set in motion, its sound is clearly audible. As the air is evacuated from the jar, the sound of the bell diminishes and finally the bell becomes inaudible.

The propagating disturbance in the sound-conducting medium is in the form of alternate compressions and rarefactions of the medium, which are initially caused by the vibrating sound source. These compressions and rarefactions are simply deviations in the density of the medium from the average value. In a gas, the variations in density are equivalent to pressure changes.

Two important characteristics of sound are *intensity*, which is determined by the magnitude of compression and rarefaction in the propagating medium, and *frequency*, which is determined by how often the compressions and

Physics in Biology and Medicine. https://doi.org/10.1016/B978-0-443-21558-2.00012-2

rarefactions take place. Frequency is measured in cycles per second, which is designated by the unit *hertz*, named after the scientist Heinrich Hertz. The symbol for this unit is Hz. (1 Hz = 1 cycle per second.)

The vibrational motion of objects can be highly complex (see Fig. 12.1), resulting in a complicated sound pattern. Still, it is useful to analyze the properties of sound in terms of simple sinusoidal vibrations such as would be set up by a vibrating tuning fork (see Fig. 12.2). The type of simple sound pattern shown in Fig. 12.2 is called a *pure tone*. When a pure tone propagates through air, the pressure variations due to the compressions and rarefactions are sinusoidal in form.

If we were to take a "snapshot" of the sound at a given instant in time, we would see pressure variations in space, which are also sinusoidal. (Such pictures can actually be obtained with special techniques.) In such a picture the distance between the nearest equal points on the sound wave is called the *wavelength*($\lambda$).

The speed of the sound wave $v$ depends on the material that propagates the sound. In air at 20°C, the speed of sound is about $3.3 \times 10^4$ cm/sec, and in water it is about $1.4 \times 10^5$ cm/sec. In general, the relationship between frequency, wavelength, and the speed of propagation is given by the following equation:

$$v = \lambda f \tag{12.1}$$

This relationship between frequency, wavelength, and speed is true for all types of wave motions.

The pressure variations due to the propagating sound are superimposed on the ambient air pressure. Thus, the total pressure in the path of a sinusoidal sound wave is of the form

$$P = P_a + P_o \sin 2\pi f t \tag{12.2}$$

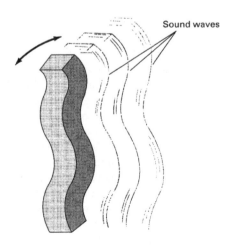

FIGURE 12.1 **A complex vibrational pattern.**

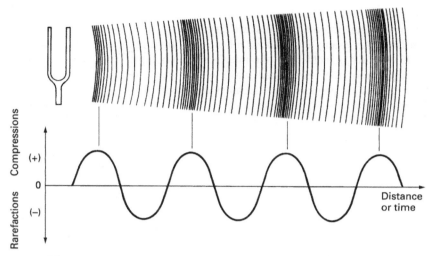

**FIGURE 12.2** Sinusoidal sound wave produced by a vibrating tuning fork.

where $P_a$ is the ambient air pressure (which at sea level at 0°C is $1.01 \times 10^5$ Pa $= 1.01 \times 10^6$ dyn/cm$^2$); $P_o$ is the maximum pressure change due to the sound wave; and $f$ is the frequency of the sound. The amount of energy transmitted by a sinusoidal sound wave per unit time through each unit area perpendicular to the direction of sound propagation is called the *intensity I* and is given by

$$I = \frac{P_o^2}{2\rho v} \qquad (12.3)$$

Here $\rho$ is the density of the medium, and v is the speed of sound propagation.

## 12.2 Some properties of waves

All waves, including sound and light, exhibit the phenomena of reflection, refraction, interference, and diffraction. These phenomena, which play an important role in both hearing and seeing, are described in detail in most basic physics texts (see for example Morgan, 1969). Here we will review them only briefly.

### 12.2.1 Reflection and refraction

When a wave enters one medium from another, part of the wave is reflected at the interface, and part of it enters the medium. If the interface between the two media is smooth on the scale of the wavelength (i.e., the irregularities of the interface surface are smaller than λ), the reflection is specular (mirror-like). If the surface has irregularities that are larger than the wavelength, the reflection is diffuse. An example of diffuse reflection is light reflected from paper.

If the wave is incident on the interface at an oblique angle, the direction of propagation of the transmitted wave in the new medium is changed (see Fig. 12.3). This phenomenon is called *refraction*. The angle of reflection is always equal to the angle of incidence, but the angle of the refracted wave is, in general, a function of the properties of the two media. The fraction of the energy transmitted from one medium to another depends again on the properties of the media and on the angle of incidence. For a sound wave incident perpendicular to the interface, the ratio of transmitted to incident intensity is given by

$$\frac{I_t}{I_i} = \frac{4\rho_1 v_1 \rho_2 v_2}{(\rho_1 v_1 + \rho_2 v_2)^2} \tag{12.4}$$

where the subscripted quantities are the velocity and density in the two media. The solution of Eq. 12.4 shows that when sound traveling in air is incident perpendicular to a water surface, only about 0.1% of the sound energy enters the water; 99.9% is reflected. The fraction of sound energy entering the water is even smaller when the angle of incidence is oblique. Water is thus an efficient barrier to sound.

### 12.2.2 Interference

When two (or more) waves travel simultaneously in the same medium, the total disturbance in the medium is at each point the vectorial sum of the

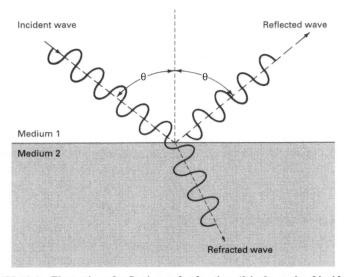

**FIGURE 12.3** **Illustration of reflection and refraction. (θ is the angle of incidence.)**

individual disturbances produced by each wave. This phenomenon is called *interference*. For example, if two waves are in phase, they add so that the wave disturbance at each point in space is increased. This is called *constructive interference* (see Fig. 12.4a). If two waves are out of phase by 180°, the wave disturbance in the propagating medium is reduced. This is called *destructive interference* (Fig. 12.4b). If the magnitudes of two out-of-phase waves are the same, the wave disturbance is completely canceled (Fig. 12.4c).

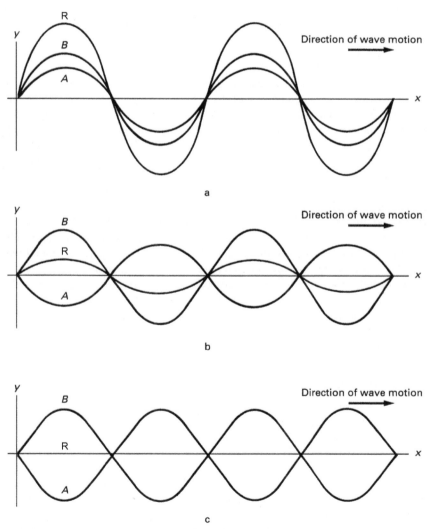

**FIGURE 12.4** (a) Constructive interference. (b, c) Destructive interference. R is the resultant of the interference of the two waves, *A* and *B*.

A special type of interference is produced by two waves of the same frequency and magnitude traveling in opposite directions. The resultant wave pattern is stationary in space and is called a *standing wave*. Such standing sound waves are formed in hollow pipes such as the flute. It can be shown that, in a given structure, standing waves can exist only at specific frequencies, which are called *resonant frequencies*. For two (or more) waves to interfere constructively or destructively, the phase of one wave with respect to the other must be well defined in time and space. In other words the phase of the disturbances produced by the waves must be correlated in time and space. Sets of waves exhibiting such correlation are referred to as *coherent*.

### 12.2.3 Diffraction

Waves have a tendency to spread as they propagate through a medium. As a result, when a wave encounters an obstacle, it spreads into the region behind the obstacle. This phenomenon is called *diffraction*. The amount of diffraction depends on the wavelength: The longer the wavelength, the greater the spreading of the wave. Significant diffraction into the region behind the obstacle occurs only if the size of the obstacle is smaller than the wavelength. For example, a person sitting behind a pillar in an auditorium hears the performer because the long wavelength sound waves spread behind the pillar. But the view of the performance is obstructed because the wavelength of light is much smaller than the pillar, and, therefore, the light does not diffract into the region behind the pillar.

Objects that are smaller than the wavelength do not produce a significant reflection. This too is due to diffraction. The wave simply diffracts around the small obstacle, much as flowing water spreads around a small stick.

Both light waves and sound waves can be focused with curved reflectors and lenses. There is, however, a limit to the size of the focused spot. It can be shown that the diameter of the focused spot cannot be smaller than about $\lambda/2$. These properties of waves have important consequences in the process of hearing and seeing.

### 12.3 Hearing and the ear

The sensation of hearing is produced by the response of the nerves in the ear to pressure variations in the sound wave. The nerves in the ear are not the only ones that respond to pressure, as most of the skin contains nerves that are pressure-sensitive. However, the ear is much more sensitive to pressure variations than any other part of the body.

Figure 12.5 is a drawing of the human ear. (The ear construction of other terrestrial vertebrates is similar.) For the purposes of description, the ear is usually divided into three main sections: The outer ear, the middle ear, and the inner ear. The sensory cells that convert sound to nerve impulses are located in the liquid-filled inner ear.

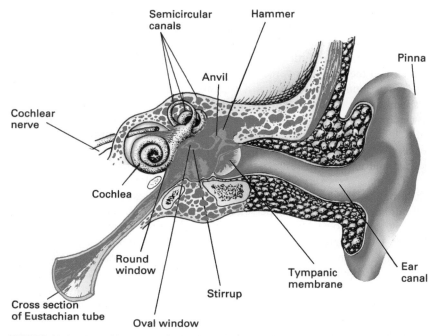

**FIGURE 12.5**  A semidiagrammatic drawing of the ear with various structures cut away and simplified to show the basic relationships more clearly. The middle ear muscles have been omitted.

The main purpose of the outer and middle ears is to conduct the sound into the inner ear.

The outer ear is composed of an external flap called the *pinna* and the ear canal, which is terminated by the *tympanic membrane* (eardrum). In many animals the pinna is large and can be rotated toward the source of the sound; this helps the animal to locate the source of sound. However, in humans the pinna is fixed and so small that it does not seem to contribute significantly to the hearing process.

The ear canal of an average adult is about 0.75 cm in diameter and 2.5 cm long, a configuration that is resonant for sound waves at frequencies around 3000 Hz. This accounts in part for the high sensitivity of the ear to sound waves in this frequency range.

For an animal to perceive sound, the sound has to be coupled from air to the sensory cells that are in the fluid environment of the inner ear. We showed earlier that direct coupling of sound waves into a fluid is inefficient because most of the sound energy is reflected at the interface. The middle ear provides an efficient conduction path for the sound waves from air into the fluid of the inner ear.

The middle ear is an air-filled cavity that contains a linkage of three bones called *ossicles* that connect the eardrum to the inner ear. The three bones are called the *hammer*, the *anvil*, and the *stirrup*. The hammer is attached to the inner surface of the eardrum, and the stirrup is connected to the oval window, which is a membrane-covered opening in the inner ear.

When sound waves produce vibrations in the eardrum, the vibrations are transmitted by the ossicles to the oval window, which in turn sets up pressure variations in the fluid of the inner ear. The ossicles are connected to the walls of the middle ear by muscles that also act as a volume control. If the sound is excessively loud, these muscles as well as the muscles around the eardrum stiffen and reduce the transmission of sound to the inner ear.

The middle ear serves yet another purpose. It isolates the inner ear from the disturbances produced by movements of the head, chewing, and the internal vibrations produced by the person's own voice. To be sure, some of the vibrations of the vocal cords are transmitted through the bones into the inner ear, but the sound is greatly attenuated. We hear ourselves talk mostly by the sound reaching our eardrums from the outside. This can be illustrated by talking with the ears plugged.

The *Eustachian tube* connects the middle ear to the upper part of the throat. Air seeps in through this tube to maintain the middle ear at atmospheric pressure. The movement of air through the Eustachian tube is aided by swallowing. A rapid change in the external air pressure such as may occur during an airplane flight causes a pressure imbalance on the two sides of the eardrum. The resulting force on the eardrum produces a painful sensation that lasts until the pressure in the middle ear is adjusted to the external pressure. The pain is especially severe and prolonged if the Eustachian tube is blocked by swelling or infection.

The conversion of sound waves into nerve impulses occurs in the *cochlea*, which is located in the inner ear. The cochlea is a spiral cavity shaped like a snail shell. The wide end of the cochlea, which contains the oval and the round windows, has an area of about 4 mm$^2$. The cochlea is formed into a spiral with about $2\frac{3}{4}$ turns. If the cochlea were uncoiled, its length would be about 35 mm.

Inside the cochlea there are three parallel ducts; these are shown in the highly simplified drawing of the uncoiled cochlea in Fig. 12.6. All three ducts are filled with a fluid. The vestibular and tympanic canals are joined at the apex of the cochlea by a narrow opening called the *helicotrema*. The cochlear duct is isolated from the two canals by membranes. One of these membranes, called the *basilar membrane*, supports the auditory nerves.

The vibrations of the oval window set up a sound wave in the fluid filling the vestibular canal. The sound wave, which travels along the vestibular canal and through the helicotrema into the tympanic canal, produces vibrations in the basilar membrane, which stimulate the auditory nerves to transmit electrical pulses to the brain (see Chapter 13). The excess energy in the sound

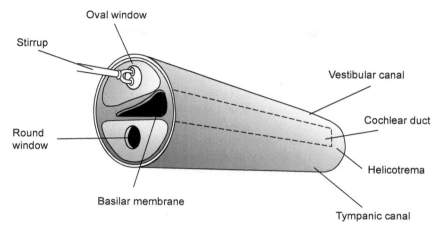

FIGURE 12.6　**An uncoiled view of the cochlea.**

wave is dissipated by the motion of the round window at the end of the tympanic canal.

The inner ear also serves a key function of providing our sense of balance. This is done by means of small calcium carbonate crystals called *otoconia* embedded in a fibrous matrix coupled to sensory cells. The small shifting motion of the otoconia under the changing force of gravity is transmitted to the sensory cells that in turn send signals to the nervous system providing information about spatial orientation of the body. The spatial information is augmented by visual signals as well as signals provided by the position of our limbs with respect to the torso.

The biosynthesis of the calcium carbonate crystals and the associated matrix occurs during fetal development. Its construction and maintenance throughout life is not yet fully understood. Several conditions manifesting as vertigo are associated with the misfunction of the otoconia system.

## 12.3.1 Performance of the ear

The nerve impulses evoke in the brain the subjective sensation of sound. *Loudness, pitch*, and *quality* are some of the terms we use to describe the sounds we hear. It is a great challenge for physiologists to relate these subjective responses with the physical properties of sound such as intensity and frequency. Some of these relationships are now well understood; others are still subjects for research.

In most cases, the sound wave patterns produced by instruments and voices are highly complex. Each sound has its own characteristic pattern. It would be impossible to evaluate the effect of sound waves on the human auditory system if the response to each sound pattern had to be analyzed separately.

Fortunately, the problem is not that complicated. About 150 years ago, J. B. J. Fourier, a French mathematician, showed that complex wave shapes can be analyzed into simple sinusoidal waves of different frequencies. In other words, a complex wave pattern can be constructed by adding together a sufficient number of sinusoidal waves at appropriate frequencies and amplitudes. Therefore, if we know the response of the ear to sinusoidal waves over a broad range of frequencies, we can evaluate the response of the ear to a wave pattern of any complexity.

An analysis of a wave shape into its sinusoidal components is shown in Fig. 12.7. The lowest frequency in the wave form is called the *fundamental*, and the higher frequencies are called *harmonics*. Figure 12.8 shows the sound pattern for a specific note played by various instruments. It is the harmonic content of the sound that differentiates one sound source from another. For a given note played by the various instruments shown in Fig. 12.8, the fundamental frequency is the same but the harmonic content of the wave is different for each instrument.

## 12.3.2 Frequency and pitch

The human ear is capable of detecting sound at frequencies between about 20 and 20,000 Hz. Within this frequency range, however, the response of the ear

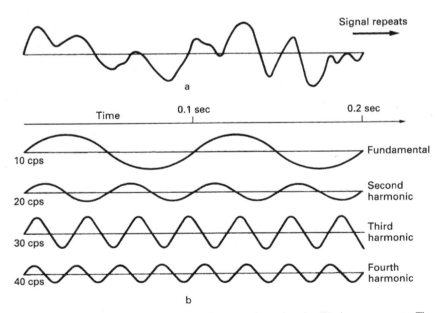

FIGURE 12.7   The analysis of (A) a complex wave shape, into its (B) sine components. The point-by-point addition of the fundamental frequency sine wave and the harmonic frequency sine waves yields the wave shape shown in (A).

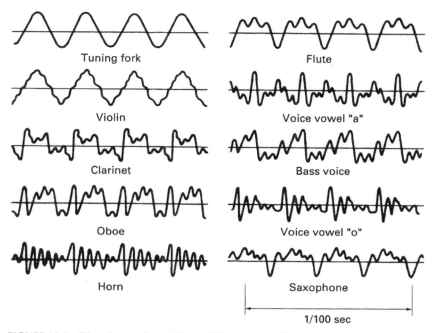

**FIGURE 12.8** **Wave forms of sound from different musical instruments sounding the same note.**

is not uniform. The ear is most sensitive to frequencies between 200 and 4000 Hz, and its response decreases toward both higher and lower frequencies. There are wide variations in the frequency response of individuals. Some people cannot hear sounds above 8000 Hz, whereas a few people can hear sounds above 20,000 Hz. Furthermore, the hearing of most people deteriorates with age.

The sensation of pitch is related to the frequency of the sound. The pitch increases with frequency. Thus, the frequency of the musical note middle C is 256 Hz, and the frequency of the A above is 440 Hz. There is, however, no simple mathematical relationship between pitch and frequency.

### 12.3.3 Intensity and loudness

The ear responds to an enormous range of intensities. At 3000 Hz, the lowest intensity that the human ear can detect is about $10^{-16}$ W/cm$^2$. The loudest tolerable sound has an intensity of about $10^{-4}$ W/cm$^2$. These two extremes of the intensity range are called the *threshold of hearing* and the *threshold of pain*, respectively. Sound intensities above the threshold of pain may cause permanent damage to the eardrum and the ossicles.

The ear does not respond linearly to sound intensity; that is, a sound that is a million times more powerful than another does not evoke a million times higher sensation of loudness. The response of the ear to intensity is closer to being logarithmic than linear.

Because of the nonlinear response of the ear and the large range of intensities involved in the process of hearing, it is convenient to express sound intensity on a logarithmic scale. On this scale, the sound intensity is measured relative to a reference level of $10^{-16}$ W/cm$^2$ (which is approximately the lowest audible sound intensity). The logarithmic intensity is measured in units of decibels (dB) and is defined as

$$\text{Logarithmic intensity} = 10 \log \frac{\text{Sound intensity in W/cm}^2}{10^{-16}\,\text{W/cm}^2} \qquad (12.5)$$

Thus, for example, the logarithmic intensity of a sound wave with a power of $10^{-12}$ W/cm$^2$ is

$$\text{Logarithmic intensity} = 10 \log \frac{10^{-12}}{10^{-16}} = 40\,\text{dB}$$

Intensities of some common sounds are listed in Table 12.1.

At one time, it was believed that the ear responded logarithmically to sound intensity. Referring to Table 12.1, a logarithmic response would imply that, for example, a busy street sounds only six times louder than the rustle of leaves, even though the power of the street sounds is a million times greater. Although it has been shown that the intensity response of the ear is not exactly logarithmic, the assumption of a logarithmic response still provides a useful guide

**TABLE 12.1 Sound levels due to various sources (representative values).**

| Source of sound | Sound level (dB) | Sound level (W/cm$^2$) |
| --- | --- | --- |
| Threshold of pain | 120 | $10^{-4}$ |
| Riveter | 90 | $10^{-7}$ |
| Busy street traffic | 70 | $10^{-9}$ |
| Ordinary conversation | 60 | $10^{-10}$ |
| Quiet automobile | 50 | $10^{-11}$ |
| Quiet radio at home | 40 | $10^{-12}$ |
| Average whisper | 20 | $10^{-14}$ |
| Rustle of leaves | 10 | $10^{-15}$ |
| Threshold of hearing | 0 | $10^{-16}$ |

for assessing the sensation of loudness produced by sounds at different intensities (see Exercises 12.1 and 12.2).

The sensitivity of the ear is remarkable. At the threshold of hearing, in the range of 2000–3000 Hz, the ear can detect a sound intensity of $10^{-16}$ W/cm$^2$. This corresponds to a pressure variation in the sound wave of only about $2.9 \times 10^{-4}$ dyn/cm$^2$ (see Exercise 12.3). Compare this to the background atmospheric pressure, which is $1.013 \times 10^6$ dyn/cm$^2$. This sensitivity appears even more remarkable when we realize that the random pressure variations in air due to the thermal motion of molecules are about $0.5 \times 10^{-4}$ dyn/cm$^2$. Thus, the sensitivity of the ear is close to the ultimate limit at which it would begin to detect the noise fluctuations in the air. The displacement of the molecules corresponding to the power at the threshold of hearing is less than the size of the molecules themselves.

The sensitivity of the ear is partly due to the mechanical construction of the ear, which amplifies the sound pressure. Most of the mechanical amplification is produced by the middle ear. The area of the eardrum is about 30 times larger than the oval window. Therefore, the pressure on the oval window is increased by the same factor (see Exercise 12.4). Furthermore, the ossicles act as a lever with a mechanical advantage of about 2. Finally, in the frequency range of around 3000 Hz, there is an increase in the pressure at the eardrum due to the resonance of the ear canal. In this frequency range, the pressure is increased by another factor of 2. Thus, the total mechanical amplification of the sound pressure in the 3000 Hz range is about $2 \times 30 \times 2 = 120$. Because the intensity is proportional to pressure squared (see Eq. 12.3), the intensity at the oval window is amplified by a factor of about 14,400.

The process of hearing cannot be fully explained by the mechanical construction of the ear. The brain itself plays an important role in our perception of sound. For example, the brain can effectively filter out ambient noise and allow us to separate meaningful sounds from a relatively loud background din. (This feature of the brain permits us to have a private conversation amid a loud party.) The brain can also completely suppress sounds that appear to be meaningless. Thus, we may lose awareness of a sound even though it still produces vibrations in our ear. The exact mechanism of interaction between the brain and the sensory organs is not yet fully understood.

## 12.4 Bats and echoes

The human auditory organs are very highly developed, yet, there are animals that can hear even better than we can. Notable among these animals are bats. They emit high-frequency sound waves and detect the reflected sounds (echoes) from surrounding objects. Their sense of hearing is so acute that they can obtain information from echoes which are in many ways as detailed as the information we can obtain with our sense of sight. The many different species of bats utilize echoes in various ways. The *Vespertilionidae* family of bats emit short chirps as they fly. The chirps last about $3 \times 10^{-3}$ seconds (3 milliseconds) with a time interval between chirps of about 70 milliseconds. Each chirp starts at a frequency of about $100 \times 10^3$ Hz and falls to about $30 \times 10^3$ Hz at the end. (The ears of bats, of course, respond to these high frequencies.) The silent interval between chirps allows the bat to detect the weak echo without interference from the primary chirp. Presumably the interval between the chirp and the return echo allows the bat to determine its distance from the object. It is also possible that differences in the frequency content of the chirp and the echo allow the bat to estimate the size of the object (see Exercise 12.5). With a spacing between chirps of 70 milliseconds, an echo from an object as far as 11.5 m can be detected before the next chirp (see Exercise 12.6). As the bat comes closer to the object (such as an obstacle or an insect), both the duration of and the spacing between chirps decrease, allowing the bat to localize the object more accurately. In the final approach to the object, the duration of the chirps is only about 0.3 milliseconds, and the spacing between them is about 5 milliseconds.

Experiments have shown that with echo location, bats can avoid wire obstacles with diameters down to about 0.1 mm, but they fail to avoid finer wires. This is in accord with our discussion of wave diffraction (see Exercise 12.7). Other animals, such as porpoises, whales, and some birds, also use echoes to locate objects, but they are not able to do so as well as bats.

## 12.5  Sounds produced by animals

Animals can make sounds in various ways. Some insects produce sounds by rubbing their wings together. The rattlesnake produces its characteristic sound by shaking its tail. In most animals, however, sound production is associated with the respiratory mechanism. In humans, the *vocal cords* are the primary source of sound. These are two reeds, shaped like lips, attached to the upper part of the trachea. During normal breathing, the cords are wide open. To produce a sound, the edges of the cords are brought together. Air from the lungs passes through the space between the edges and sets the cords into vibration. The frequency of the sounds is determined by the tension on the vocal cords. The fundamental frequency of the average voice is about 140 Hz for males and about 230 Hz for females. The sound produced by the vocal cords is substantially modified as it travels through the passages of the mouth and throat. The tongue also plays an important role in the final sound. Many voice sounds are produced outside the vocal cords (e.g., the consonant *s*). The sounds in a whispering talk are also produced outside the vocal cords.

## 12.6  Acoustic traps

Electronically generated sounds that mimic those of animals and insects are increasingly being used as lures to trap the creatures. Electronic fishing lures are now commercially available. One such device mimics the distress call of a mackerel and attracts marlin and other larger fish to the fishhook.

To obtain baseline data on bat populations, often the bats have to be captured and examined. In one such study, the social call of a rare Bechstein's bat that inhabits the woodlands of southeast England was synthesized, luring the bats into the net. (The bats were released after examination.)

The Mediterranean fruit fly, commonly called medfly, is a pest that infests fruits and other crops, causing on the order of $1 billion in damages worldwide. At present, spraying of pesticides is the most common way of controlling the medfly. An environmentally more friendly way of controlling the pest has been sought for many years. Sound traps under development may provide a viable alternative. The male medfly produces with its wings a vibration at a fundamental frequency of about 350 Hz accompanied by complex harmonics. The female medflies are attracted to this courtship call and can be lured into a trap.

## 12.7 Clinical uses of sound

### 12.7.1 The stethoscope

The most familiar *clinical use of sound* is in the analysis of body sounds with a *stethoscope*. This instrument consists of a small bell-shaped cavity attached to a hollow flexible tube. The bell is placed on the skin over the source of the body sound (such as the heart, intestines, or lungs). The sound is then conducted by the pipe to the ears of the examiner, who evaluates the functioning of the organ (see Exercise 12.8). The importance and significance of sounds emanating from the various organs within the body have been known since antiquity. Physicians listened to the sounds of the various organs by placing their ear in contact with the skin over the organ of interest. The stethoscope was developed in 1816 by a French physician, Rene Laennec, reportedly because he was uncomfortable placing his ear directly on the chests of women. Many stethoscope designs are currently in use, several of them for specific purposes. One modified version of the stethoscope consists of two bells that are placed on different parts of the body. The sound picked up by one bell is conducted to one ear, and the sound from the other bell is conducted to the other ear. The two sounds are then compared. With this device, it is possible, for example, to listen simultaneously to the heartbeats of the fetus and of the pregnant mother.

Other methods, most of them utilizing ultrasound, now provide a more accurate and detailed diagnosis of organ functions than does the stethoscope. However, it is likely that the stethoscope, because of its simplicity and its significance as a symbol of the medical profession, will remain a first-line tool in a physical examination.

## 12.7.2 Ultrasonic waves

With special electronically driven crystals, it is possible to produce mechanical waves at very high frequencies, up to about 20 MHz. These waves, which are simply the extension of sound to high frequencies, are called *ultrasound (or ultrasonic) waves*. Because of their short wavelength, ultrasonic waves can be focused onto small areas and can be imaged much as visible light (see Exercise 12.9).

## 12.7.3 Ultrasonic imaging

Ultrasonic waves penetrate tissue and are scattered and absorbed within it. The scattered and reflected ultrasound contains information about the form and structure of the tissue. This ultrasound signal is converted into a visible image using specialized computer techniques. The technique is called *ultrasound imaging*. Therefore, structures within living organisms can be examined with ultrasound, as with X-rays. By capturing rapid sequential images of the organs, motions of the internal organs can be obtained. Ultrasonic examinations are safer than X-rays but do not provide the same information obtainable with X-rays. For some applications, such as in the examination of a fetus, the ultrasonic method is preferred by most physicians.

## 12.7.4 Ultrasonic flow meter

The frequency of sound detected by an observer depends on the relative motion between the source and the observer. This phenomenon is called the *Doppler effect*. It can be shown (see Exercise 12.10) that if the observer is stationary and the source is in motion, the frequency of the sound f′ detected by the observer is given by

$$f' = f \frac{v}{v \mp v_s} \tag{12.6}$$

where f is the frequency in the absence of motion, $v$ is the speed of sound, and $v_s$ is the speed of the source. The minus sign in the denominator is to be used when the source is approaching the observer, and the plus sign is when the source is receding.

Using the Doppler effect, it is possible to measure motions within a body. One device utilizing the Doppler effect is the ultrasonic flow meter, which produces ultrasonic waves that are scattered by blood cells flowing in the blood vessels. The frequency of the scattered sound is altered by the Doppler effect. The velocity of blood flow is obtained by comparing the incident frequency with the frequency of the scattered ultrasound.

### 12.7.5 Echocardiography

Echocardiography is an ultrasound imaging technique that can provide key information about the functioning of the heart. Several echocardiographic techniques can be used. The simplest of these is the transthoracic *echocardiogram*. Here the ultrasound is aimed at the heart and the receiver (called transducer) positioned on the chest of the patient records the reflections (echoes) from the heart. The reflections are converted into visible images of the heart and the motions of components within the heart. In some cases, a contrast agent is injected into the heart to enhance parts of the image.

If the image produced using a standard transthoracic echocardiogram is not clear enough, a transesophageal echocardiogram is obtained. Here the transducer in a flexible tube is guided down the throat and the esophagus bringing the detector closer to the heart. Significantly more detail and clearer images can be obtained with this technique.

Both transthoracic and transesophageal echocardiography can be used in conjunction with measurement of the Doppler shift. With this combination, information can be obtained about blood flow within the heart and in and out of the heart. Narrowing of arteries, leaky heart valves, and weaknesses in the separation of the heart chambers can be detected.

### 12.7.6 Therapeutic use of ultrasound

Within the tissue, vibrational energy of the ultrasonic wave is converted to heat. In ultrasound imaging, the ultrasound power level used is about 0.1 W (when focused on the object about $2-15$ W/cm$^2$), and the imaging is intermittent at pulse length on the order 1 to 10 microseconds. At these power levels, tissue heating is negligible. At a power level about 10 times that used for imaging, and the power applied continuously rather than in the pulsed mode, the ultrasound energy is sufficient to heat selected parts of a patient's body more efficiently and evenly than can be done with conventional heat lamps. This type of treatment, called *diathermy*, is used to relieve pain and promote the healing of injuries.

### 12.7.7 Focused ultrasound surgery

At a sufficiently high ultrasound intensity, it is possible to raise the temperature of tissue cells high enough to destroy the tissue. Increasing the temperature of cells to about 56°C from the normal tissue temperature of about 37°C for a period of a few seconds (typically between 1 and 10 seconds) will destroy them. Figure 12.9 illustrates the basic method of focused ultrasound surgery (FUS). A specially designed vibrating crystal source produces an ultrasound beam that is focused into the volume of cells to be destroyed inside the tissue. As is shown in Exercise 12.11, a focused exposure of 12 W of ultrasound into a 0.15 cm$^3$

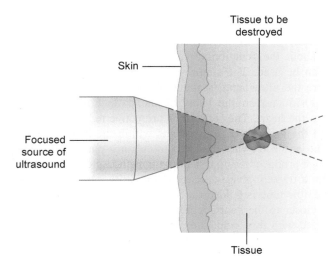

**FIGURE 12.9 Focused ultrasound surgery.** A specially designed vibrating crystal source produces an ultrasound beam focused on the volume of cells to be destroyed inside the tissue.

volume of tissue for 1 second will raise its temperature to about 56°C, sufficient to destroy the tissue without harming the surrounding cells. Tissue volume of that size is typical of controlled FUS exposure. The FUS technique has been used to destroy tumors and to reduce the size of enlarged prostate. The dead cells are removed by the natural clearing mechanisms of the body. FUS has the advantage of a minimally invasive technique that can reach relatively deep below the skin to affect the destruction of tissue. On the other hand, the technique is slow because to destroy or cut a significant-sized region of tissue, many pulses are required, each lasting a second or more.

## 12.7.8 Elastography

*Elastography* is an imaging technique that maps the elastic properties of soft tissue such as liver, prostate, or lymph nodes. Hardening of the tissue (i.e., reduction of elasticity) usually indicates a medical problem such as a tumor that is usually harder and therefore less elastic than normal tissue. A reduction of liver elasticity may indicate cirrhosis or liver cancer.

Elastography is directly related to manual palpation, an ancient medical technique described already in Egyptian papyrus documents dating back to 1500 BCE. Manual palpation continues to be used as a first line in a physical examination of many organs accessible to touch, most often in the examination of breasts, prostate, and abdomen. Elastography provides a more accurate and quantitative evaluation of the tissue elasticity and makes possible examination of organs deeper in the body not accessible to manual palpation.

Medical examination by manual percussion is likewise an ancient technique, similar to palpation, except that the practitioner uses sound rather than touch to obtain information about the organs of interest. The practitioner places his or her hand with fingers spread on the patient contacting the area over the organ of interest. The practitioner then taps sequentially the regions between the spread fingers and listens, with a stethoscope, to the quality of the sound being produced. Typically, regions in the body filled with air provide a resonant higher frequency sound. Regions filled with tissue yield a duller, less resonant sound depending on the nature of the tissue. Percussion is most commonly used in examining large organs mostly lungs, intestines, liver, and spleen.

In current methods of elastographic image production, the tissue to be examined is distorted by one means or another, and the subsequent response of the tissue is recorded. Often the distortion is produced simply by an externally applied compression. An ultrasound image is taken before and after the applied compression. The parts of the image that show the smallest compression are the least elastic and are most likely regions of diseased tissue.

*Magnetic resonance elastography* (MRE) is a relatively new (in use since 2009) technique that provides detailed three-dimensional information about tissue. The technique does not utilize ultrasound and is at present most usefully used in liver examinations. The tissue is distorted by a low-frequency pulsating force while images are taken using magnetic resonance imaging techniques (see Chapter 17, Section 17.2). The technique provides detail and accuracy that were previously obtainable only through biopsy. Further, while biopsy provides information only about relatively small selected regions of the organ, at times possibly missing important regions of pathology, MRE yields an image of the whole organ.

## Exercises

**12.1.** The intensity of a sound produced by a point source decreases as the square of the distance from the source. Consider a riveter as a point source of sound and assume that the intensities listed in Table 12.1 are measured at a distance 1 m away from the source. What is the maximum distance at which the riveter is still audible? (Neglect losses due to energy absorption in the air.)

**12.2.** Referring to Table 12.1, approximately how much louder does busy street traffic sound than a quiet radio?

**12.3.** Calculate the pressure variation corresponding to a sound intensity of $10^{-16}$ W/cm$^2$. (The density of air at 0°C and 1 atm pressure is $1.29 \times 10^{-3}$ g/cm$^3$; for the speed of sound, use the value $3.3 \times 10^4$ cm/sec.)

**12.4.** Explain why the relative sizes of the eardrum and the oval window result in pressure magnification in the inner ear.

**12.5.** Explain how a bat might use the differences in the frequency content of its chirp and echo to estimate the size of an object.

**12.6.** With a 70-millisecond space between chirps, what is the farthest distance at which a bat can detect an object?

**12.7.** In terms of diffraction theory, discuss the limitations on the size of the object that a bat can detect with its echolocation.

**12.8.** A stethoscope tube has an inner diameter of 0.7 cm, and its length is 70 cm. Sound traveling along the tube attenuates 3 dB/m. What is the intensity of the faintest sound emanating from the surface of the body that the examiner can detect? Use data in Table 12.1.

**12.9.** Estimate the smallest size of objects in soft tissue that can be detected with ultrasound at a frequency of $5 \times 10^6$ Hz. The speed of sound in soft tissue is 1540 m/sec. Assume that the optimum spatial resolution is approximately the wavelength of the probing ultrasound.

**12.10.** With the help of a basic physics textbook, explain the Doppler effect, and derive Eq. 12.6.

**12.11.** Estimate the increase in tissue temperature when 0.15 cm$^3$ volume of tissue is exposed to 12 W of ultrasound for 1 second. Assume that the density of the tissue is 1 g/cm$^3$, the specific heat (or heat capacity) is 1 cal/g C°, and the amount of heat conducted away from the heated area during the 1-second pulse interval is negligible.

# Chapter 13

# Electricity

The word *electricity* usually evokes the image of a manmade technology because we usually associate electricity with devices such as amplifiers, televisions, and computers. This technology has certainly played an important role in our understanding of living systems, as it has provided the major tools for the study of life processes. However, many life processes themselves involve electrical phenomena. The nervous system of animals and the control of muscle movement, for example, are both governed by electrical interactions. Even plants rely on electrical forces for some of their functions. In this chapter, we will describe some of the electrical phenomena in living organisms, and in Chapter 14 we will discuss the applications of electrical technology in biology and medicine. A brief review of electricity in Appendix B summarizes the concepts, definitions, and equations used in the text.

## 13.1 The nervous system

The most remarkable use of electrical phenomena in living organisms is found in the nervous system of animals. Specialized cells called *neurons* form a complex network within the body that receives, processes, and transmits information from one part of the body to another. The center of this network is located in the brain, which has the ability to store and analyze information. Based on this information, the nervous system controls various parts of the body. The nervous system is very complex. The human nervous system, for example, consists of about $10^{10}$ interconnected neurons. It is, therefore, not surprising that, although the nervous system has been studied for more than a hundred years, its functioning as a whole is still poorly understood. It is not known how information is stored and processed by the nervous system, nor is it known how the neurons grow into patterns specific to their functions. Yet some aspects of the nervous system are now well known. Specifically, during the past 40 years, the method of signal propagation through the nervous system has been firmly established. The messages are electrical pulses transmitted by the neurons. When a neuron receives an appropriate stimulus, it produces electrical pulses that are propagated along its cablelike structure. The pulses are constant in magnitude and duration, independent of the intensity of the stimulus. The strength of the stimulus is conveyed by the number of pulses

Physics in Biology and Medicine. https://doi.org/10.1016/B978-0-443-21558-2.00013-4

produced. When the pulses reach the end of the "cable," they activate other neurons or muscle cells.

### 13.1.1 The neuron

The neurons, which are the basic units of the nervous system, can be divided into three classes: *Sensory* neurons, *motor* neurons, and *interneurons.* The sensory neurons receive stimuli from sensory organs that monitor the external and internal environment of the body. Depending on their specialized functions, the sensory neurons convey messages about factors such as heat, light, pressure, muscle tension, and odor to higher centers in the nervous system for processing. The motor neurons carry messages that control the muscle cells. These messages are based on information provided by the sensory neurons and by the central nervous system located in the brain. The interneurons transmit information between neurons.

Each neuron consists of a cell body to which are attached input ends called *dendrites* and a long tail called the *axon,* which propagates the signal away from the cell (see Fig. 13.1). The far end of the axon branches into nerve endings that transmit the signal across small gaps to other neurons or muscle cells. A simple sensory-motor neuron circuit is shown in Fig. 13.2. A stimulus from a muscle produces nerve impulses that travel to the spine. Here the signal is transmitted to a motor neuron, which in turn sends impulses to control the muscle. Such simple circuits are often associated with reflex actions. Most nervous connections are far more complex.

The axon, which is an extension of the neuron cell, conducts the electrical impulses away from the cell body. Some axons are long indeed—in people, for example, the axons connecting the spine with the fingers and toes are more than a meter in length. Some of the axons are covered with a segmented sheath of fatty material called *myelin.* The segments are about 2 mm long, separated by gaps called the *nodes of Ranvier.* We will show later that the myelin sheath increases the speed of pulse propagation along the axon.

Although each axon propagates its own signal independently, many axons often share a common path within the body. These axons are usually grouped into nerve bundles.

The ability of the neuron to transmit messages is due to the special electrical characteristics of the axon. Most of the data about the electrical and chemical properties of the axon is obtained by inserting small needlelike probes into the axon. With such probes, it is possible to measure currents flowing in the axon and to sample its chemical composition. Such experiments are usually difficult as the diameter of most axons is very small. Even the largest axons in the human nervous system have a diameter of only about 20 μm $(20 \times 10^{-4}$ cm$)$. The squid, however, has a giant axon with a diameter of about 500 μm (0.5 mm), which is large enough for the

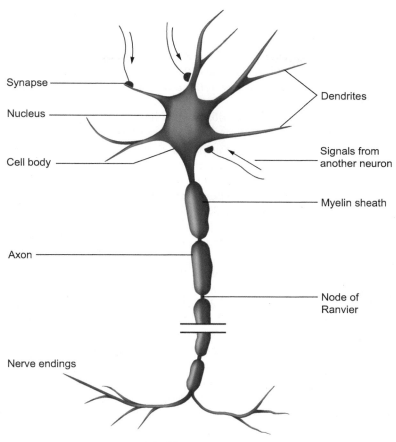

Synapse

Nucleus

Cell body

Dendrites

Signals from
another neuron

Myelin sheath

Axon

Node of
Ranvier

Nerve endings

FIGURE 13.1   **A neuron.**

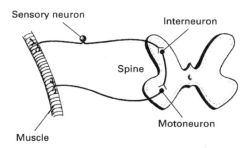

Sensory neuron

Interneuron

Spine

Muscle

Motoneuron

FIGURE 13.2   **A simple neural circuit.**

convenient insertion of probes. Much of the information about signal transmission in the nervous system has been obtained from experiments with the squid axon.

## 13.1.2 Electrical potentials in the axon

In the aqueous environment of the body, salt and various other molecules dissociate into positive and negative ions. As a result, body fluids are relatively good conductors of electricity. Still, these fluids are not nearly as conductive as metals; their resistivity is about 100 million times greater than that of copper, for example.

The inside of the axon is filled with an ionic fluid that is separated from the surrounding body fluid by a thin membrane (Fig. 13.3). The axon membrane, which is only about $50 - 100$ Å thick, is a relatively good but not perfect electrical insulator. Therefore, some current can leak through it.

The electrical resistivities of the internal and the external fluids are about the same, but their chemical compositions are substantially different. The external fluid is similar to sea water. Its ionic solutes are mostly positive sodium ions and negative chlorine ions. Inside the axon, the positive ions are mostly potassium ions, and the negative ions are mostly large negatively charged organic molecules.

Because there is a large concentration of sodium ions outside the axon and a large concentration of potassium ions inside the axon, we may ask why the concentrations are not equalized by diffusion. In other words, why don't the sodium ions leak into the axon and the potassium ions leak out of it? The answer lies in the properties of the axon membrane.

In the resting condition, when the axon is not conducting an electrical pulse, the axon membrane is highly permeable to potassium and only slightly permeable to sodium ions. The membrane is impermeable to the large organic ions. Thus, while sodium ions cannot easily leak in, potassium ions can certainly leak out of the axon. However, as the potassium ions leak out of the axon, they leave behind the large negative ions, which cannot follow them through the membrane. As a result, a negative potential is produced inside the axon with respect to the outside. This negative potential, which has been measured to be about 70 mV, holds back the outflow of potassium so that in equilibrium, the concentration of ions is as we have stated. Some sodium ions

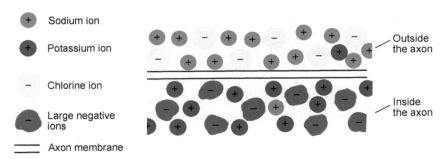

FIGURE 13.3    **The axon membrane and surroundings.**

do in fact leak into the axon, but they are continuously removed by a metabolic mechanism called the *sodium pump*. This pumping process, which is not yet fully understood, transports sodium ions out of the cell and brings in an equal number of potassium ions.

### 13.1.3 Action potential

The description of the axon that we have so far given applies to other types of cells as well. Most cells contain an excess concentration of potassium ions and are at a negative potential with respect to their surroundings. The special property of the neuron is its ability to conduct electrical impulses.

Physiologists have studied the properties of nerve impulses by inserting a probe into the axon and measuring the changes in the axon voltage with respect to the surrounding fluid. The nerve impulse is elicited by some stimulus on the neuron or the axon itself. The stimulus may be an injected chemical, mechanical pressure, or an applied voltage. In most experiments the stimulus is an externally applied voltage, as shown in Fig. 13.4.

A nerve impulse is produced only if the stimulus exceeds a certain threshold value. When this value is exceeded, an impulse is generated at the point of stimulation and propagates down the axon. Such a propagating impulse is called an *action potential*. An action potential as a function of time at one point on the axon is shown in Fig. 13.5. The scales of time and voltage are typical of most neurons. The arrival of the impulse is marked by a sudden rise of the potential inside the axon from its negative resting value to about + 30 mV. The potential then rapidly decreases to about −90 mV and returns more slowly back to the initial resting state. The whole pulse passes a given point in a few milliseconds. The speed of pulse propagation depends on the type of axon. Fast-acting axons propagate the pulse at speeds up to 100 m/sec. The mechanism for the action potential is discussed in a following section.

Impulses produced by a given neuron are always of the same size and propagate down the axon without attenuation. The nerve impulses are produced at a rate proportional to the intensity of the stimulus. There is, however,

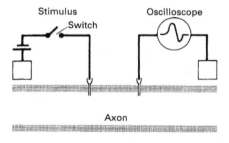

FIGURE 13.4   **Measuring the electrical response of the axon.**

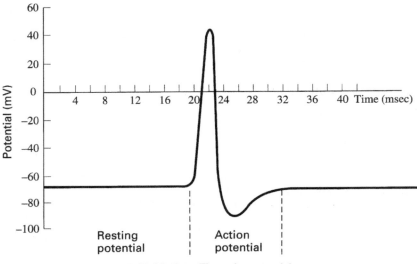

FIGURE 13.5   The action potential.

an upper limit to the frequency of impulses because a new impulse cannot begin before the previous one is completed.

### 13.1.4 Axon as an electric cable

In the analysis of the electrical properties of the axon, we will use some of the techniques of electrical engineering. This treatment is more complex than the methods used in the other sections of the text. The added complexity, however, is necessary for the quantitative understanding of the nervous system.

Although the axon is often compared to an electrical cable, there are profound differences between the two. Still, it is possible to gain some insight into the functioning of the axon by analyzing it as an insulated electric cable submerged in a conducting fluid. In such an analysis, we must take into account the resistance of the fluids both inside and outside the axon and the electrical properties of the axon membrane. Because the membrane is a leaky insulator, it is characterized by both capacitance and resistance. Thus, we need four electrical parameters to specify the cable properties of the axon.

The capacitance and the resistance of the axon are distributed continuously along the length of the cable. It is, therefore, not possible to represent the whole axon (or any other cable) with only four circuit components. We must consider the axon as a series of very small electrical-circuit sections joined together. When a potential difference is set up between the inside and the outside of the axon, four currents can be identified: The current outside the axon, the current inside the axon, the current through the resistive component of the membrane, and the current through the capacitive component of the

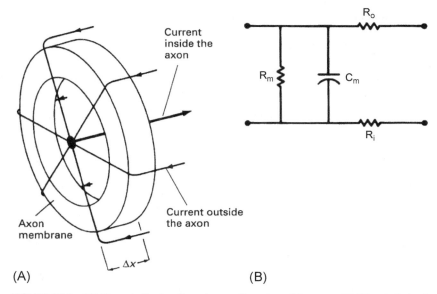

**FIGURE 13.6** (A) Currents flowing through a small section of the axon. (B) Electrical circuit representing a small section of the axon.

membrane (see Fig. 13.6). The electrical circuit representing a small axon section of length $\Delta x$ is shown in Fig. 13.6. In this small section, the resistances of the outside and of the inside fluids are $R_o$ and $R_i$, respectively. The membrane capacitance and resistance are shown as $C_m$ and $R_m$. The whole axon is just a long series of these subunits joined together. This is shown in Fig. 13.7. Sample values of the circuit parameters for both a myelinated and a nonmyelinated axon of radius $5 \times 10^{-6}$ m are listed in Table 13.1. (These values were obtained from Delchar (1997).) Note that the values in Table 13.1 are quoted for a 1-m length of the axon. The unit *mho* for the conductivity of the axon membrane is defined in Appendix B.

An examination of the axon performance shows immediately that the circuit in Fig. 13.7 does not explain some of the most striking features of the axon. An electrical signal along such a circuit propagates at nearly the speed of light $\left(3 \times 10^8 \text{ m/sec}\right)$, whereas a pulse along an axon travels at a speed that is at most about 100 m/sec. Furthermore, as we will show, the circuit in Fig. 13.7 dissipates an electrical signal very quickly, yet we know that action potentials propagate along the axon without any attenuation. Therefore, we must conclude that an electrical signal along the axon does not propagate by a simple passive process.

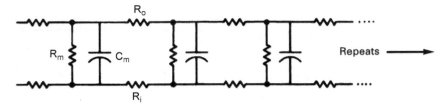

FIGURE 13.7   **The axon represented as an electrical cable.**

**TABLE 13.1** Properties of sample axons.

| Property | Nonmyelinated axon | Myelinated axon |
|---|---|---|
| Axon radius | $5 \times 10^{-6}$ m | $5 \times 10^{-6}$ m |
| Resistance per unit length of fluid both inside and outside axon (r) | $6.37 \times 10^{9}$ $\Omega$/m | $6.37 \times 10^{9}$ $\Omega$/m |
| Conductivity per unit length of axon membrane (gm) | $1.25 \times 10^{-4}$ mho/m | $3 \times 10^{-7}$ mho/m |
| Capacitance per unit length of axon (c) | $3 \times 10^{-7}$ F/m | $8 \times 10^{-10}$ F/m |

## 13.1.5  Propagation of the action potential

After many years of research, the propagation of an impulse along the axon is now reasonably well understood. (See Fig. 13.8.) When the magnitude of the voltage across a portion of the membrane is reduced below a threshold value, the permeability of the axon membrane to sodium ions increases rapidly. As a result, sodium ions rush into the axon, cancel out the local negative charges, and, in fact, drive the potential inside the axon to a positive value. This process produces the initial sharp rise of the action potential pulse. The sharp positive spike in one portion of the axon increases the permeability to the sodium immediately ahead of it, which in turn produces a spike in that region. In this way the disturbance is sequentially propagated down the axon, much as a flame is propagated down a fuse.

The axon, unlike a fuse, renews itself. At the peak of the action potential, the axon membrane closes its gates to sodium and opens them wide to po-tassium ions. The potassium ions now rush out, and, as a result, the axon potential drops to a negative value slightly below the resting potential. After a few milliseconds, the axon potential returns to its resting state and that portion of the axon is ready to receive another pulse.

The number of ions that flow in and out of the axon during the pulse is so small that the ion densities in the axon are not changed appreciably. The

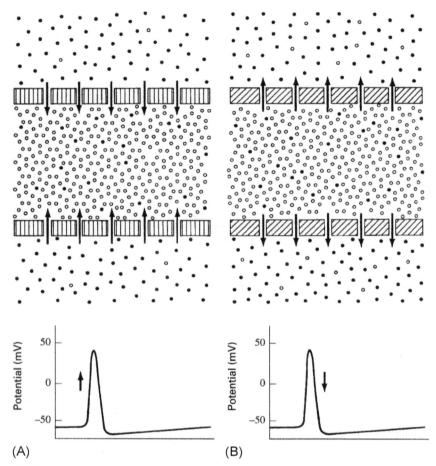

**FIGURE 13.8** **The action potential.** (A) The action potential begins with the axon membrane becoming highly permeable to sodium ions (closed circles) which enter the axon, making it positive. (B) The sodium gates then close and potassium ions (open circles) leave the axon making the interior negative again.

cumulative effect of many pulses is balanced by metabolic pumps that keep the ion concentrations at the appropriate levels. Using Eq. B.5 in Appendix B, we can estimate the number of sodium ions that enter the axon during the rising phase of the action potential. The initial inrush of sodium ions changes the amount of electrical charge inside the axon. We can express this change in charge $\Delta Q$ in terms of the change in the voltage $\Delta V$ across the membrane capacitor C, that is,

$$\Delta Q = C\Delta V \qquad (13.1)$$

In the resting state, the axon voltage is $-70\,mV$. During the pulse, the voltage changes to about $+30\,mV$, resulting in a net voltage change across the membrane of $100\,mV$. Therefore, $\Delta V$ to be used in Eq. 13.1 is $100\,mV$.

The calculations outlined in Exercise 13.1 show that, in the case of the nonmyelinated axon described in Table 13.1, during each pulse, $1.87\times 10^{11}$ sodium ions enter per meter of axon length. The same number of potassium ions leaves during the following part of the action potential. (Measurements show that actually the ion flow is about three times higher than our simple estimate.) The exercise also shows that, in the resting state, the number of sodium ions inside a meter length of the axon is about $7\times 10^{14}$ and the number of potassium ions is $7\times 10^{15}$. Thus, the inflow and the outflow of ions during the action potential is negligibly small compared to the equilibrium density.

Another simple calculation using Eq. B.6 yields an estimate of the minimum energy required to propagate the impulse along the axon. During the propagation of one pulse, the whole axon capacitance is successively discharged and then must be recharged again. The energy required to recharge a meter length of the nonmyelinated axon is

$$E = \frac{1}{2}C(\Delta V)^2 = \frac{1}{2}\times 3\times 10^{-7}\times (0.1)^2 = 1.5\times 10^{-9}\,J/m \qquad (13.2)$$

where C is the capacitance per meter of the axon. Because the duration of each pulse is about $10^{-2}$ seconds, and an axon can propagate at most 100 pulses/sec, even at peak operation, the axon requires only $1.5\times 10^{-7}\,W/m$ to recharge its capacitance.

## 13.1.6 An analysis of the axon circuit

The circuit in Fig. 13.7 does not contain the pulse-conducting mechanism of the axon. It is possible to incorporate this mechanism into the circuit by connecting small signal generators along the circuit. However, the analysis of such a complex circuit is outside the scope of this text. Even the circuit in Fig. 13.7 cannot be fully analyzed without calculus. We will simplify this circuit by neglecting the capacitance of the axon membrane. The circuit is then as shown in Fig. 13.9a. This representation is valid when the capacitors are fully charged so that the capacitive current is zero. With this model, we will be able to calculate the voltage attenuation along the cable when a steady voltage is applied at one end. The simplified model, however, cannot make predictions about the time-dependent behavior of the axon.

The problem then is to calculate the voltage $V(x)$ at point x when a voltage $V_a$ is applied at point $x_0$ (see Fig. 13.9a). The approach is to calculate first the voltage drop across a small incremental cable section of length $\Delta x$ cut by lines

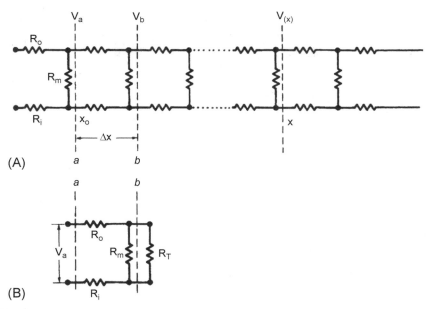

**FIGURE 13.9** (A) Approximation to the circuit in Fig. 13.7 with the capacitances neglected. (B) The resistances to the right of line *b* replaced by the equivalent resistor $R_T$.

*a* and *b* (see Hobbie (1973, 1997)). We assume that the cable is infinite in length and that the total cable resistance to the right of line *b* is $R_T$. Thus, the whole cable to the right of line *b* is replaced by $R_T$, as shown in Fig. 13.9b. Because the cable is infinite, the resistance to the right of any vertical cut equivalent to line *b* is also $R_T$. Specifically, the resistance to the right of line *a* is $R_T$. We can, therefore, calculate $R_T$ by equating the resistance to the right of line *a* in Fig. 13.9b to $R_T$, that is,

$$R_T = R_o + R_i + \frac{R_T R_m}{R_T + R_m} \tag{13.3}$$

Measurements show that the resistivities inside and outside the axon are about the same. Therefore, $R_i = R_o = R$ and Eq. 13.3 simplifies to

$$R_T = 2R + \frac{R_T R_m}{R_T + R_m} \tag{13.4}$$

The solution of Eq. 13.4 (see Exercise 13.2) yields

$$R_T = R + \left[R^2 + 2R\, R_m\right]^{1/2} \tag{13.5}$$

A simple circuit analysis (see Exercise 13.3) shows that

$$V_b = \frac{V_a}{1 + \left[\dfrac{(2R)(R_T + R_m)}{R_T R_m}\right]} = \frac{V_a}{1 + \beta} \tag{13.6}$$

where $\beta$ is the quantity in the square brackets.

We can calculate $\beta$ from the measured parameters shown in Table 13.1. The resistances R and $R_m$ are the values for a small axon section of length $\Delta x$. Therefore,

$$R = r\Delta x \ and \ \frac{1}{R_m} = g_m \Delta x \quad or \quad R_m = \frac{1}{g_m \Delta x}$$

From Eq. 13.5, it can be shown (see Exercise 13.4) that if $\Delta x$ is very small, then

$$R_T = \left(\frac{2r}{g_m}\right)^{1/2} \tag{13.7}$$

and

$$\beta = (2rg_m)^{1/2}\Delta x = \frac{\Delta x}{\lambda} \tag{13.8}$$

where

$$\lambda = \left(\frac{1}{2rg_m}\right)^{1/2} \tag{13.9}$$

Now returning to Eq. 13.6, since $\Delta x$ is vanishingly small, $\beta$ is also very small. Therefore, the term $1/(1+\beta)$ is approximately equal to $1-\beta$ (see Exercise 13.5). Consequently, the voltage $V_b$ at $b$, a distance $\Delta x$ away from $a$, is

$$V_b = V_a\left[1 - \frac{\Delta x}{\lambda}\right] \tag{13.10}$$

To obtain the voltage at a distance x away from line $a$, we divide this distance into increments of $\Delta x$ such that $n\Delta x = x$. We can then apply Eq. 13.10 successively down the cable and obtain the voltage at x (see Exercise 13.6) as

$$V(x) = V_a\left[1 - \frac{\Delta x}{\lambda}\right]^n \tag{13.11}$$

It can be shown that, for small $\Delta x$ and large n, Eq. 13.11 can be written as (see Exercise 13.7)

$$V(x) = V_a e^{-x/\lambda} \tag{13.12}$$

Eq. 13.12 states that if a steady voltage $V_a$ is applied across one point in the axon membrane, the voltage decreases exponentially down the axon. From Table 13.1, for a nonmyelinated axon $\lambda$ is about 0.8 mm. Therefore, at a distance 0.8 mm from the point of application, the voltage decreases to 37% of its value at the point of application.

Myelinated axons, because of their outer sheath, have a much smaller membrane conductance than axons without myelin. As a result, the value of $\lambda$ is larger. Using the values given in Table 13.1, we can show that $\lambda$ is 16 mm for a myelinated axon. This result helps to explain the faster pulse conduction along myelinated axons. As we mentioned earlier, the myelin sheath is in 2-mm-long segments. The action potential is generated only at the nodes between the segments. The pulse propagates through the myelinated segments as a fast conventional electrical signal. Because $\lambda$ is 16 mm, the pulse decreases by only 13% as it traverses one segment, and it is still sufficiently intense to generate an action potential at the next node.

### 13.1.7 Synaptic transmission

So far, we have considered the propagation of an electrical impulse down the axon. Now we shall briefly describe how the pulse is transmitted from the axon to other neurons or muscle cells.

At the far end, the axon branches into nerve endings which extend to the cells that are to be activated. Through these nerve endings, the axon transmits signals, usually to a number of cells. In some cases the action potential is transmitted from the nerve endings to the cells by electrical conduction. In the vertebrate nervous system, however, the signal is usually transmitted by a chemical substance. The nerve endings are actually not in contact with the cells. There is a gap, about a nanometer wide $\left(1 \text{ nm} = 10^{-9} \text{ m} = 10^{-7} \text{ cm}\right)$ between the nerve ending and the cell body. These regions of interaction between the nerve ending and the target cell are called *synapses* (see Fig. 13.10). When the impulse reaches the synapse, a chemical substance is released at the nerve ending, which quickly diffuses across the gap and stimulates the adjacent cell. The chemical is released in bundles of discrete size.

Usually a neuron is in synaptic contact with many sources. Often a number of synapses must be activated simultaneously to start the action potential in the target cell. The action potential produced by a neuron is always of the same magnitude. The neuron operates in an all-or-none mode: It either produces an action potential of the standard size or does not fire at all. In some cases the chemicals released at the synapse do not stimulate the cell but inhibit its response to impulses arriving along a different channel. Presumably, these types of interactions permit decisions to be made on a cellular level. The details of these processes are not yet fully understood.

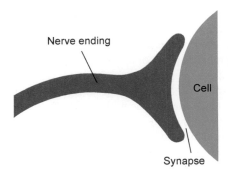

FIGURE 13.10   **Synapse.**

### 13.1.8 Action potentials in muscles

Muscle fibers produce and propagate electrical impulses in much the same way as neurons. The action potential in the muscle fiber is initiated by the impulses arriving from motor neurons. This stimulation causes a reduction of the potential across the fiber membrane which initiates the ionic process involved in the pulse propagation. The shape of the action potential is the same as in the neuron, except that its duration is usually longer. In skeletal muscles, the action potential lasts about 20 milliseconds, whereas in heart muscles it may last a quarter of a second.

After the action potential passes through the muscle fiber, the fiber contracts. In Chapter 5, we briefly discussed some aspects of muscle contraction. The details of this process are not yet fully understood.

Within the skeletal muscle fibers, mechanoreceptor organs called *muscle spindles* continuously transmit information on the state of muscle contraction. This information is relayed via neurons for processing and further action. In this way, the movement of muscles is continuously under control.

It is possible to stimulate muscle fibers by an external application of an electric current. This effect was first observed in 1780 by Luigi Galvani, who noted that a frog's leg twitched when an electric current passed through it. (Galvani's initial interpretation of this effect was wrong.) External muscle stimulation is a useful clinical technique for maintaining muscle tone in cases of temporary muscle paralysis resulting from nerve disorders.

### 13.1.9 Surface potentials

The voltages and currents associated with the electrical activities in neurons, muscle fibers, and other cells extend to regions outside the cells. As an example, consider the propagation of the action potential along the axon. As the voltage at one point on the axon drops suddenly, a voltage difference is produced between this point and the adjacent regions. Consequently, current

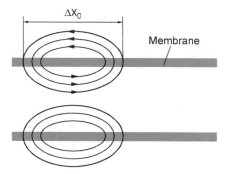

**FIGURE 13.11**   The action potential produces currents both inside and outside the axon.

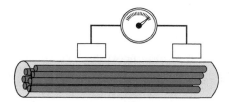

**FIGURE 13.12   Surface potential along a nerve bundle.**

flows both inside and outside the axon (see Fig. 13.11). As a result, a voltage drop develops along the outer surface length of the axon.

Experiments are sometimes performed on a whole nerve consisting of many axons. As shown in Fig. 13.12, two electrodes are placed along the nerve bundle, and the voltage between them is recorded. This measured voltage is the sum of the surface potentials produced by the individual axons and yields some information about the collective behavior of the axons.

Electric fields associated with the activities in cells extend all the way to the surface of the animal body. Thus, along the surface of the skin, we can measure electric potentials representing the collective cell activities associated with certain processes in the body. Based on this effect, clinical techniques have been developed to obtain, from the skin surface, information about the activities of the heart (*electrocardiography*) and the brain (*electroencepha-lography*). The measurement of these surface signals will be discussed in Chapter 14.

Surface signals are associated with many other activities, such as movement of the eye, contractions of the gastrointestinal tract, and movement of muscles. Using a technique called *electromyography* (EMG), measurement of action potentials and their speed of propagation along muscles can provide information about muscle and nerve disorders. (see Hobbie (1973)). Surface potentials associated with metabolic activities have also been observed in plants and bones, as discussed in the following sections.

### 13.1.10 Role of myelin in learning

As we mentioned in Section 13.1.1, some axons are covered by a *myelin sheath* that significantly speeds up the propagation and reduces attenuation of the electrical signal (i.e., the action potential pulses) along the axon that connects the neurons. The importance of the myelin sheath is clear for signal propagation through long axons such as those connecting neurons at the base of the spinal cord to the toes (a length of about 1 meter). Recent experiments indicate that this attribute of myelin, namely that of greatly improving electrical conduction of the action potential, is also key to learning physical skills.

In the late 1940s, a Canadian neurologist, *Donald Hebb*, formulated a hypothesis of learning that was subsequently encapsulated in a simple, catchy phrase: "Neurons that fire together, wire together." A simplified version of this learning model can be described as follows: In the brain, many thousands of neurons are interconnected to form networks. The connections between neurons are provided by the axons that emanate from each neuron. A specific network of neurons is associated with a specific action, such as throwing a ball, for example, or responding to a specific stimulus. Learning a motor skill requires repetition of the action. According to Hebb's hypotheses, with repetition of the action, the connections between the neurons in the network speed up and strengthen, resulting in improvement of the specific skill. This strengthening of the neuronal connections is by analogy to electrical systems referred to as being wired together as a result of repeated collective neuron network stimulation (i.e., being fired together).

Initially, it was not known how this is accomplished. Gradually, it emerged that myelin is the key. (Myelin is the white matter in the brain.) With repeated usage, the axons that interconnect the neurons in the motor skill–related network are wrapped in myelin that speeds up specific neuronal connections while leaving axons unrelated to the specific network unaltered.

This process was clearly confirmed experimentally in 2014 by a UCLA research group. They demonstrated that mice with ability to manufacture myelin easily learned a skill such as running fast on a complex wheel, while mice with blocked ability to make myelin were unable to learn the task. (For details, see Exercise 13.9.) This experiment relates specifically to motor skills. However, it is likely that a similar process governs all learning.

## 13.2 Electricity in plants

The type of propagating electrical impulses we have discussed in connection with neurons and muscle fibers have also been found in certain plant cells. The shape of the action potential is the same in both cases, but the duration of the action potential in plant cells is a thousand times longer, lasting about 10 seconds. The speed of propagation of these plant action potentials is also rather slow, only a few centimeters per second. In plant cells, as in neurons, the action potential is elicited by various types of electrical, chemical, or mechanical stimulation. However, the initial rise in the plant cell potential is produced by an inflow of calcium ions rather than sodium ions.

The role of action potentials in plants is not yet known. It is possible that they coordinate the growth and the metabolic processes of the plant and perhaps control the long-term movements exhibited by some plants.

## 13.3 Electricity in the bone

When certain types of crystals are mechanically deformed, the charges in them are displaced; as a result, they develop voltages along the surface. This phenomenon is called the *piezoelectric effect* (Fig. 13.13). Bone is composed of a crystalline material that exhibits the piezoelectric effect. It has been suggested that these piezoelectric voltages play a role in the formation and nourishment of the bone.

The body has mechanisms for both building and destroying bone. New bone is formed by cells called *osteoblasts* and is dissolved by cells called *osteoclasts*. It has been known for some time that a living bone will adapt its structure to a long-term mechanical load. For example, if a compressive force bends a bone, after a time, the bone will assume a new shape. Some portions of the bone gain substance and others lose it in such a way as to strengthen the bone in its new position. It has been suggested that the appropriate addition and removal of bone tissue is guided by the piezoelectric potentials produced by the deformation (see Fig. 13.14).

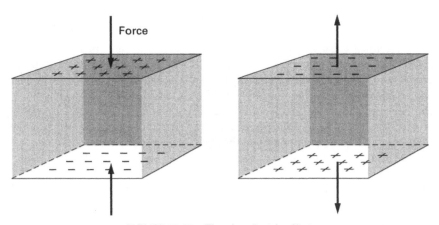

FIGURE 13.13   **The piezoelectric effect.**

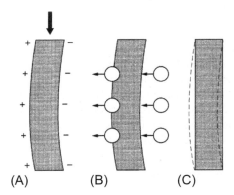

**FIGURE 13.14** (A) Charges are produced along the bone as a result of deformation. (B) Guided by the charge distribution, substance is added and removed from the bone. (C) Reconstructed bone.

Piezoelectricity in the bone may have yet another function. All tissue including bone has to be nourished by fluids. The nourishing fluids move into the bone through very narrow canals. Without a pumping mechanism, the flow of this fluid would be too slow to provide the bone with the necessary nutrients. It has been suggested that the piezoelectric voltages produced by forces due to normal body movement act on the ions in the nutrient fluid and pump it in and out of the bone.

## 13.4 Electric fish

Most animals do not possess sensory organs that are specifically designed to detect external electric fields, but sharks and rays are exceptions. These fish have small organs along their skin which are remarkably sensitive to electric fields in water. A shark responds to an electric field as small as 1 $\mu V/m$, which is in the range of fields found along the skin of animals. (A flashlight battery with terminals separated by 1500 km produces a field of this strength.) The shark uses these electrical organs to locate animals buried in sand and perhaps even to communicate with other sharks. Sharks are also known to bite boat propellers, probably in response to the electric field generated in the proximity of the metal.

An equally remarkable use of electricity is found in the electric eel, which can generate along its skin electric pulses up to 500 V with currents reaching 80 mA. The eel uses this ability as a weapon. When it comes in contact with its prey, the high-voltage pulse passes through the victim and stuns it.

The electric organ of the eel consists of specialized muscle fibers that are connected together electrically. The high voltage is produced by a series interconnection of many cells, and the large current is obtained by connecting the series chains in parallel (see Exercise 13.8). A number of other fish possess similar electric organs.

## Exercises

**13.1.** (a) Using Eq. 13.1 and the data in Table 13.1, calculate the number of ions entering the axon during the action potential per meter of nonmyelinated axon length. (The charge on the ion is $1.6 \times 10^{-19}$ coulomb.) (b) During the resting state of the axon, typical concentrations of sodium and potassium ions inside the axon are 15 and 150 mmol/L, respectively. From the data in Table 13.1, calculate the number of ions per meter length of the axon.

$$1 \text{ mol/L} = 6.02 \times 10^{20} \, \frac{\text{particles (ions, atoms, etc.)}}{\text{cm}^3}$$

**13.2.** From Eq. 13.4, obtain a solution for $R_T$. (Remember that $R_T$ must be positive.)

**13.3.** Verify Eq. 13.6.

**13.4.** Show that when $\Delta x$ is very small, $R_T$ is given by Eq. 13.7.

**13.5.** Show that if $\beta$ is small, $1/(1 + \beta) \approx 1 - \beta$. (Refer to tables of series expansion.)

**13.6.** Verify Eq. 13.11.

**13.7.** Using the binomial theorem, show that Eq. 13.11 can be written as

$$V(x) = V_a\left[1 - \frac{n\Delta x}{\lambda} + \frac{n(n-1)}{2!}\left(\frac{\Delta x}{\lambda}\right)^2\right.$$
$$\left. - \frac{n(n-1)(n-2)}{3!}\left(\frac{\Delta x}{\lambda}\right)^3 + \cdots\right]$$

Since $\Delta x$ is vanishingly small, n must be very large. Show that the above equation approaches the expansion for an exponential function. (Refer to tables of series expansion.)

**13.8.** (a) From the data provided in the text, estimate the number of cells that must be connected in series to provide the 500 V observed at the skin of the electric eel. (b) Estimate the number of chains that must be connected in parallel to provide the observed currents. Assume that the size of the cell is $10^{-5}$ m, the pulse produced by a single cell is 0.1 V, and the duration of pulse is $10^{-2}$ seconds. Use the data in the text and in Exercise 13.1 to estimate the current flowing into a single cell during the action potential.

**13.9.** Describe the 2014 UCLA experiments that provided a clear demonstration that myelin coating of the interconnecting axons is required for learning physical skills. (Use the *Science* article by McKensey et al. (2014))

# Chapter 14

# Electrical technology

Electrical technology was developed by applying some of the basic principles of physics to problems in communications and industry. Although this technology was directed primarily toward industrial and military applications, it has made great contributions to the life sciences. Electrical technology has provided tools for the observation of biological phenomena that would have been otherwise inaccessible. It yielded most of the modern clinical and diagnostic equipment used in medicine. Even the techniques developed for the analysis of electrical devices have been useful in the study of living systems. In this chapter, we will describe some of the applications of electrical technology in these areas.

## 14.1 Electrical technology in biological research

Our understanding of the world would be greatly limited if it were based only on observations made by our unaided senses. As well developed as our senses are, their responses are limited. We cannot hear sound at frequencies above 20,000 Hz. We cannot see electromagnetic radiation outside the limited wavelength region between about 400 nm and 700 nm $(1 \text{ nm} = 10^{-9} \text{ m})$. Even in this visible range, we cannot detect variations in light intensity that occur at a rate faster than about 20 Hz. Although many of the vital processes within our bodies are electrical, our senses cannot detect small electric fields directly. Electrical technology has provided the means for translating information from many areas into the domain of our senses.

Electrical technology is a vast subject that we cannot possibly cover in this short chapter. Here we will simply outline the general techniques used in observing life processes. A description of the various electrical components can be found in for example, Cromwell et al. (1973), and Davidovits (1972).

A diagram of a typical experimental setup in biology is shown in Fig. 14.1. The various subunits of the experiment are shown as blocks of specialized functions. We start with the phenomenon we want to observe but which we cannot detect with our senses. This may be, for example, a high-frequency sound emitted by a bat, the electrochemical activity of a cell, the subtle movement of a muscle, or the light emitted by a fluorescent dye. These phenomena are first translated into electrical signals, which then carry information about the intensity and time variations of the original event. Specialized

Physics in Biology and Medicine. https://doi.org/10.1016/B978-0-443-21558-2.00014-6

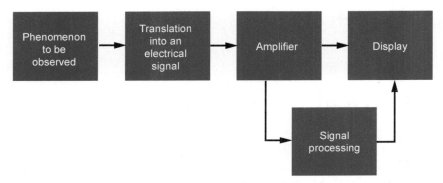

FIGURE 14.1 An experiment in biology.

devices are required to perform this task. Some of these devices are commonly found in our everyday technology; others are rather esoteric. Sound, for example, is translated into electrical signals by microphones. Light can be translated into an electrical current by photomultipliers.

The electrical signals generated in this way are usually too weak to drive the final instrument that displays the signals for our observation, so the power and amplitude of the signal are increased by a device called an *amplifier*. The amplified signal then drives the display unit.

The display unit must be matched to the type of signal that is being observed. A slowly varying signal can be displayed on a voltmeter, which has a pointer that moves in accord with the current. Somewhat faster signals are often displayed on a pen recorder, which draws the shape of the signal on a chart. Very fast signals are recorded by a device called an *oscilloscope*. This instrument is similar to a television set. A beam of electrons generated inside the device hits a fluorescent screen that emits lights at the point of impact. The motion and intensity of the beam are controlled by electrical signals applied to the oscilloscope. The resulting picture on the screen displays the information content of the signal. More modern oscilloscopes (as well as television sets) use light-emitting diodes to generate the picture. Speakers can also translate into sound electrical signals that have a time variation in the audible range of frequencies.

Often the experimental signal is very noisy. In addition to the desired information, it contains spurious signals due to various sources extraneous to the main phenomenon. Techniques have been developed for analyzing such signals and extracting the relevant information from the noise. In modern experiments, this processing is often performed by a computer.

## 14.2 Diagnostic equipment

Most of the diagnostic equipment in medicine utilizes electrical technology in one form or another. Even the traditional stethoscope is now available with electronic modifications that increase its sensitivity. We will describe here only two of the many diagnostic instruments found in a modern clinical facility: The *electrocardiograph (ECG)* and the *electroencephalograph (EEG).*

As a result of the ionic currents associated with electrical activities in the cells, potential differences are produced along the surface of the body. (See Chapter 13.) By measuring these potential differences between appropriate points on the surface of the body, it is possible to obtain information about the functioning of specific organs. The surface potentials are usually very small and, therefore, must be amplified before they can be displayed for examination.

## 14.2.1 The electrocardiograph

The ECG is an instrument that records surface potentials associated with the electrical activity of the heart. The surface potentials are conducted to the instrument by metal contacts called *electrodes* which are fixed to various parts of the body. Usually the electrodes are attached to the four limbs and over the heart. Voltages are measured between two electrodes at a time. (See Fig. 14.2.)

A typical normal signal recorded between two electrodes is shown in Fig. 14.3. The main features of this waveform are identified by the letters *P, Q, R, S,* and *T.* The shape of these features varies with the location of the electrodes. A trained observer can diagnose abnormalities by recognizing deviations from normal patterns.

The wave shape in Fig. 14.3 is explained in terms of the pumping action of the heart described in Chapter 8. The rhythmic contraction of the heart is initiated by the *pacemaker,* which is a specialized group of muscle cells located near the top of the right atrium. Immediately after the pacemaker fires,

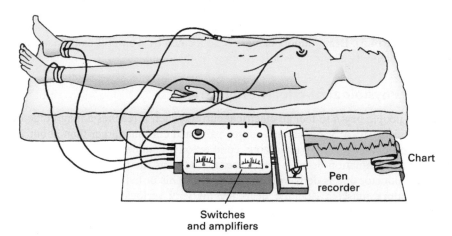

FIGURE 14.2 **Electrocardiography.**

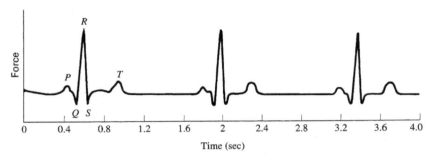

FIGURE 14.3  **An electrocardiogram.**

the action potential propagates through the two atria. The *P* wave is associated with the electrical activity that results in the contraction of the atria. The *QRS* wave is produced by the action potential associated with the contraction of the ventricles. The *T* wave is caused by currents that bring about the recovery of the ventricle for the next cycle.

### 14.2.2 The electroencephalograph

The EEG measures potentials along the surface of the scalp. Here again electrodes are attached to the skin at various positions along the scalp. The instrument records the voltages between pairs of electrodes. The EEG signals are much more complex and difficult to interpret than those produced by the ECG. The EEG signals are certainly the result of collective neural activity in the brain. However, so far, it has not been possible to relate unambiguously the EEG potentials to specific brain functions. Nevertheless, certain types of patterns are known to be related to specific activities, as illustrated in Fig. 14.4.

EEGs have been useful in diagnosing various brain disorders. Epileptic seizures, for example, are characterized by pronounced EEG abnormalities (see Fig. 14.5). Brain tumors can often be located by a careful examination of EEG potentials along the whole contour of the scalp.

## 14.3 Physiological effects of electricity

Most people begin to feel an electrical current when it reaches a magnitude of about 500 µA. A 5 mA current causes pain, and currents larger than about 10 mA produce sustained tetanizing contraction of some muscles. This is a dangerous situation because under these conditions the person cannot release the conductor that is delivering the current into his or her body. The brain, the respiratory muscles, and the heart are all very seriously affected by large electric currents. Currents in the range of a few hundred milliamperes flowing across the head produce convulsions resembling epilepsy. Currents in this range are used in electric shock therapy to treat certain mental disorders.

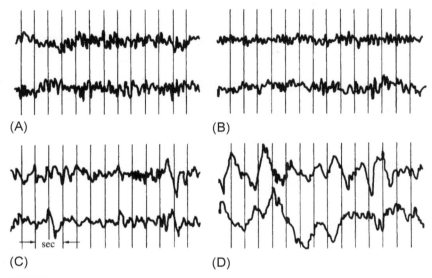

(A)  (B)  (C)  (D)

**FIGURE 14.4    EEG potentials between two pairs of electrodes.** (A) Subject alert, (B) subject drowsy, (C) light sleep, (D) deep sleep.

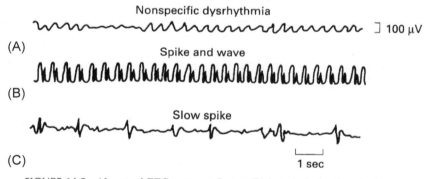

Nonspecific dysrhythmia

(A)

] 100 µV

Spike and wave

(B)

Slow spike

(C)

1 sec

**FIGURE 14.5    Abnormal EEG patterns.** Pattern (B) is typical of petit mal seizures.

Currents in the range of a few amperes flowing in the region of the heart can cause death within a few minutes. In this connection, a large current of about 10 A is often less dangerous than a 1 A current. When the smaller current passes through the heart, it may tetanize only part of the heart, thereby causing a desynchronization of the heart action; this condition is called fibrillation. The movements of the heart become erratic and ineffective in pumping blood. Usually fibrillation does not stop when the current source is removed. A large current tetanizes the whole heart, and when the current is discontinued, the heart may resume its normal rhythmic activity.

## 14.4 Bioelectronic medicine

As was pointed out in Section 14.3, electrical currents affect organs of the body in a variety of ways. The effect of large currents can be deadly, but depending on the organ, smaller currents may have therapeutic benefits. Several devices utilizing the biological effects of electricity have been used in medical practice for many decades. We will begin this section by describing briefly four such more commonly used applications of electrical techniques. We will then discuss more recently developed techniques for using electrical phenomena in medicine. The hope of researchers developing these new, more advanced electrical techniques is to replace drug-based medications with much less intrusive and more effective electrical methods. This new field has been named bioelectronic medicine.

### 14.4.1 Defibrillators

Heart fibrillations, which are desynchronized heart contractions, often occur during a heart attack and during cardiac surgery. The tetanizing effect of large currents can be used to synchronize the heart. A clinical device designed for this purpose is called a defibrillator. A capacitor in this device is charged to about 6000 V and stores about 200 J of energy. Two electrodes connected to the capacitor through a switch are placed on the chest. When the switch is closed, the capacitor rapidly discharges through the heart. The current pulse lasts about 5 milliseconds. During this period, the heart is tetanized, that is, stopped (see Exercise 14.1). The hope is that after the pulse, the heart will resume its normal beat. Often, the heart must be shocked a few times before it resynchronizes.

### 14.4.2 Heart pacemakers and implantable cardioverter-defibrillators

In some heart diseases, the pacemaker cells in the heart that control the timing of the heartbeat cease to function properly. In such cases, electronic pacemakers have been very useful. A pacemaker applies controlled, properly timed electrical pulses to the heart, overriding erratic signals that may be naturally generated. The device is in most cases implanted in the region of the heart below the skin. Pacemakers have been unambiguously successful in regulating erratic heartbeat and allowing people to resume normal activities. In the United States about 200,000 pacemakers are installed annually. About 1 million are installed worldwide.

For high-risk patients that suffer both from heart rhythm irregularities and periodic fibrillations, a newer device called an implantable cardioverter-defibrillator (ICD) is often implanted into the chest. This device is essentially a combination of a pacemaker and a defibrillator. Such a device should certainly be expected to reduce sudden cardiac death. However, some studies suggest that the device does not significantly decrease the death rate of patients (see Exercise 14.5).

### 14.4.3 Electrical muscle stimulation

An electrical pulse applied to a muscle via electrodes attached externally to the skin causes the muscle to contract much the same way as a neurally generated action potential does. As was discussed in Section 13.1.8, such electric stimulation of paralyzed skeletal muscles is useful for maintaining their tone and retarding muscle atrophy while the nerves regrow. However, electric muscle stimulation is effective in retarding atrophy only for a limited time.

Muscle stimulators are available commercially, with manufacturers claiming that electrical stimulation of muscles will build strength without exercise. Evidence for these claims is not convincing.

### 14.4.4 Electroconvulsive therapy

In electroconvulsive therapy (briefly mentioned in Section 14.3), two electrodes are in electrical contact with the outer skin of the skull and a pulse of electric current in the range of several hundred milliamperes is passed through the brain. The current triggers a limited-duration seizure. The procedure is used to treat severe depression and catatonia, where other treatments have failed or cannot be used. The biochemistry or neurological effects of the treatment are not known and its efficacy has not been clearly established. Some claim that the procedure is effective in 80% of the cases. Others point out that the treatment has beneficial effects only in about 50% of cases and seldom lasts longer than a few months (see Exercise 14.6).

### 14.4.5 Vagus nerve stimulation

The vagus nerve is a long nerve that emanates from the brain and, at the neck, splits into two branches that travel down each side of the body connecting the brain to several organs. The vagus nerve contains on the order 100,000 fibers that send as well as receives signals to and from the throat, heart, lungs, and abdomen, including indirectly the liver and the pancreas, and reaches all the way down to the colon. The vagus nerve affects several physiological functions, among them blood flow to various organs, blood pressure, digestion, and breathing.

Because the vagus nerve interacts with and controls many organs with a wide range of physiological functions, it has been conjectured that electrical stimulation of this nerve is likely to have therapeutic effects. In 1997, based on a set of studies, the US Food and Drug Administration (FDA) approved the use of vagus nerve stimulation for people with severe ill-controlled epileptic seizures. In 2005, the procedure was approved for debilitating drug-resistant depressive disorder. The success of vagus nerve stimulation is limited but significant. In 30% of the cases, electrical stimulation of the vagus nerve decreased the number of seizures by half or more. However, for about 50% of the patients, the treatment had no effect at all. In a few cases, vagus nerve stimulation made the seizure disorder worse.

For drug-resistant depressive disorder, 27% of people receiving vagus nerve stimulation reported significant improvement in their condition. By comparison, only 9% in the control group reported an improvement.

So far, four conditions have received FDA approval for treatment by vagus nerve stimulation: Seizures, depression, and most recently, obesity and migraine headaches. However, experiments exploring the possible benefits of vagus nerve stimulation are being explored for other conditions, among them metabolic syndrome, heart disease, tinnitus, and several autoimmune diseases.

In vagus nerve stimulation, a small-pulse generator a few centimeters in diameter with associated batteries is implanted through an incision usually in the left side of the chest under the clavicle. The vagus nerve is accessed through a second incision in the neck and the electrical leads are wrapped around the nerve. A few weeks after surgery, and over several visits, the duration, frequency, and strength of the electrical pulses delivered to the vagus nerve are adjusted and programmed for optimum results. Migraine headaches are being treated with external electrical stimulation rather than with implanted devices.

### 14.4.6 Vagus nerve and the immune system

In 1998, Kevin Tracey and his coworkers discovered a connection between the vagus nerve and the immune system that may become an important component of bioelectronic medicine.

The immune system protects the body from invading pathogens such as bacteria, viruses, and other foreign organisms. Such organisms trigger the immune response that, depending on the nature of the invading organism, involves several organs, among them bone marrow, lymphatic system, thymus gland, and spleen. The immune system produces molecules that destroy or inactivate harmful foreign pathogens. Inflammation is part of the defense mechanism that destroys the pathogens. A protein called tumor necrosis factor (TNF) has a key role in triggering the inflammation response by mobilizing the various components of the immune system to respond appropriately to the harmful intrusion. However, in some cases, the immune system overreacts or malfunctions producing an excess of TNF that triggers the destruction of the body's own tissue. Such self-destruction by the immune system results in several so-called autoimmune diseases, among them type 1 diabetes, rheumatoid arthritis, multiple sclerosis, and Crohn's disease.

Through a series of animal studies, Tracey and his colleagues showed that electrical stimulation of the vagus nerve reduces the production of the inflammation-inducing TNF protein. Further studies showed that only a subset of about 10,000 of the 100,000 or so fibers carried by the vagus nerve affects the production of the TNF protein, and this subset of fibers is activated by a significantly lower electrical current than the other vagus nerve fibers. This finding was highly significant, indicating that electrical stimulation could selectively turn off the excess production of TNF without affecting other functions of the vagus nerve.

Animal experiments, mostly with rats, have demonstrated successful blocking of TNF and a reduction of excessive inflammatory response. Small-scale human experiments are in progress, studying the effect of vagus nerve stimulation on rheumatoid arthritis and Crohn's disease. So far, the results are promising but not as fully successful as was hoped. It seems that some people do not respond to vagus nerve stimulation. Only about 50% of the patients in the experimental studies have experienced significant improvement in their condition.

At this time (2023), the mechanism for the therapeutic effect of electrical stimulation of the vagus nerve is not known. Nor is it known why some patients benefit from the procedure while others are unaffected. Most experts in this field agree that the full potential of the technique will not be realized until the relevant parts of the nervous system are adequately mapped and the processes involved in vagus nerve stimulation are explained.

### 14.4.7 Nanomedicine

Miniaturizations and increased functionality of electronic devices have been so rapid that it is now possible to envision realistic devices on the scale of nanometers that could be implanted in the body to stimulate individual cells or nerve fibers and to monitor, as well as alter, if required, a variety of physical functions.

The hope and expectations for the use of nanoelectronics in medicine far exceed the capabilities of currently used bioelectronics devices. One can envision a variety of health conditions that implanted nanoelectronic devices could alleviate. In a diabetic person, the device could monitor blood sugar and then trigger insulin production if needed. A nanodevice could monitor blood pressure and then electrically stimulate organs that would reduce blood pressure as needed. An implanted nanocontrol device could regulate breathing in an asthma attack.

One nanosized bioelectronic device is already in development. Promising new techniques are being explored to restore vision. It is estimated that 250 million people worldwide, 15 million in the United States, are severely impaired visually due to macular degeneration and genetic disorders such as retinitis pigmentosa. In these conditions, the optic nerve that sends signals to the brain is usually intact. The problem resides with the damaged light receptors (rods and cones) in the retina. Several research groups are working on the design and construction of artificial retinas. Typically, these are sheets composed of nanosized components that generate an electrical pulse in response to impinging photons, much the way in which a biological retina does. In one design, the electrical pulse is conducted to the neurons of the optic nerve via neuron-adhering carbon nanotubes. The ultimate aim is to construct a retina-like structure that could be implanted in the eye and restore vision.

In the development of the artificial retina, the neural connections to the optic nerve are relatively easy to trace. This is not the case for most planned nanobioelectronic applications. Here success will depend on detailed mapping of the neural system because the nanoelectronic device will have to be connected to specific neuron or neural fiber, and the connection will have to be more accurate than in the larger-scale application such as vagus nerve stimulation.

### 14.4.8 Concerns

Most applications we discussed involve implantation of a device into the body. Safety concerns therefore mandate that the devices be biocompatible as well as electrically and mechanically safe. The safety of implantable devices is continually improving, but at this time (2023), several safety issues remain unresolved. An article dated January 14, 2018, issue of *The New York Times* points out that currently the medical device industry is inadequately controlled. Many implantable devices presently in use have not been thoroughly clinically tested, and the oversight of their applications is inadequate.

These implanted medical devices have improved and saved lives of many people, but they have also caused much unnecessary pain and damage. For example, as pointed out in this and other articles, thousands of metal hip replacements have failed causing much harm to patients. Likewise, thousands of heart defibrillators had to be removed because of faulty wiring. Such removal causes serious complications in 15% of patients. In the case of another device, the implanted vagus nerve stimulators, it was noted that the death rate of patients with this implant was higher than expected. Yet the device was granted conditional approval. Clearly, the testing and oversight of implantable devices has to be improved. Till then, the decisions about the safety and benefits of these devices must be carefully evaluated by the patients and their physicians.

### 14.5 Control systems

Many of the processes in living systems must be controlled to meet the requirements of the organism. We have already encountered a few examples of controlled processes in our earlier discussions. Temperature control in the body and the growth of bones were two cases where various processes had to be regulated in order to achieve the desired condition. In this section, we will describe briefly a useful general method of analyzing such control systems.

Features common to all control systems are shown in Fig. 14.6. Each block represents an identifiable function within the control system. The control process consists of the following:

1. The parameter to be controlled. This may be the temperature of the skin, the movement of muscles, the rate of heartbeat, the size of the bone, and so on.
2. A means of monitoring the parameter and transmitting information about its state to some decision-making center. This task is usually performed by the sensory neurons.
3. Some reference value to which the controlled parameters are required to comply. The reference value may be in the central nervous system in the form of a decision, for example, about the position of the hand. In this case, the reference value is changeable and is set by the central nervous system. Many references for body functions are autonomous, however, not under the cognitive control of the brain.
4. A method for comparing the state of the parameter with the reference value and for transmitting instructions to bring the two into accord. The instructions may be transmitted by nerve impulses or in some cases by

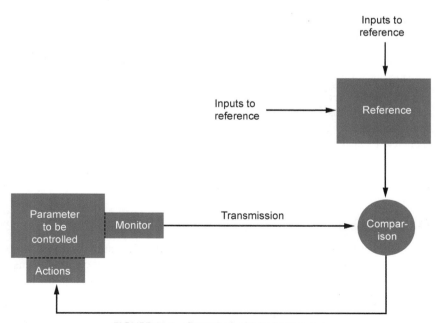

**FIGURE 14.6  Control of a biological process.**

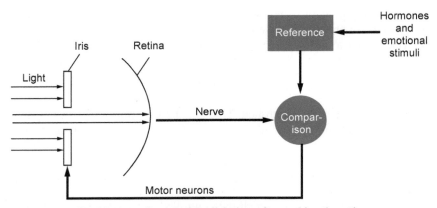

FIGURE 14.7   **Control of the light intensity reaching the retina.**

chemical messengers called *hormones*, which diffuse through the body and control various metabolic functions.

5. A mechanism for translating the messages into actions that alter the state of the controlled parameter. In the case of the hand position, for example, this is the contraction of a set of muscle fibers.

We will now illustrate these concepts with a concrete example of the control of the light intensity reaching the retina of the eye (see Fig. 14.7). Light enters the eye through the *pupil*, which is the dark opening in the center of the *iris*. (The iris is the colored disk in the eyeball.) The size of the opening decreases as the light intensity increases. Thus, the iris acts somewhat like the automatic diaphragm in a camera. Clearly this action must be governed by a control system.

Light reaching the retina is converted to neural impulses, which are generated at a frequency proportional to the light intensity. At some place along the nervous system of vision, this information is interpreted and compared to a preset reference value stored probably in the brain. The reference itself can be altered by hormones and various emotional stimuli. The result of this comparison is transmitted by means of nerve impulses to the muscles of the iris, which then adjust the size of the opening in response to this signal.

## 14.6 Feedback

For many years engineers have studied mechanical and electrical systems that have the general characteristics of control systems in biological organisms. Voltage regulators, speed controls, and thermostatic heat regulators all have features in common with biological control systems. Engineers have developed techniques for analyzing and predicting the behavior of control systems. These techniques have also been useful in the study of biological systems.

An engineering analysis of such systems is usually done in terms of input and output. In the light-intensity control example, the input is the light reaching the retina, and the output is the response of the retina to light. The system itself is that which yields an output in response to the input. In our case, this is the retina and the associated nerve circuits. The aim of the iris control system is to maintain the output as constant as possible.

The most significant point to note about control systems such as the one shown in Fig. 14.7 is that the output affects the input itself. Such systems are called *feedback systems* (because information about the output is fed back to the input). The system is said to have *negative feedback* if it opposes a change in the input and *positive feedback* if it augments a change in the input. The light control in Fig. 14.7 is a negative feedback because an increase in the light intensity causes a decrease in the iris opening and a corresponding reduction of the light reaching the retina. Regulation of body temperature by sweating or shivering is another example of negative feedback, whereas sexual arousal is an example of positive feedback. In general, negative feedback keeps the system response at a relatively constant level. Therefore, most biological feedback systems are in fact negative.

We will illustrate the method of system analysis with an example from electrical engineering. We will analyze in these terms a voltage amplifier that has part of its output fed back to the input. Let us first consider a simple amplifier without feedback (see Fig. 14.8). The amplifier is an electric device that increases the input voltage $(V_{in})$ by a factor A; that is, the output voltage $V_{out}$ is

$$V_{out} = AV_{in} \tag{14.1}$$

It is evident from this equation that the amplification A is simply determined by the ratio of the output and input voltages; that is,

$$A = \frac{V_{out}}{V_{in}} \tag{14.2}$$

Now let us introduce feedback (Fig. 14.9). Part of the output $(\beta \times V_{out})$ is added back to the amplifier input so that the voltage at the input terminal of the amplifier $(V'_{in})$ is

$$V'_{in} = V_{in} + \beta \times V_{out} \tag{14.3}$$

Here $V_{in}$ is the externally applied voltage. The amplification of the total feedback system is

FIGURE 14.8 **An amplifier without feedback.**

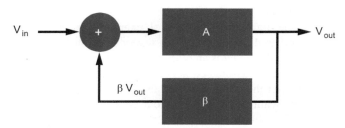

FIGURE 14.9  An amplifier with feedback.

$$A_{feedback} = \frac{V_{out}}{V_{in}} \tag{14.4}$$

Using the fact that $V_{out} = AV'_{in}$, we can show that (see Exercise 14.2)

$$A_{feedback} = \frac{A}{1 - A\beta} \tag{14.5}$$

Now if $\beta$ is a negative number, the amplification with feedback is smaller than the amplification without feedback (i.e., $A_{feedback}$ is smaller than A). A negative $\beta$ implies that the voltage is added out of phase with the external input voltage. This is negative feedback. With positive $\beta$, we have positive feedback and increased amplification.

This type of analysis has the advantage that we can learn about the system without a detailed knowledge of the individual system components. We can vary the frequency, magnitude, and duration of the input voltage and measure the corresponding output voltage. From these measurements, we can obtain some information about the amplifier and the feedback component without knowing anything about the transistors, resistors, capacitors, and other components that make up the device. We could, of course, obtain this information and much more through a detailed analysis of the device in terms of its basic components, but this would involve much more work.

In the study of complex biological functions, the systems approach is often very useful because the details of the various component processes are unknown. For example, in the iris control system, very little is known about the processing of the visual signals, the mechanism of comparing these signals to the reference, or the nature of the reference itself. Yet by shining light at various intensities, wavelengths, and durations into the eye and by measuring the corresponding changes in the iris opening, we can obtain significant information about the system as a whole and even about the various subunits. Here the techniques developed by the engineers are useful in analyzing the system (see Exercises 14.3 and 14.4). However, many biological systems are so complicated with many inputs, outputs, and feedbacks that even the simplified systems approach cannot yield a tractable formulation.

## 14.7 Sensory aids

Sight and hearing are the two principal pathways through which our brain receives information about the external world. The two organs, eyes and ears, that transmit light and sound information into the brain are often damaged and their function needs to be supplemented. Eyeglasses came into use in the 1200s. At first, these visual aids provided only simple magnified images of objects. Gradually a sophisticated technology evolved to produce eyeglasses to compensate for a wide range of visual problems (see Chapter 15).

Ear horns, in one form or another, have been used to aid hearing for thousands of years. These devices improve hearing by collecting sound from an area significantly larger than the pinna (see Chapter 12).

Electrical technology has led to the development of devices that greatly enhance hearing and in some cases even restore hearing. The restoration of vision is far more challenging, and while several avenues of research are being pursued, the final goal seems far in the future.

### 14.7.1 Hearing aids

The basic principle of hearing aids is simple. A microphone converts sound to an electrical signal. The electrical signal is amplified and converted back into sound using a speaker-type device. The net result is an amplification of the sound that enters the ear.

The first hearing aids became commercially available in the 1930s. They were relatively large, cumbersome devices using battery-powered vacuum tube amplifiers. The batteries had to be replaced daily.

The much smaller transistor amplifiers that became available in the 1950s made hearing aids truly practicable. Transistorized hearing aids were now small enough to be placed in the ear. The application of digital computer technology to hearing aids was another major improvement that allowed individual tailoring of the device to compensate for the specific hearing deficits of the user. Using various feedback networks, modern hearing aids automatically adjust the volume of the sound so that quiet sounds can be heard and loud sounds are not painfully overwhelming.

### 14.7.2 Cochlear implant

A cochlear implant functions differently from a hearing aid. A hearing aid simply amplifies incoming sound compensating for the diminished functioning of the ear. A cochlear implant converts sound to electrical signals of the type produced by the inner ear in response to sound that enters the ear. The electrical signal is wirelessly transmitted to electrodes surgically implanted in the inner ear. The signals applied to the electrodes stimulate the auditory nerve to produce the sensation of sound. Thus the cochlear implant actually mimics the functions of the ear and can restore partial hearing to the deaf.

A sketch of a cochlear implant system is shown in Fig. 14.10. The external part of the system is small enough to be placed behind the ear. It consists of a microphone, a signal processor, and a transmitter. The internal part consists of the receiver and an array of electrodes implanted and wound through the cochlea.

The microphone converts the sound to an electrical signal. Such electrical signals as are produced by the microphone could themselves stimulate the auditory nerve, but the neural signals produced by such stimulation would not be interpreted by the brain as sound. In the normal ear, the fluid-filled cochlea processes the sound signal according to frequency such that the various frequency components of the incoming sound stimulate nerve endings along

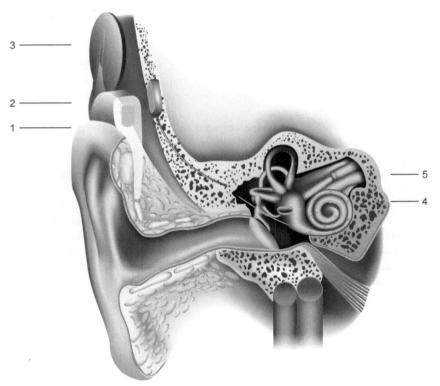

**FIGURE 14.10  Cochlear implant.** (1) Sounds are picked up by the microphone. (2) The signal is then "coded" (turned into a special pattern of electrical pulses). (3) These pulses are sent to the coil and are then transmitted across the skin to the implant. (4) The implant sends a pattern of electrical pulses to the electrodes in the cochlea. (5) The auditory nerve picks up these electrical pulses and sends them to the brain. The brain recognizes these signals as sound.

different parts of the basilar membrane (see Chapter 12). This type of a frequency-selective stimulation of the neural network provided by the cochlea is essential if the signal is to be interpreted by the brain as sound.

One of the main challenges in the design of cochlear implants was the development of signal-processing techniques that duplicated the action of a normal cochlea. Much of the work in this area was done in the 1950s and 1960s. First experiments with human implants began in the mid-1960s and continued through the 1970s. In 1984, FDA approved implantation into adults and, shortly after, into children.

Usually a person receiving an implant is not immediately able to hear sounds properly. A period of training and speech therapy is needed before the full benefits of the device are realized (see Exercise 14.7).

## Exercises

**14.1.** From the data in the text, compute the capacitance of the capacitor in the defibrillator and calculate the magnitude of the average current flowing during the pulse.

**14.2.** Verify Eq. 14.5.

**14.3.** Draw a block diagram for the following control systems. (a) Control of the body temperature in a person. (b) Control of the hand in drawing a line. (c) Control of the reflex action when the hand draws away from a painful stimulus. Include here the type of control that the brain may exercise on this movement. (d) Control of bone growth in response to pressure.

**14.4.** For each of the control systems in Exercise 14.3, discuss how the system could be studied experimentally.

**14.5.** Evaluate the effectiveness of implantable cardioverter-defibrillators (ICDs) in prolonging the lifespan of patients. (Use the internet for appropriate references.)

**14.6.** Evaluate the effectiveness of electroconvulsive therapy. (Use the internet for appropriate references.)

**14.7.** Discuss the controversy surrounding cochlear implants.

Chapter 15

# Optics

Light is the electromagnetic radiation in the wavelength region between about 400 and 700 nm $\left(1 \text{ nm} = 10^{-9} \text{ m}\right)$. Although light is only a tiny part of the electromagnetic spectrum, it has been the subject of much research in both physics and biology. The importance of light is due to its fundamental role in living systems. Most of the electromagnetic radiation from the sun that reaches the Earth's surface is in this region of the spectrum, and life has evolved to utilize it. In photosynthesis, plants use light to convert carbon dioxide and water into organic materials, which are the building blocks of living organisms. Animals have evolved light-sensitive organs, which are their main source of information about their surroundings. Some bacteria and insects can even produce light through chemical reactions.

Optics, which is the study of light, is one of the oldest branches of physics. It includes topics such as microscopes, telescopes, vision, color, pigments, illumination, spectroscopy, and lasers, all of which have applications in the life sciences. In this chapter, we will discuss four of these topics: Vision, telescopes, microscopes, and optical fibers. Background information needed to understand this chapter is reviewed in Appendix C.

## 15.1 Vision

Vision is our most important source of information about the external world. It has been estimated that about 70% of a person's sensory input is obtained through the eye. The three components of vision are the stimulus, which is light; the optical components of the eye, which image the light; and the nervous system, which processes and interprets the visual images.

## 15.2 Nature of light

Experiments performed during the 19th century showed conclusively that light exhibits all the properties of wave motion, which were discussed in Chapter 12. At the beginning of this century, however, it was shown that wave concepts alone do not explain completely the properties of light. In some cases, light and other electromagnetic radiation behave as if composed of small packets (quanta) of energy. These packets of energy are called *photons*.

Physics in Biology and Medicine. https://doi.org/10.1016/B978-0-443-21558-2.00015-8

For a given frequency f of the radiation, each photon has a fixed amount of energy E, which is

$$E = hf \tag{15.1}$$

where h is Planck's constant, equal to $6.63 \times 10^{-27}$ erg sec.

In our discussion of vision, we must be aware of both of these properties of light. The wave properties explain all phenomena associated with the propagation of light through bulk matter, and the quantum nature of light must be invoked to understand the effect of light on the photoreceptors in the retina.

## 15.3 Structure of the eye

A diagram of the human eye is given in Fig. 15.1. The eye is roughly a sphere, approximately 2.4 cm in diameter. All vertebrate eyes are similar in structure but vary in size. Light enters the eye through the cornea, which is a transparent section in the outer cover of the eyeball. The light is focused by the lens system of the eye into an inverted image at the photosensitive retina, which covers the back surface of the eye. Here the light produces nerve impulses that convey information to the brain.

The focusing of the light into an image at the retina is produced by the curved surface of the cornea and by the crystalline lens inside the eye. The focusing power of the cornea is fixed. The focus of the crystalline lens, however, is alterable, allowing the eye to view objects over a wide range of distances.

In front of the lens is the iris, which controls the size of the pupil, or entrance aperture into the eye (see Chapter 14). Depending on the intensity of

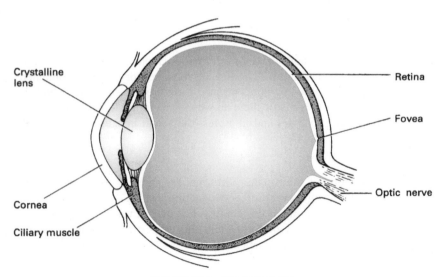

FIGURE 15.1   **The human eye.**

the light, the diameter of the aperture ranges from 2 to 8 mm. The cavity of the eye is filled with two types of fluid, both of which have a refractive index about the same as water. The front of the eye, between the lens and the cornea, is filled with a watery fluid called the *aqueous humor*. The space between the lens and the retina is filled with the gelatinous *vitreous humor*.

## 15.4 Accommodation

The focusing of the eye is controlled by the ciliary muscle, which can change the thickness and curvature of the lens. This process of focusing is called *accommodation*. When the ciliary muscle is relaxed, the crystalline lens is fairly flat, and the focusing power of the eye is at its minimum. Under these conditions, a parallel beam of light is focused at the retina. Because light from distant objects is nearly parallel, the relaxed eye is focused to view distant objects. In this connection, "distant" is about 6 m and beyond (Exercise 15.1).

The viewing of closer objects requires greater focusing power. The light from nearby objects is divergent as it enters the eye; therefore, it must be focused more strongly to form an image at the retina. There is, however, a limit to the focusing power of the crystalline lens. With the maximum contraction of the ciliary muscle, a normal eye of a young adult can focus on objects about 15 cm from the eye. Closer objects appear blurred. The minimum distance of sharp focus is called the *near point of the eye*.

The focusing range of the crystalline lens decreases with age. The near point for a 10-year-old child is about 7 cm, but by the age of 40, the near point shifts to about 22 cm. After that, the deterioration is rapid. At age 60, the near point is shifted to about 100 cm. This decrease in the accommodation of the eye with age is called *presbyopia*.

## 15.5 Eye and the camera

Although the designers of the photographic camera did not consciously imitate the structure of the eye, many of the features in the two are remarkably similar (see Fig. 15.2). Both consist of a lens system that focuses a real, inverted image onto a photosensitive surface. In the eye, as in the camera, the diameter of the light entrance is controlled by a diaphragm that is adjusted in accordance with the available light intensity. In a camera, the image is focused by moving the lens with respect to the film. In the eye, the distance between the retina and the lens is fixed; the image is focused by changing the thickness of the lens.

Even the photosensitive surfaces are somewhat similar. Both photographic film and the retina consist of discrete light-sensitive units, microscopic in size, which undergo chemical changes when they are illuminated.[1] In fact, under special circumstances, the retina can be "developed," like film, to show the

---

1. Light detection in digital cameras utilizes charge storage rather than chemical reactions.

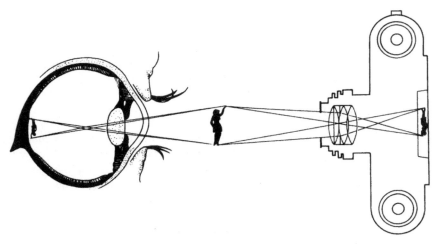

FIGURE 15.2   **The eye and the camera.**

image that was projected on it. This was first demonstrated in the 1870s by the German physiologist W. Kuhne. He exposed the eye of a living rabbit to light coming through a barred window. After 3 minutes of exposure to light, the rabbit was killed and its retina was immersed in an alum solution, which fixed the retinal reaction. The barred window was clearly visible on the retina. A few years later, Kuhne fixed the retina "of the eye from the head of a guillotined criminal." He observed an image, but he could not interpret it in terms of anything that the man had seen before he was beheaded.

The analogy between the eye and the camera, however, is not complete. As we will describe later, the eye goes far beyond the camera in processing the images that are projected on the retina.

### 15.5.1 Aperture and depth of field

The iris is the optical aperture of the eye, and its size varies in accordance with the available light. If there is adequate light, the quality of the image is best with the smallest possible aperture. This is true for both the eye and the camera.

There are two main reasons for the improved image with reduced aperture. Imperfections in lenses tend to be most pronounced around the edges. A small aperture restricts the light path to the center of the lens and eliminates the distortions and aberrations produced by the periphery.

A smaller aperture also improves the image quality of objects that are not located at the point on which the eye or the camera is focused. An image is in sharp focus at the retina (or film) only for objects at a specific distance from the lens system. Images of objects not at this specific plane are blurred at the retina (see Fig. 15.3); in other words, a point that is not in exact focus appears as a disk on the retina. The amount of blurring depends on the size of the aperture. As shown in Fig. 15.3, a small aperture reduces the diameter of the blurred spot and allows the formation of a relatively clear image from objects that are not on the plane to which the eye is focused. The range of object distances over which a good image is formed for a given setting of the focus is called the *depth of field*. Clearly a small aperture increases the depth of field. It can be shown that the depth of field is inversely proportional to the diameter of the aperture (see Exercise 15.2).

## 15.6 Lens system of the eye

The focusing of the light into a real, inverted image at the retina is produced by refraction at the cornea and at the crystalline lens (see Fig. 15.4). The focusing or refractive power of the cornea and the lens can be calculated using Eq. C.9 (Appendix C). The data required for the calculation are shown in Table 15.1.

The largest part of the focusing, about two-thirds, occurs at the cornea. The power of the crystalline lens is small because its index of refraction is only slightly greater than that of the surrounding fluid. In Exercise 15.3, it is shown that the refractive power of the cornea is 42 diopters, and the refractive power of the crystalline lens is variable between about 19 and 24 diopters. (For a definition of the unit *diopter*, see Appendix C.)

The refractive power of the cornea is greatly reduced when it is in contact with water (see Exercise 15.4). Because the crystalline lens in the human eye cannot compensate for the diminished power of the cornea, the human eye under water is not able to form a clear image at the retina and vision is

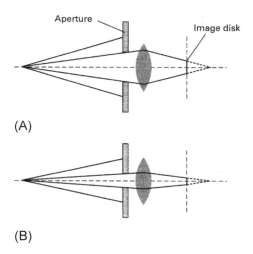

**FIGURE 15.3   Size of image disk.** (A) With large aperture, (B) with small aperture.

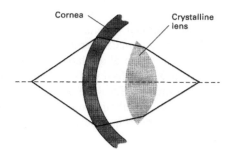

**FIGURE 15.4**  **Focusing by the cornea and the crystalline lens (not to scale).**

**TABLE 15.1** Parameters for the eye.

| | Radius (mm) | | |
| --- | --- | --- | --- |
| | Front | Back | Index of refraction |
| Cornea | 7.8 | 7.3 | 1.38 |
| Lens, min. power | 10.00 | −6.0 | |
| Lens, max. power | 6.0 | −5.5 | 1.40 |
| Aqueous and vitreous humor | | | 1.33 |

blurred. In fish eyes, which have evolved for seeing under water, the lens is intended to do most of the focusing. The lens is nearly spherical and has a much greater focusing power than the lens in the eyes of terrestrial animals (see Exercise 15.5).

## 15.7 Reduced eye

To trace accurately the path of a light ray through the eye, we must calculate the refraction at four surfaces (two at the cornea and two at the lens). It is possible to simplify this laborious procedure with a model called the *reduced eye*, shown in Fig. 15.5. Here all the refraction is assumed to occur at the front surface of the cornea, which is constructed to have a diameter of 5 mm. The eye is assumed to be homogeneous, with an index of refraction of 1.333 (the same as water). The retina is located 2 cm behind the cornea. The nodal point n is the center of corneal curvature located 5 mm behind the cornea.

This model represents most closely the relaxed eye which focuses parallel light at the retina, as can be confirmed using Eq. C.9. For the reduced eye, the second term on the right-hand side of the equation vanishes because the light is focused within the reduced eye so that $n_L = n_2$. Eq. C.9, therefore, simplifies to

$$\frac{n_1}{p} + \frac{n_L}{q} = \frac{n_L - n_1}{R}$$ (15.2)

where $n_1 = 1, n_L = 1.333$, and $R = 0.5$ cm. Because the incoming light is parallel, its source is considered to be at infinity (i.e., $p = \infty$). Therefore, the distance q at which parallel light is focused is given by

$$\frac{1.333}{q} = \frac{1.333 - 1}{5}$$

or

$$q = \frac{1.333 \times 5}{0.333} = 20 \text{ mm}$$

The anterior focal point F for the reduced eye is located 15 mm in front of the cornea. This is the point at which parallel light originating inside the eye is focused when it emerges from the eye (see Exercise 15.6).

Although the reduced eye does not contain explicitly the mechanism of accommodation, we can use the model to determine the size of the image formed on the retina. The construction of such an image is shown in Fig. 15.6. Rays from the limiting points of the object *A* and *B* are projected through the nodal point to the retina. The limiting points of the image at the retina are *a* and *b*. This construction assumes that all the rays from points *A* and *B* that

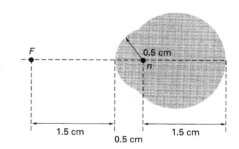

FIGURE 15.5    The reduced eye.

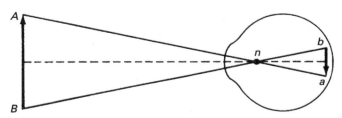

FIGURE 15.6    Determination of the image size on the retina.

enter the eye are focused on the retina at points *a* and *b*, respectively. Rays from all other points on the object are focused correspondingly between these limits. The triangles *AnB* and *anb* are similar; therefore, the relation of object to image size is given by

$$\frac{\text{Object size}}{\text{Image size}} = \frac{\text{Distance of object from nodal point}}{\text{Distance of image from nodal point}} \qquad (15.3)$$

or

$$\frac{AB}{ab} = \frac{An}{an}$$

Consider as an example the image of a person 180 cm tall standing 2 m from the eye. The height of the full image at the retina is

$$\text{Height of image} = 180 \times \frac{1.5}{205} = 1.32 \text{ cm}$$

The size of the face in the image is about 1.8 mm, and the nose is about 0.4 mm.

## 15.8 Retina

The retina consists of photoreceptor cells in contact with a complex network of neurons and nerve fibers which are connected to the brain via the optic nerve (see Fig. 15.7). Light absorbed by the photoreceptors produces nerve impulses

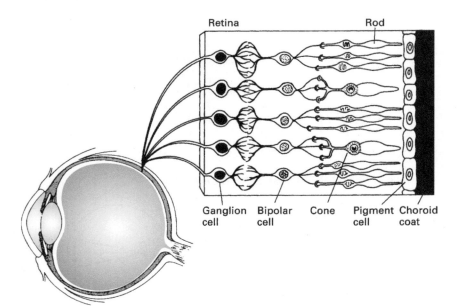

**FIGURE 15.7 The retina.**

that travel along the neural network and then through the optic nerve into the brain. The photoreceptors are located behind the neural network, so the light must pass through this cell layer before it reaches the photoreceptors.

There are two types of photoreceptor cells in the retina: *Cones* and *rods*. The cones are responsible for sharp color vision in daylight. The rods provide vision in dim light.

Near the center of the retina is a small depression about 0.3 mm in diameter which is called the *fovea*. It consists entirely of cones packed closely together. Each cone is about 0.002 mm (2 μm) in diameter. Most detailed vision is obtained on the part of the image that is projected on the fovea. When the eye scans a scene, it projects the region of greatest interest onto the fovea.

The region around the fovea contains both cones and rods. The structure of the retina becomes more coarse away from the fovea. The proportion of cones decreases until, near the edge, the retina is composed entirely of rods. In the fovea, each cone has its own path to the optic nerve. This allows the perception of details in the image projected on the fovea. Away from the fovea, a number of receptors are attached to the same nerve path. Here the resolution decreases, but the sensitivity to light and movement increases.

With the structure of the retina in mind, let us examine how we view a scene from a distance of about 2 m. From this distance, at any one instant, we can see most distinctly an object only about 4 cm in diameter. An object of this size is projected into an image about the size of the fovea.

Objects about 20 cm in diameter are seen clearly but not with complete sharpness. The periphery of large objects appears progressively less distinct. Thus, for example, if we focus on a person's face 2 m away, we can see clearly the facial details, but we can pick out most clearly only a subsection about the size of the mouth. At the same time, we are aware of the person's arms and legs, but we cannot detect, for example, details about the person's shoes.

## 15.9 Resolving power of the eye

So far in our discussion of image formation, we have used geometric optics, which neglects the diffraction of light. Geometric optics assumes that light from a point source is focused into a point image. This is not the case. When light passes through an aperture such as the iris, diffraction occurs, and the wave spreads around the edges of the aperture.[2] As a result, light is not focused into a sharp point but into a diffraction pattern consisting of a disk surrounded by rings of diminishing intensity.

If light originates from two point sources that are close together, their image diffraction disks may overlap, making it impossible to distinguish the two points. An optical system can resolve two points if their corresponding

---

2. If there are no smaller apertures in the optical path, the lens itself must be considered as the aperture.

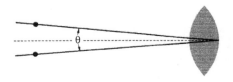

FIGURE 15.8   **Two points being resolvable if the angle θ is greater than 1.22λ/d.**

diffraction patterns are distinguishable. This criterion alone predicts that two points are resolvable (see Fig. 15.8) if the angular separation between the lines joining the points to the center of the lens is equal to or greater than a critical value given by

$$\theta = \frac{1.22\lambda}{d} \tag{15.4}$$

where $\lambda$ is the wavelength of light and $d$ is the diameter of the aperture. The angle $\theta$ is given in radians (1 rad = 57.3°). With green light ($\lambda = 500$ nm) and an iris diameter of 0.5 cm, this angle is $1.22 \times 10^{-4}$ rad.

Experiments have shown that the eye does not perform this well. Most people cannot resolve two points with an angular separation of less than $5 \times 10^{-4}$ rad. Clearly there are other factors that limit the resolution of the eye. Imperfections in the lens system of the eye certainly impede the resolution. But perhaps even more important are the limitations imposed by the structure of the retina.

The cones in the closely packed fovea are about 2 μm in diameter. To resolve two points, the light from each point must be focused on a different cone and the excited cones must be separated from each other by at least one cone that is not excited. Thus at the retina, the images of two resolved points are separated by at least 4 μm. A single unexcited cone between points of excitation implies an angular resolution of about $3 \times 10^{-4}$ rad (see Exercise 15.7a). Some people with acute vision do resolve points with this separation, but most people do not. We can explain the limits of resolution demonstrated by most normal eyes if we assume that, to perceive distinct point images, there must be three unexcited cones between the areas of excitation. The angular resolution is then, as observed, $5 \times 10^{-4}$ rad (see Exercise 15.7b).

Let us now calculate the size of the smallest detail that the unaided eye can resolve. To observe the smallest detail, the object must be brought to the closest point on which the eye can focus. Assuming that this distance is 20 cm from the eye, the angle subtended by two points separated by a distance x is given by (see Fig. 15.9)

$$\tan^{-1}\frac{\theta}{2} = \frac{x/2}{20} \tag{15.5}$$

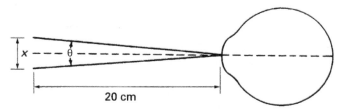

FIGURE 15.9   **Resolution of the eye.**

If θ is very small, as is the case in our problem, the tangent of the angle is equal to the angle itself and

$$\theta = \frac{x}{20}$$

Because the smallest resolvable angle is $5 \times 10^{-4}$ rad, the smallest resolvable detail x is

$$x = 5 \times 10^{-4} \times 20 = 100 \ \mu m = 0.1 \ mm$$

Using the same criterion, we can show (see Exercise 15.8) that the facial features such as the whites of the eye are resolvable from as far as 20 m.

## 15.10 Threshold of vision

The sensation of vision occurs when light is absorbed by the photosensitive rods and cones. At low levels of light, the main photoreceptors are the rods. Light produces chemical changes in the photoreceptors, which reduce their sensitivity. For maximum sensitivity, the eye must be kept in the dark (dark adapted) for about 30 minutes to restore the composition of the photoreceptors.

Under optimum conditions, the eye is a very sensitive detector of light. The human eye, for example, responds to light from a candle as far away as 20 km. At the threshold of vision, the light intensity is so small that we must describe it in terms of photons. Experiments indicate that an individual photoreceptor (rod) is sensitive to 1 quantum of light. This, however, does not mean that the eye can see a single photon incident on the cornea. At such low levels of light, the process of vision is statistical.

In fact, measurements show that about 60 quanta must arrive at the cornea for the eye to perceive a flash. Approximately half the light is absorbed or reflected by the ocular medium. The 30 or so photons reaching the retina are spread over an area containing about 500 rods. It is estimated that only five of these photons are actually absorbed by the rods. It seems, therefore, that at least five photoreceptors must be stimulated to perceive light.

The energy in a single photon is very small. For green light at 500 nm, it is

$$E = hf = \frac{hc}{\lambda} = \frac{6.63 \times 10^{-27} \times 3 \times 10^{10}}{5 \times 10^{-5}} = 3.98 \times 10^{-12} \text{ erg}$$

This amount of energy, however, is sufficient to initiate a chemical change in a single molecule which then triggers the sequence of events that leads to the generation of the nervous impulse.

## 15.11 Vision and the nervous system

Vision cannot be explained entirely by the physical optics of the eye. There are many more photoreceptors in the retina than fibers in the optic nerve. It is, therefore, evident that the image projected on the retina is not simply transmitted point by point to the brain. A considerable amount of signal processing occurs in the neural network of the retina before the signals are transmitted to the brain. The neural network "decides" which aspects of the image are most important and stresses the transmission of those features. In a frog, for example, the neurons in the retina are organized for most active response to movements of small objects. A fly moving across the frog's field of vision will produce an intense neural response, and if the fly is close enough, the frog will lash out its tongue to capture the fly. On the other hand, a large object, clearly not food for the frog, moving in the same vision field will not elicit a neural response. Evidently the optical processing system of the frog enhances its ability to catch small insects while reducing the likelihood of being noticed by larger, possibly dangerous creatures passing through the neighborhood.

The human eye also possesses important processing mechanisms. It has been shown that movement of the image is necessary for human vision as well. In the process of viewing an object, the eye executes small rapid movements, 30 to 70 per second, which alter slightly the position of the image on the retina. Under experimental conditions, it is possible to counteract the movement of the eye and stabilize the position of the retinal image. It has been found that, under these conditions, the image perceived by the person gradually fades.

## 15.12 Defects in vision

There are three common defects in vision associated with the focusing system of the eye: *Myopia* (nearsightedness), *hyperopia* (farsightedness), and *astigmatism*. The first two of these defects are best explained by examining the imaging of parallel light by the eye. Figs. 15.11b and 15.12b show correction for myopia and hyperopia using external lenses.

The relaxed normal eye focuses parallel light onto the retina (Fig. 15.10). In the myopic eye the lens system focuses the parallel light in front of the retina (Fig. 15.11a). This misfocusing is usually caused by an elongated eyeball or an excessive curvature of the cornea. In hyperopia the problem is reversed (see Fig. 15.12a). Parallel light is focused behind the retina. The

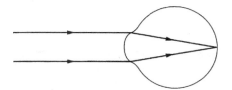

FIGURE 15.10   **The normal eye.**

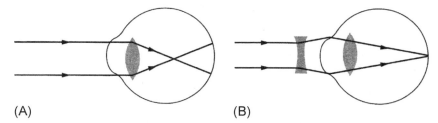

(A)                                                    (B)

FIGURE 15.11   (A) Myopia. (B) Its correction.

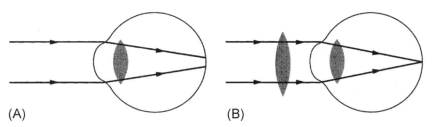

(A)                                                    (B)

FIGURE 15.12   (A) Hyperopia. (B) Its correction.

problem here is caused by an eyeball that is shorter than normal or by the inadequate focusing power of the eye. The hyperopic eye can accommodate objects at infinity, but its near point is farther away than is normal. Hyperopia is, thus, similar to presbyopia. These two defects can be summarized as follows: The myopic eye converges light too much, and the hyperopic eye not enough.

Astigmatism is a defect caused by a nonspherical cornea. An oval-shaped cornea, for example, is more sharply curved along one plane than another; therefore, it cannot form simultaneously sharp images of two perpendicular lines. One of the lines is always out of focus, resulting in distorted vision.

All three of these defects can be corrected by lenses placed in front of the eye. Myopia requires a diverging lens to compensate for the excess refraction in the eye. Hyperopia is corrected by a converging lens, which adds to the focusing power of the eye. The uneven corneal curvature in astigmatism is compensated for by a cylindrical lens (Fig. 15.13), which focuses light along one axis but not along the other.

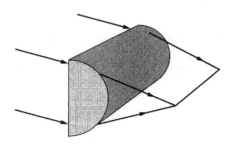

FIGURE 15.13   **Cylindrical lens for astigmatism.**

## 15.13 Lens for myopia

Let us assume that the farthest object a certain myopic eye can properly focus is 2 m from the eye. This is called the *far point of the eye*. Light from objects farther away than this is focused in front of the retina (Fig. 15.11a). Here the purpose of the corrective lens is to make parallel light appear to come from the far point of the eye (in this case, 2 m). With such a corrective lens, the eye is able to form images of objects all the way to infinity.

The focal length of the lens is obtained by using Eq. C.6, which is

$$\frac{1}{p} + \frac{1}{q} = \frac{1}{f}$$

Here p is infinity, as this is the effective distance for sources of parallel light. The desired location $q$ for the virtual image is $-200$ cm. The focal length of the diverging lens (see Eq. C.4) is, therefore,

$$\frac{1}{f} = \frac{1}{\infty} + \frac{1}{-200} \text{ or } f = -200 \text{ cm} = -5 \text{ diopters}$$

## 15.14 Lens for presbyopia and hyperopia

In these disorders, the eye cannot focus properly on close objects. The near point is too far from the eye. The purpose of the lens is to make light from close objects appear to come from the near point of the unaided eye. Let us assume that a given hyperopic eye has a near point at 150 cm. The desired lens is to allow the eye to view objects at 25 cm. The focal length of the lens is again obtained from Eq. C.6:

$$\frac{1}{p} + \frac{1}{q} = \frac{1}{f}$$

Here p is the object distance at 25 cm and q is $-150$ cm, which is the distance of the virtual image at the near point. The focal length f for the converging lens is given by

$$\frac{1}{f} = \frac{1}{25 \text{ cm}} - \frac{1}{150 \text{ cm}} \text{ or } f = 30 \text{ cm} = 33.3 \text{ diopters}$$

## 15.15 Extension of vision

The range of vision of the eye is limited. Details on distant objects cannot be seen because their images on the retina are too small. The retinal image of a 20-m-high tree at a distance of 500 m is only 0.6 mm high. The leaves on this tree cannot be resolved by the unaided eye (see Exercise 15.9). Observation of small objects is limited by the accommodation power of the eye. We have already shown that because the average eye cannot focus light from objects closer than about 20 cm, its resolution is limited to approximately 100 μm.

Over the past 300 years, two types of optical instruments have been developed to extend the range of vision: The telescope and the microscope. The telescope is designed for the observation of distant objects. The microscope is used to observe small objects that cannot be seen clearly by the naked eye. Both of these instruments are based on the magnifying properties of lenses. A third more recent aid to vision is the fiberscope, which utilizes total internal reflection to allow the visualization of objects normally hidden from view.

### 15.15.1 Telescope

A drawing of a simple telescope is shown in Fig. 15.14. Parallel light from a distant object enters the first lens, called the *objective lens* or *objective*, which forms a real, inverted image of the distant object. Because light from the distant object is nearly parallel, the image is formed at the focal plane of the objective. (The drawing shows the light rays from only a single point on

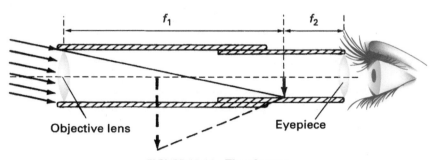

FIGURE 15.14   **The telescope.**

the object.) The second lens, called the *eyepiece*, magnifies the real image. The telescope is adjusted so that the real image formed by the objective falls just within the focal plane of the eyepiece. The eye views the magnified virtual image formed by the eyepiece. The total magnification—the ratio of image to object size—is given by

$$\text{Magnification} = -\frac{f_1}{f_2} \tag{15.6}$$

where $f_1$ and $f_2$ are the focal lengths of the objective and the eyepiece, respectively. As can be seen from Eq. 15.6, greatest magnification is obtained with a long focal length objective and a short focal length eyepiece.

### 15.15.2 Microscope

A simple microscope consists of a single lens that magnifies the object (Fig. 15.15). Better results can be obtained, however, with a two-lens system compound microscope, shown in Fig. 15.16. The compound microscope, like the telescope, consists of an objective lens and an eyepiece, but the objective of the microscope has a short focal length. It forms a real image $I_1$ of the object; the eye views the final magnified image $I_2$ formed by the eyepiece.

The microscope is an important tool in the life sciences. Its invention in the 1600s marked the beginning of the study of life on the cellular level. The early microscope produced highly distorted images, but years of development have

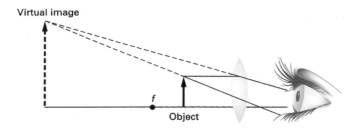

**FIGURE 15.15    Simple magnifier.**

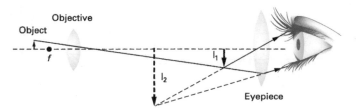

**FIGURE 15.16    Schematic diagram of a compound microscope.**

perfected the device nearly to its theoretical optimum. The resolution of the best modern microscopes is determined by the diffraction properties of light, which limit the resolution to about half the wavelength of light. In other words, with a good modern microscope, we can observe objects as small as half the wavelength of the illuminating light.

We will not present here the details of microscopy. These can be found in many basic physics texts (see, for example Ackerman, 1962). We will, however, describe a special-purpose scanning confocal microscope designed in the laboratory of Paul Davidovits and M. David Egger.

### 15.15.3 Confocal microscopy

With conventional microscopes, it is not possible to observe small objects embedded in translucent materials. For example, cells located beneath the surface of tissue, such as buried brain cells in living animals, cannot be satisfactorily observed with conventional microscopes.

Light can certainly penetrate through tissue. This can be demonstrated simply by inserting a flashlight into the mouth and observing the light passing through the cheeks. In principle, therefore, we should be able to form a magnified image of a cell inside the tissue. This could be done by shining light into the tissue and collecting the light reflected from the cell. Unfortunately there is a problem associated with the straightforward use of this technique. Light is reflected and scattered not only by the cell of interest but also by the surface of the tissue and by the cells in front and behind the cell being examined. This spurious light is also intercepted by the microscope and masks the image of the single cell layer within the tissue (see Fig. 15.17).

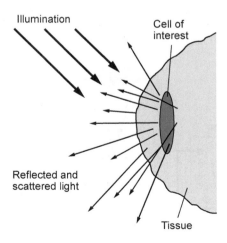

**FIGURE 15.17  Light scattered and reflected from tissue.**

Over the years, a number of microscopes have been designed that have attempted to solve this problem. The most successful of these is the *confocal microscope*. The principle of confocal microscopy was first described by Marvin Minsky in 1957. In the 1960s, Davidovits and Egger modified the Minsky design and built the first successful confocal microscope for observation of cells within living tissue.

The confocal microscope is designed to accept light only from a thin slice within the tissue and to reject light reflected and scattered from other regions. A schematic diagram of the Davidovits—Egger microscope is shown in Fig. 15.18. Although the device does not resemble a conventional microscope,

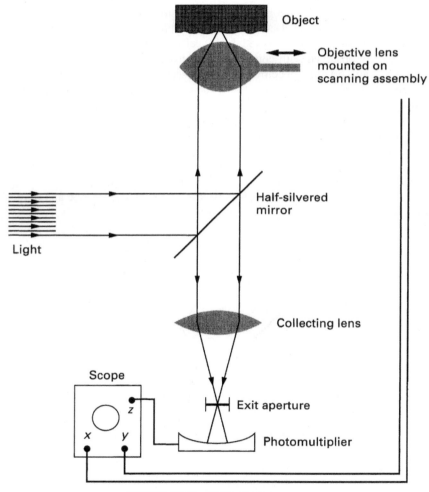

FIGURE 15.18 **Confocal microscope.**

it certainly does produce magnified images. This microscope requires a parallel beam of light for illumination of the object. As the source of parallel light, Davidovits and Egger used a laser with a power output that is relatively low so that it does not damage the tissue under observation. The laser beam is reflected by a half-silvered mirror into the objective lens, which focuses the beam to a point inside the tissue. Because the light is parallel, the beam is brought to a point at the principal focus of the lens. The depth of this point in the tissue can be changed by altering the distance between the lens and the tissue.

Light is scattered and reflected from all points in the path of the entering light, and part of this returning light is intercepted by the objective lens. However, only light originating from the focal point emerges from the lens as a parallel beam; light from all other points either converges toward or diverges from the lens axis. The returning light passes through the half-silvered mirror and is intercepted by the collecting lens. Only the parallel component of the light is focused into the small exit aperture that is placed at the principal focal point of the collecting lens. Nonparallel light is defocused at the exit aperture. A photomultiplier placed behind the exit aperture produces a voltage proportional to the light intensity transmitted through the exit aperture. This voltage is then used to control the intensity of an electron beam in the oscilloscope.

So far, we have one spot on the screen of the oscilloscope that glows with a brightness proportional to the reflectivity of one point inside the tissue. In order to see a whole cell or region of cells, we must scan the region point by point. This is done by moving the lens in its own plane so that the focal point scans an area inside the tissue. The motion of the lens does not affect the parallelism of the light originating at the focal point of the objective lens. Therefore, at every instant, the output of the photomultiplier and the corresponding brightness of the spot on the screen are proportional to the reflectivity of the point being scanned. While the object is scanned, the electron beam in the oscilloscope is moved in synchrony with the motion of the objective lens. Thus, the screen shows a picture of a very thin section within the tissue.

The magnification of this microscope is simply the ratio of the electron beam excursion on the oscilloscope face to the excursion of the scanning lens. For a 0.1-mm excursion of the lens, the electron beam may be adjusted to move 5 cm. The magnification is then 500. The resolution of the device is determined by the size of the spot focused by the objective. The diffraction properties of light limit the minimum spot size to about half the wavelength of light. The optimum resolution is, therefore, about the same as in conventional microscopes.

The first biologically significant observations with the confocal microscope were those of endothelial cells on the inside of the cornea in live frogs. Such observations cannot be made with conventional microscopes because the light

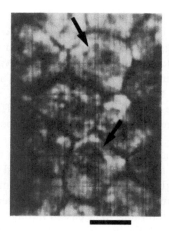

FIGURE 15.19 **Corneal endothelial cells in an intact eye of a living bullfrog.** Arrows indicate outlines of the nuclei in two of the cells. Calibration mark, 25 μm.

reflected from the front surface of the cornea masks the weak reflections from the endothelial cells. The picture of these cells shown in Fig. 15.19 was obtained by photographing the image on the oscilloscope screen. The confocal microscope is now a major observational tool in most biology laboratories. In the more recent versions of the instrument, the object is scanned with moving mirrors and the image is processed by computers. Image improvement obtained with modern confocal microscopy is illustrated in Fig. 15.20.

### 15.15.4 Fiber optics

Fiber-optic devices are now used in a wide range of medical applications. The principle of their operation is simple. As discussed in Appendix C in connection with Snell's law, light traveling in a material with a high index of refraction is totally reflected back into the material if it strikes the boundary of the material with lower refractive index at an angle greater than the critical angle $\theta_c$. In this way, light can be confined to travel within a glass cylinder, as shown in Fig. 15.21. This phenomenon has been well known since the early days of optics. However, major breakthroughs in materials technology were necessary before the phenomenon could be widely utilized.

Optical fiber technology, developed in the 1960s and 1970s, made it possible to manufacture low-loss, thin, highly flexible glass fibers that can carry light over long distances. A typical optical fiber is about 10 μm in diameter and is made of high-purity silica glass. The fiber is coated with a cladding to increase light trapping. Such fibers can carry light over tortuously twisting paths for several kilometers without significant loss.

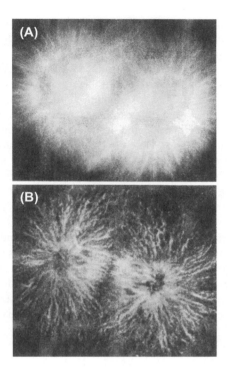

**FIGURE 15.20**  Microscope images of sea urchin embryos obtained with (A) a conventional microscope showing out-of-focus blur and (B) a modern confocal microscope. *(Part (A) from Matsumoto B. Meth Cell Biol 1993;38:22. Part (B) from Wright JS, et al. J Cell Sci 1989;94:617—24, with permission from the Company of Biologists Ltd.)*

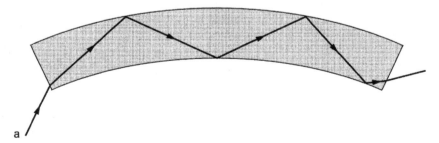

**FIGURE 15.21**  **Light confined to travel inside a glass cylinder by total reflection.**

*Fiberscopes* or *endoscopes* are the simplest of the fiber-optic medical devices. They are used to visualize and examine internal organs such as the stomach, heart, and bowels. A fiberscope consists of two bundles of optical fibers tied into one flexible unit. Each bundle is typically a millimeter in diameter, consisting of about 10,000 fibers. For some applications, the bundles are thicker, up to about 1.5 cm in diameter. Depending on their use, the bundles vary in length from 0.3 to 1.2 m.

The two bundles as a unit are introduced into the body through orifices, veins, or arteries and are threaded toward the organ to be examined. Light from a high-intensity source, such as a xenon arc lamp, is focused into one bundle, which carries the light to the organ to be examined. Each of the fibers in the other bundle collects light reflected from a small region of the organ and carries it back to the observer. Here the light is focused into an image that can be viewed by the eye or displayed on a cathode ray or some other type of electronic screen. In the usual arrangement, the illuminating bundle surrounds the light-collecting bundle. Most endoscopes now utilize attached miniature video cameras to form images of the internal organs for display on TV monitors.

The use of fiber-optic devices has been greatly expanded by attaching to the endoscope remotely controlled miniature instruments to perform surgical operations without major surgical incisions. More recent applications of fiber optics include measurement of pressure in arteries, bladder, and uterus using optical sensors and laser surgery, where powerful laser light is directed through one of the bundles to the tissue which is selectively destroyed (see Chapter 16).

## 15.16 Physiological and psychological effects of light on people

Vision is only one of many ways (although certainly the central one) that light interacts with humans and most other animals. Perhaps second in importance is the light-induced production of vitamin D by the skin that is initiated by exposure of the skin to light in the UVB range (about 280 nm to 320 nm). Vitamin D is essential for the absorption of calcium to maintain bone health.

Vitamin D deficiency in children results in soft bones, a condition called rickets, while in adults, inadequate supply of vitamin D causes bone pain and a decreased ability of the bones to carry the weight of the body. In sunny climates, vitamin D requirements of the body can be supplied by light-induced skin production. However, in northern regions with short daylight hours and much of the time spent indoors, a diet rich in vitamin D and a periodic check of vitamin D levels in the body are recommended.

Vitamin D deficiency has been linked to a range of diseases, such as several types of cancer, diabetes, and high blood pressure, among others. However, the connection between vitamin D deficiency and maladies other than bone disease has not been firmly established.

A variety of effects of light interacting with the visual system but not directly with image production are well documented. Results of both objective and subjective tests indicate that higher illumination almost immediately increases a subject's alertness.

Light impinging on the eye provides information to other parts of the brain that set the sleep—wake and other time (circadian) rhythms of the body. It appears that optimum well-being requires a person to be in synchrony with the night—day circadian rhythm of the surroundings. Several studies indicate that nightshift workers suffer in numbers higher than the population average, from diseases such as cardiovascular problems, gastrointestinal difficulties, metabolic syndrome, and some types of cancer.

The seasonal effect of low-light levels on mood is well documented. Many people suffer low levels of depression during winter months. The phenomenon has been designated as *seasonal affective disorder* (SAD). The hypothesis has been put forth that for people suffering from this syndrome, the production of the neurotransmitter serotonin is reduced in diminished light. Exposure for several hours each day to artificial light mimicking sunlight seems to alleviate the symptoms of SAD. The effects of light on the biochemistry of the brain are not well understood and remain an active area of research.

## Exercises

**15.1.** Compute the change in the position of the image formed by a lens with a focal length of 1.5 cm as the light source is moved from its position at 6 m from the lens to infinity.

**15.2.** A point source of light that is not exactly in focus produces a disk image at the retina. Assume that the image is acceptable provided the image diameter of the defocused point source is less than a. Show that the depth of field is inversely proportional to the diameter of the aperture.

**15.3.** Using data presented in the text, calculate the focusing power of the cornea and of the crystalline lens.

**15.4.** Calculate the refractive power of the cornea when it is in contact with water. The index of refraction for water is 1.33.

**15.5.** Calculate the focusing power of the lens in the fish eye. Assume that the lens is spherical with a diameter of 2 mm. (The indices of refraction are as in Table 15.1.) The index of refraction for water is 1.33.

**15.6.** Calculate the distance of the point in front of the cornea at which parallel light originating inside the reduced eye is focused.

**15.7.** Using the dimensions of the reduced eye (Fig. 15.5), calculate the angular resolution of the eye (use Fig. 15.6 as an aid) (a) with a single unexcited cone between points of excitation and (b) with four unexcited cones between areas of excitation.

**15.8.** Calculate the distance from which a person with good vision can see the whites of another person's eyes. Use data in the text and assume the size of the eye is 1 cm.

**15.9.** Calculate the size of the retinal image of a 10-cm leaf from a distance of 500 m.

**15.10.** The 1928 Nobel Prize for Chemistry was awarded to Adolf Windaus for his work with vitamin D. This work was conducted in connection with attempts to cure rickets. Describe these experiments. (Relevant articles are found on the web.)

# Chapter 16

# Atomic physics

Modern atomic and nuclear physics are among the most impressive scientific achievements of this century. There is hardly an area of science or technology that does not draw on the concepts and techniques developed in these fields. Both the theories and techniques of atomic and nuclear physics have played an important role in the life sciences. The theories have provided a solid foundation for understanding the structure and interaction of organic molecules, and the techniques have provided many tools for both experimental and clinical work. Contributions from this field have been so numerous and influential that it is impossible to do them justice in a single chapter. Of necessity, therefore, our discussion will be restricted to a survey of the subject. We will present a brief description of the atom and the nucleus, which will lead into a discussion of the applications of atomic and nuclear physics to the life sciences.

## 16.1 The atom

By 1912, through the work of J. J. Thompson, E. Rutherford, and their colleagues, a number of important facts had been discovered about atoms that make up matter. It was found that atoms contain small negatively charged electrons and relatively heavier positively charged protons. The proton is about 2000 times heavier than the electron, but the magnitude of the charge on the two is the same. There are as many positively charged protons in an atom as negatively charged electrons. The atom as a whole is, therefore, electrically neutral. The identity of an atom is determined by the number of protons it has. For example, hydrogen has 1 proton, carbon has 6 protons, and silver has 47 protons. Through a series of ingenious experiments, Rutherford showed that most of the atomic mass is concentrated in the nucleus consisting of protons and that the electrons are somehow situated outside the nucleus. It was subsequently discovered that the nucleus also contains another particle, the neutron, which has approximately the same mass as the proton but is electrically neutral.

Although the nucleus contains most of the atomic mass, it occupies only a small part of the total atomic volume. The diameter of the whole atom is on the order of $10^{-8}$ cm, but the diameter of the nucleus is only about $10^{-13}$ cm. The configuration of the electrons around the nucleus was not known at that time.

Physics in Biology and Medicine. https://doi.org/10.1016/B978-0-443-21558-2.00016-X

In 1913, the Danish physicist Niels Bohr proposed a model for the atom, which explained many observations that were puzzling scientists at that time. When Bohr first became acquainted with atomic physics, the subject was in a state of confusion. A number of theories had been proposed for the structure of the atom, but none explained satisfactorily the existing experimental results. The most surprising observed property of atoms was the light emitted by them.[1] When an element is put into a flame, it emits light at sharply defined wavelengths, called *spectral lines*. Each element emits its own characteristic spectrum of light. This is in contrast to a glowing filament in a light bulb, for example, which emits light over a continuous range of wavelengths.

Prior to Bohr, scientists could not explain why these colors were emitted by atoms. Bohr's model of the atom explained the reason for the sharp spectra. Bohr started with the model of the atom as proposed by Rutherford. At the center of the atom is the positive nucleus made up of protons (and neutrons). The electrons orbit around the nucleus much as the planets orbit around the sun. They are maintained in orbit by the electrostatic attraction of the nucleus. And here is the major feature of the Bohr model: So that the model would explain the emission of spectral lines, Bohr had to postulate that the electrons are restricted to distinct orbits around the nucleus. In other words, the electrons can be found only in certain allowed orbits. Bohr was able to calculate the radii of these allowed orbits and show that the spectral lines are emitted as a consequence of the orbital restrictions. Bohr's calculations are found in most elementary physics texts.

The orbital restrictions are most easily illustrated with the simplest atom, hydrogen, which has a single-proton nucleus and one electron orbiting around it (Fig. 16.1). Unless energy is added to the atom, the electron is found in the allowed orbit closest to the nucleus. If energy is added to the atom, the electron may "jump" to one of the higher allowed orbits farther away from the nucleus, but the electron can never occupy the regions between the allowed orbits.

The Bohr model was very successful in explaining many of the experimental observations for the simple hydrogen atom. But to describe the behavior of atoms with more than one electron, it was necessary to impose an additional restriction on the structure of the atom: The number of electrons in a given orbit cannot be greater than $2n^2$, where $n$ is the order of the orbit from the nucleus. Thus, the maximum number of electrons in the first allowed orbit is $2 \times (1)^2 = 2$; in the second allowed orbit, it is $2 \times (2)^2 = 8$; in the third orbit, it is $2 \times (3)^2 = 18$, and so on.

The atoms are found to be constructed in accordance with these restrictions. Helium has two electrons, and, therefore, its first orbit is filled.

---

1. In atomic physics, the word *light* is not restricted solely to the visible part of the electromagnetic spectrum. Radiation at shorter wavelength (ultraviolet) and longer wavelength (infrared) is also often referred to as light.

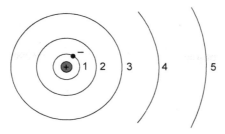

**FIGURE 16.1** **Bohr model for the hydrogen atom.** The electron orbits about the nucleus and can occupy only discrete orbits with radii 1, 2, 3, and so on.

Lithium has three electrons, two of which fill the first orbit; the third electron, therefore, must be in the second orbit. This simple sequence is not completely applicable to the very complex atoms, but basically, this is the way the elements are constructed.

A specific amount of energy is associated with each allowed orbital configuration of the electron. Therefore, instead of speaking of the electron as being in a certain orbit, we can refer to it as having a corresponding amount of energy. Each of these allowed values of energy is called an *energy level*. An energy level diagram for an atom is shown in Fig. 16.2. Note that every element has its own characteristic energy level structure. The electrons in the atom can occupy only specific energy states; that is, in a given atom the electron can have an energy $E_1$, $E_2$, $E_3$, and so on, but it cannot have an energy between these two values. This is a direct consequence of the restrictions on the allowed electron orbital configurations.

The lowest energy level that an electron can occupy is called the *ground state*. This state is associated with the orbital configuration closest to the nucleus. The higher allowed energy levels, called *excited states*, are associated with larger orbits and different orbital shapes. Normally the electron occupies the lowest energy level, but it can be excited into a higher energy state by adding energy to the atom.

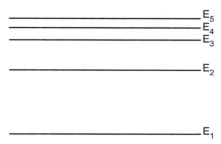

**FIGURE 16.2** **Energy levels for an atom.**

An atom can be excited from a lower to a higher energy state in a number of different ways. The two most common methods of excitation are electron impact and absorption of electromagnetic radiation. Excitation by electron impact occurs most frequently in a gas discharge. If a current is passed through a gas of atoms, the colliding electron is slowed down and the electron in the atom is promoted to a higher energy configuration. When the excited atoms fall back into the lower energy states, the excess energy is given off as electromagnetic radiation. Each atom releases its excess energy in a single photon. Therefore, the energy of the photon is simply the difference between the energies of the initial state $E_i$ and the final state $E_f$ of the atom. The frequency f of the emitted radiation is given by

$$f = \frac{\text{Energy of photon}}{\text{Planck constant}} = \frac{E_i - E_f}{h} \qquad (16.1)$$

Transition between each pair of energy levels results in the emission of light at a specific frequency, called transition or resonance frequency. Therefore, a group of highly excited atoms of a given element emit light at a number of well-defined frequencies which constitute the optical spectrum for that element.

An atom in a given energy level can also be excited to a higher level by light at a specific frequency. The frequency must be such that each photon has just the right amount of energy to promote the atom to one of its higher allowed energy states. Atoms, therefore, absorb light only at the specific transition frequencies, given by Eq. 16.1. Light at other frequencies is not absorbed. If a beam of white light (containing all the frequencies) is passed through a group of atoms of a given species, the spectrum of the transmitted light shows gaps corresponding to the absorption of the specific frequencies by the atoms. This is called the *absorption spectrum* of the atom. In their undisturbed state, most of the atoms are in the ground state. The absorption spectrum, therefore, usually contains only lines associated with transitions from the ground state to higher allowed states (Fig. 16.3).

Optical spectra are produced by the outer electrons of the atom. The inner electrons, those closer to the nucleus, are bound more tightly and are consequently more difficult to excite. However, in a highly energetic collision with another particle, an inner electron may be excited. When in such an excited atom an electron returns to the inner orbit, the excess energy is again released as a quantum of electromagnetic radiation. Because the binding energy here is about a thousand times greater than for the outer electrons, the frequency of the emitted radiation is correspondingly higher. Electromagnetic radiation in this frequency range is called *X-rays*.

The Bohr model also explained qualitatively the formation of chemical bonds. The formation of chemical compounds and matter in bulk is due to the distribution of electrons in the atomic orbits. When an orbit is not filled to capacity (which is the case for most atoms), the electrons of one atom can

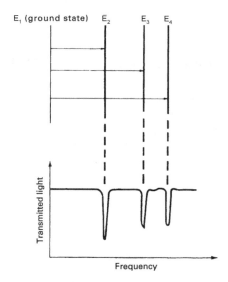

FIGURE 16.3   **The absorption spectrum.**

partially occupy the orbit of another. This sharing of orbits draws the atoms together and produces bonding between atoms. As an example, we show in Fig. 16.4 the formation of a hydrogen molecule from two hydrogen atoms. In the orbit of each of the hydrogen atoms there is room for another electron. A completely filled orbit is the most stable configuration; therefore, when two hydrogen atoms are close together, they share each other's electrons, and, in this way, the orbit of each atom is completely filled part of the time. This shared orbit can be pictured as a rubber band pulling the two atoms together. Therefore, the sharing of the electrons binds the atoms into a molecule. While the sharing of electrons pulls the atoms together, the coulomb repulsion of the

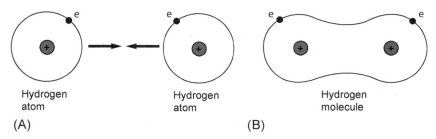

FIGURE 16.4   **A schematic representation for the formation of a hydrogen molecule.** (A) Two separate hydrogen atoms. (B) When the two atoms are close together, the electrons share each other's orbit, which results in the binding of the two atoms into a molecule.

nuclei tends to keep them apart. The equilibrium separation between atoms in a molecule is determined by these two counterforces. In a similar way, more complex molecules, and ultimately bulk matter, are formed.

Atoms with completely filled orbits (these are atoms of the so-called *noble gases*—helium, neon, argon, krypton, and xenon) cannot share electrons with other elements and are, therefore, chemically most inert.

Molecules also have characteristic spectra both in emission and in absorption. Because molecules are more complicated than atoms, their spectra are correspondingly more complex. In addition to the electronic configuration, these spectra also depend on the motion of the nuclei. Still the spectra can be interpreted and are unique for each type of molecule.

## 16.2 Spectroscopy

The absorption and emission spectra of atoms and molecules are unique for each species. They can serve as fingerprints in identifying atoms and molecules in various substances. Spectroscopic techniques were first used in basic experiments with atoms and molecules, but they were soon adopted in many other areas, including the life sciences.

In biochemistry, spectroscopy is used to identify the products of complex chemical reactions. In medicine, spectroscopy is used routinely to determine the concentration of certain atoms and molecules in the body. From a spectroscopic analysis of urine, for example, one can determine the level of mercury in the body. Blood-sugar level is measured by first producing a chemical reaction in the blood sample, which results in a colored product. The concentration of this colored product, which is proportional to the blood-sugar level, is then measured by absorption spectroscopy.

The basic principles of spectroscopy are simple. In emission spectroscopy the sample under investigation is excited by an electric current or a flame. The emitted light is then examined and identified. In absorption spectroscopy, the sample is placed in the path of a beam of white light. Examination of the transmitted light reveals the missing wavelengths which identify the components in the substance. Both the absorption and the emission spectra can also provide information about the concentration of the various components in the substance. In the case of emission, the intensity of the emitted light in the spectrum is proportional to the number of atoms or molecules of the given species. In absorption spectroscopy, the amount of absorption can be related to the concentration. The instrument used to analyze the spectra is called a *spectrometer*. This device records the intensity of light as a function of wavelength.

A spectrometer, in its simplest form, consists of a focusing system, a prism, and a detector for light (see Fig. 16.5). The focusing system forms a parallel beam of light that passes through the sample under study. The light then passes through the prism. The prism, which can be rotated, breaks up the beam into

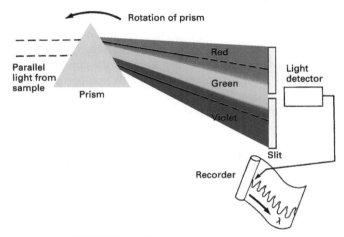

**FIGURE 16.5   The measurement of spectra.**

its component wavelengths. At this point, the fanned-out spectrum can be photographed and identified. Usually, however, the spectrum is detected a small section at a time. This is accomplished with the narrow exit slit, which intercepts only a portion of the spectrum. As the prism is rotated, the whole spectrum is swept sequentially past the slit. The position of the prism is calibrated to correspond with the wavelength impinging on the slit. The light that passes through the slit is detected by a photodetector which produces an electrical signal proportional to the light intensity. The intensity of the signal as a function of wavelength can be displayed on a chart recorder.

Spectrometers used in routine clinical work are automated and can be operated by relatively unskilled personnel. The identification and interpretation of the spectra produced by less well-known molecules, however, require considerable training and skill. In addition to identifying the molecule, such spectra also yield information about the molecular structure. The use of spectrometers is further explored in Exercise 16.1.

## 16.3  Quantum mechanics

Although the Bohr model explained many observations, from the very beginning, the theory appeared contrived. Certainly the concept of stable allowed orbits with a specific number of electrons seemed arbitrary. The model, however, was a daring step in a new direction that eventually led to the development of quantum mechanics.

In the quantum mechanical description of the atom, it is not possible to assign exact orbits or trajectories to electrons. Electrons possess wavelike properties and behave as clouds of specific shape around the nucleus. The artificial postulates in Bohr's theory are a natural consequence of the quantum mechanical approach to the atom. Furthermore, quantum mechanics explains many phenomena outside the scope of the Bohr model. The shape of simple molecules, for example, can be shown to be the direct consequence of the interaction between the electron configurations in the component atoms.

The concept that particles may exhibit wavelike properties was introduced in 1924 by Louis de Broglie. This suggestion grew out of an analogy with light which was then known to have both wave- and particle-like properties. De Broglie suggested by analogy that particles may exhibit wavelike properties. He showed that the wavelength λ of the matter waves would be

$$\lambda = \frac{h}{mv} \tag{16.2}$$

where *m* and *v* are the mass and velocity of the particle, respectively, and *h* is the Planck's constant.

In 1925, De Broglie's hypothesis was confirmed by experiments that showed that electrons passing through crystals form wavelike diffraction patterns with a configuration corresponding to a wavelength given by Eq. 16.2.

## 16.4 Electron microscope

In Chapter 15, we pointed out that the size of the smallest object observable by a microscope is about half the wavelength of the illuminating radiation. In light microscopes, this limits the resolution to about 200 nm $\left(2000 \, \text{Å}\right)$. Because of the wave properties of electrons, it is possible to construct microscopes with a resolution nearly 1000 times smaller than this value.

It is relatively easy to accelerate electrons in an evacuated chamber to high velocities so that their wavelength is much less than $10^{-10}$ m $\left(1 \, \text{Å}\right)$. Furthermore, the direction of motion of the electrons can be altered by electric and magnetic fields. Thus, suitably shaped fields can act as lenses for the electrons. The short wavelength of electrons coupled with the possibility of focusing them has led to the development of electron microscopes that can observe objects 1000 times smaller than are visible with light microscopes. The basic construction of an electron microscope is shown in Fig. 16.6. The

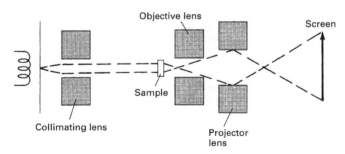

**FIGURE 16.6  The electron microscope.**

similarities between the electron and the light microscope are evident: Both have the same basic configuration of two lenses which produce two-stage magnification. Electrons are emitted from a heated filament and are then accelerated and collimated into a beam. The beam passes through the thin sample under examination, which diffracts the electrons in much the same way as light is diffracted in an optical microscope. But because of their short wavelength, the electrons are influenced by much smaller structures within the sample. The transmitted electrons are focused into a real image by the objective lens. This image is then further magnified by the projector lens, which projects the final image onto film or a fluorescent screen. At present, (2023) the best resolution of electron microscopes is about $5 \times 10^{-10}$ m (0.5 Å), limited by the current quality of electron optics.

Because electrons are scattered by air, the microscope must be contained in an evacuated chamber. Furthermore, the samples under examination must be dry and thin. These conditions, of course, present some limitations in the study of biological materials. The samples have to be specially prepared for electron microscopic examination. They must be dry, thin, and in some cases coated. Nevertheless, electron microscopes have yielded beautiful pictures showing details in cell structure, biological processes, and recently even large molecules such as DNA in the process of replication (see Fig. 16.7).

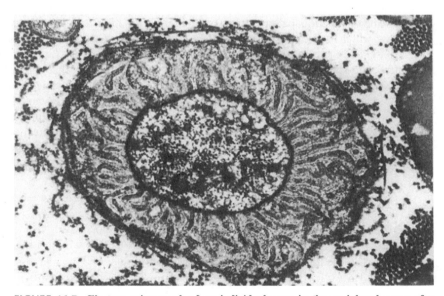

**FIGURE 16.7 Electron micrograph of an individual axon in the peripheral nerve of a mouse.** The cross-section of the axon at the level of the node of Ranvier is about 2.5 μm in width. Surrounding the axon is a differentiated region of the myelin sheath. *(Photograph courtesy of Professor Dan Kirschner, Biology Department, Boston College, and Dr. Bela Kosaras, Primate Center, Southborough, MA.)*

## 16.5 X-rays

In 1895, Wilhelm Conrad Roentgen announced his discovery of X-rays. He had found that when high-energy electrons hit a material such as glass, the material emitted radiation that penetrated objects which are opaque to light. He called this radiation X-rays. It was shown subsequently that X-rays are short-wavelength electromagnetic radiation emitted by highly excited atoms. Roentgen showed that X-rays could expose film and produce images of objects in opaque containers. Such pictures are possible if the container transmits X-rays more readily than the object inside. A film exposed by the X-rays shows the shadow cast by the object.

Within 3 weeks of Roentgen's announcement, two French physicians, Oudin and Barthélemy, obtained X-rays of bones in a hand. Since then, the use of X-rays has become one of the most important diagnostic tools in medicine. With current techniques, it is even possible to view internal body organs that are quite transparent to X-rays. This is done by injecting into the organ a fluid opaque to X-rays. The walls of the organ then show up clearly by contrast.

X-rays have also provided valuable information about the structure of biologically important molecules. The technique used here is called *crystallography*. The wavelength of X-rays is on the order of $10^{-10}$ m, about the same as the distance between atoms in a molecule or crystal. Therefore, if a beam of X-rays is passed through a crystal, the transmitted rays produce a diffraction pattern that contains information about the structure and composition of the crystal. The diffraction pattern consists of regions of high and low X-ray intensity, which when photographed show spots of varying brightness (Fig. 16.8).

Diffraction studies are most successfully done with molecules that can be formed into a regular periodic crystalline array. Many biological molecules can in fact be crystallized under the proper conditions. It should be noted, however, that the diffraction pattern is not a unique, unambiguous picture of the molecules in the crystal. The pattern is a mapping of the collective effect of the arrayed molecules on the X-rays that pass through the crystal. The

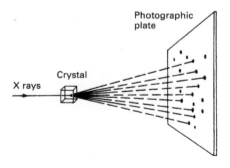

FIGURE 16.8 **Arrangement for detecting diffraction of X-rays by a crystal.**

structure of the individual molecule must be deduced from the indirect evidence provided by the diffraction pattern.

If the crystal has a simple structure—such as sodium chloride—the X-ray diffraction pattern is also simple and relatively easy to interpret. Complicated crystals, however, such as those synthesized from organic molecules, produce very complex diffraction patterns. But, even in this case, it is possible to obtain some information about the structure of the molecules forming the crystal (for details, see Ackerman, 1962). To resolve the three-dimensional features of the molecules, diffraction patterns must be formed from thousands of different angles. The patterns are then analyzed with the aid of a computer. These types of studies provided critical information for the determination of the structure of penicillin, vitamin $B_{12}$, DNA, and many other biologically important molecules.

## 16.6 X-ray computerized tomography

The usual X-ray picture does not provide depth information. The image represents the total attenuation as the X-ray beam passes through the object in its path. For example, a conventional X-ray of the lung may reveal the existence of a tumor, but it will not show how deep in the lung the tumor is located. Several *tomographic techniques* (CT scans) have been developed to produce slice images within the body which provide depth information. (Tomography is from the Greek word *tomos*, meaning section.) Presently the most commonly used of these is X-ray computerized tomography (CT scan), developed in the 1960s. The basic principle of the technique in its simplest

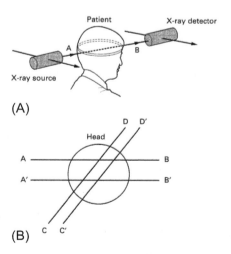

**FIGURE 16.9** (A) Basic principle of computed axial tomography. (B) Rotation of the source—detector combination provides information about the X-ray transmission properties of each point within the plane of the object to be studied.

form is illustrated in Fig. 16.9a and 16.9b. A thin beam of X-rays passes through the plane we want to visualize and is detected by a diametrically opposing detector. For a given angle with respect to the object (in this case the head), the X-ray source—detector combination is moved laterally, scanning the region of interest as shown by the arrow in Fig. 16.9a. At each position, the detected signal carries integrated information about X-ray transmission properties of the full path, in this case $A - B$. The angle is then changed by a small amount (about $1°$) and the process is repeated full circle around the object. As indicated in Fig. 16.9b, by rotating the source—detector combination, information can be obtained about the intersection points of the X-ray beams.

In Fig. 16.9b, we show schematically the scanning beam at two angles with two lateral positions at each angle. While at each position, the detected signal carries integrated information about the full path, two paths that intersect contain common information about the one point of intersection. In the figure, four such points are shown at the intersection of the beams $A - B, A' - B', C - D,$ and $C' - D'$. The multiple images obtained by translation and rotation contain information about the X-ray transmission properties of each point within the plane of the object to be studied. These signals are stored, and by a rather complex computer analysis, a point-by-point image is constructed of the thin slice scanned within the body.

The visualized slices within the body obtained in this way are typically about 2 mm thick. In the more recent versions of the instrument, a fan rather than a beam of X-rays scans the object, and an array of multiple detectors are used to record the signal. Data acquisition is sped up in this way, yielding an image in a few seconds.

## 16.7 Lasers

As was pointed out in Section 16.1, when light at the frequency corresponding to the transition between two energy levels of atoms (or molecules) is passed through a collection of these atoms, photons are absorbed from the light beam by atoms in the lower energy level, raising them to the higher (excited) level. Atoms in an excited level can return to the lower state by emitting a photon at the corresponding resonance frequency (see Eq. 16.1). This type of emission is called *spontaneous emission*. However, atoms in an excited state can emit photons also in another way.

In 1916, Albert Einstein analyzed the interaction of electromagnetic radiation with matter using quantum mechanics and equilibrium considerations. His results showed that while light interacting with atoms in a lower energy state is absorbed, there is a parallel interaction of light with atoms in the excited energy state. The light at the resonance frequency interacts with the excited atoms by stimulating them to make a transition back into the lower energy state. In the process, each stimulated atom emits a photon at the resonance frequency and in phase with the stimulating light. This type of light emission is called *stimulated emission*.

In a collection of atoms or molecules under equilibrium conditions, more atoms are in a lower energy state than in a higher one. When a beam of light at resonance frequency passes through a collection of atoms in equilibrium, more photons are taken out of the beam by absorption than are added to it by stimulated emission and the light beam is attenuated. However, through a variety of techniques, it is possible to reverse the normal situation and cause more atoms to occupy a higher than a lower energy state. A collection of atoms, with more atoms occupying the higher state, is said to have an *inverted population* distribution. When light at resonance frequency passes through atoms with inverted population distribution, more photons are added to the beam by stimulated emission than are taken out of the beam by absorption. As a result, the intensity of the light beam increases. In other words, the light is amplified. A medium with an inverted population can be made into a special type of light source called a *laser* (*l*ight *a*mplification by *s*timulated *e*mission of *r*adiation) (see Exercises 16.3 and 16.4).

Light emitted by a laser has some unique properties. The light across the laser beam is *coherent*. That is, the phase of the wave at all points across the laser light beam is correlated in time and space. As a result, the emitted light can be formed into a highly parallel beam that can be subsequently focused into a very small area, typically on the order of the wavelength of light. In this way a large amount of energy can be delivered into a small region with high degree of positional precision. Further, the light emitted by a laser is monochromatic (single color), with the wavelength determined by the amplifying medium.

The first laser was built in 1960. Since then, many different types of laser have been developed, operating over a wide range of energies and wavelengths covering the full spectrum from infrared to ultraviolet. Some lasers produce short-duration, highly intense light pulses, while others operate in a continuous mode. Lasers are now widely used in science, technology, and increasingly also in medicine. Fig. 16.10 shows an argon-ion laser that emits green or blue light (depending on settings) and is one of the lasers frequently used in medical applications.

FIGURE 16.10 **Argon-ion laser.** *(From http://www.nationallaser.com.)*

## 16.7.1 Lasers surgery

It was evident soon after the development of the first laser that the device would be very useful as a surgical tool. Intense laser light focused onto a small area could burn off and vaporize selected tissue without damage to neighboring areas. Bleeding and pain during such a procedure would be minimized because blood vessels are cauterized and nerve endings are sealed. Infections would likewise be reduced because the cutting tool is not in physical contact with the tissue.

Before lasers could be successfully used in surgical procedures, a wide range of studies had to be conducted to understand the effect of intense light on various types of tissue. Further, technology had to be developed for precise control of light intensity and duration and for accurate positioning of the focal point. While the surgical use of lasers is growing in many areas of medicine and dentistry, the positional accuracy of laser tissue removal is particularly important in neurosurgery and eye surgery, where a fraction of a millimeter offset can make the difference between success and failure.

Ophthalmologists were among the first to use lasers for a wide range of procedures. The repair of retinal detachments and retinal tears is one such application. As a result of trauma or disease, the retina may detach from the back of the eye or may develop tears. Left untreated, this condition leads to a loss of vision. Laser procedures have been very successful in arresting retinal degeneration and restoring normal vision. Laser light is focused through the iris onto the boundary of the detached or torn region of the retina. The tissue is burned and the subsequent scarring "welds" the retina to the underlying tissue.

In another ophthalmological application, lasers are used to treat diabetic retinopathy. Diabetes often causes disorders in blood circulation including leaks in the retinal blood vessels. Such a condition can cause serious damage to the retina and the optic nerve. Laser light focused on the damaged blood vessel seals the leak and halts further retinal deterioration. Unfortunately, the course of the disease is not halted and new leaks develop that require repeated treatments. Fig. 16.11 shows a typical eye surgery setup.

A relatively recent but now widely used application of lasers in ophthalmology is the LASIK technique (Laser-Assisted In Situ Keratomileusis). This is a laser surgical procedure that reshapes the cornea with the aim of correcting focusing problems associated with myopia, hyperopia, and astigmatism. In this procedure, the computer that controls the laser is first programmed for the amount and location of the corneal tissue to be removed. Then using a cutting instrument called a microkeratome, a flap is cut in the front part of the cornea and the flap is folded back. The midpart of the cornea is reshaped by the computer-controlled laser pulses that deliver the correct amount of energy to evaporate the corneal tissue at the set locations. As a result of this procedure, the need for eyeglasses is often eliminated.

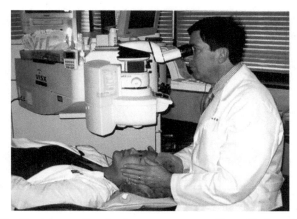

**FIGURE 16.11** **Physician performing laser eye surgery.** *(From Chabner, Davi-Ellen, The language of medicine. Elsevier, 2011.)*

## 16.7.2 Lasers in medical imaging

As was discussed in connection with confocal microscopy (Section 15.15.3), light in the red and near-infrared region of the spectrum penetrates effectively into biological tissue. However, with conventional optical techniques (see Appendix C), the light emerging from the tissue cannot be formed into a useful image of a tissue layer within the bulk because most of the light that is reflected (or transmitted) by the tissue is multiply scattered by cells in front of and behind the layer of interest. To form a useful image of a specific tissue layer, the imaging system must pick out the small amount of light originating from the layer of interest and eliminate the effect of the light scattered from other parts of the tissue within which the signal of interest is buried.

The confocal microscope is one system designed to perform this task (see Section 15.15.3). In the early 1990s, another technique was developed, referred to as *optical coherence tomography (OCT)*. The typical resolution of an OCT instrument is about 10 μm compared to the significantly higher 1 $\mu$m resolution of a confocal microscope. However, the OCT instrument can form images of cells up to 2 or 3 mm inside the tissue, whereas the depth penetration of a confocal microscope is typically less than 1 mm.

A simplified schematic diagram of an OCT apparatus is shown in Fig. 16.12. A near-infrared laser beam (wavelength ～800 nm) is split into two beams by a partially silvered mirror. One beam shines into the tissue to be examined. The other beam is reflected from another mirror and provides a reference for the detection of the signal reflected from the tissue. The reference beam and the light reflected from the tissue are merged at the detector. The light that has been multiply scattered within the tissue has lost its phase correlation with the illuminating laser beam and therefore also with the

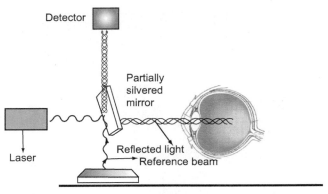

**FIGURE 16.12** **Schematic diagram of optical coherence tomography (OCT) instrument, providing details of retinal structure.**

reference laser beam. This uncorrelated scattered light does not form an interference pattern with reference beam. The small fraction of the reflected beam that is singly scattered by the cells in the tissue retains a phase relationship with the reference beam and forms an interference pattern at the detector. The interference pattern contains information about the position and reflectivity of the source cell. As the reference beam mirror is scanned in a three-dimensional raster, a changing interference pattern is formed that contains information about the light reflected layer by layer from the cells within the sample. The interference pattern is processed by a computer, yielding a conventional, easily interpretable image. The OCT apparatus has so far been most useful in providing microscopic details of the retinal structures, as is illustrated in the figure. Other applications particularly in dermatology are being developed.

### 16.7.3 Lasers in medical diagnostics

An example of a laser-based noninvasive diagnostic instrument is a recently developed infrared optical scanning device designed to detect intracranial bleeding often caused by concussion or hemorrhagic stroke. Failure to detect such bleeding promptly, within an hour or so, can cause irreversible brain damage or death. A computed axial tomography (CAT) scan is the usual method of diagnosing brain bleeding (hematoma). However, CAT scan instrumentation is expensive and is found only at major medical facilities. The newly developed device costs about 1% of a CT scanner, it is portable, about the size of a book, and can perform the diagnosis in just 2 minutes. Further, the patient is not exposed to the high doses of radiation required to produce a CT image.

The operation of the instrument is based on the difference in the optical properties of blood and brain tissue. Absorption by blood of light in the near-infrared region of the spectrum is much greater than absorption of the light by brain tissue. In one version of the device, light from an 808-nm diode laser illuminates a part of the cranium and the reflected as well as the transmitted light is detected. Symmetric regions of the cranium, such as the left and right sides, are illuminated and the measured light intensities are compared. Because blood absorbs more of the light than brain tissue, both the reflected and transmitted light is reduced from the part of the cranium that is affected by bleeding. A comparison of measured light intensities from different regions of the cranium identifies the presence and location of the hematoma.

## 16.8 Atomic force microscopy

Over the past 30 years, several *scanning probe microscope* imaging techniques have been developed and perfected to form high-resolution images of surfaces. In these instruments probes positioned close to the surface scan the object in a raster fashion, producing images that allow visualization of surface atoms and molecules. Of the several instruments in this category, the *atomic force microscope (AFM)* is at this point the most useful one for biological applications. The resolution of the AFM instrument is comparable to that of the electron microscope with a very important advantage. The AFM samples can be studied in air or in biologically native aqueous environments. By contrast, samples studied by electron microscopy must be contained in an evacuated chamber.

A schematic diagram of the AFM is shown in Fig. 16.13. A sharp tip probe made of silicon nitride with a diameter on the order of a few nanometers is

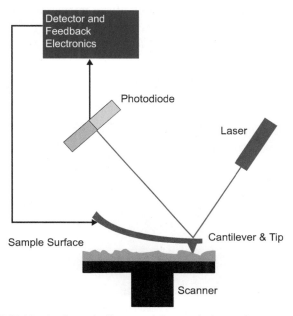

**FIGURE 16.13   A schematic diagram of the atomic force microscope (AFM).**

attached to a cantilever spring. The sample to be examined is placed on the scanner platform with positional control in three dimensions (x, y define the horizontal plane, z the vertical position). The probe tip is positioned close (between 1 and 10 nm) to the sample surface. The inherent charge distributions on the sample and the tip set up an attractive coulomb force that pulls the probe tip toward the sample surface. The closer the surface is to the tip, the larger the bending force exerted on the cantilever. As the sample is moved in the scanning pattern, the distance between the tip and the sample surface changes in accordance with the molecular configuration of the surface. The corresponding change in the force alters the position of the cantilever that is monitored by the laser beam reflected from the cantilever surface. In one frequently used arrangement, a feedback mechanism controlled by the reflected laser beam moves the scanner platform up or down so as to keep the distance between the tip and the sample constant. While the scanner moves the platform in the predetermined raster plane, the feedback signal is at all times proportional to the height variations of the sample surface under the scanning tip. The feedback signal is recorded as a function of the scanning tip position. The feedback signal together with the information about the scanning tip position is used to produce an image of the sample surface. An example of such an AMF-produced image of bacteria (*Bacillus cereus*) is shown in Fig. 16.14.

Scanning probe microscopes such as the AFM can also be used to apply a force to individual surface atoms and molecules and, thereby, in a controlled way, to alter their position. In this way a pattern can be constructed atom by atom. Some quantitative aspects of atomic force microscopy are presented in Exercise 16.5.

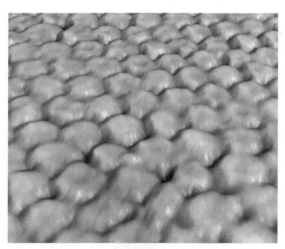

**FIGURE 16.14**  **AFM-produced image of bacteria *(Bacillus cereus)*.** *(Source: Nanomedicine: Nanotechnology, biology, and medicine. Elsevier; 2007.)*

## 16.9 Cryoelectron microscopy

A highly promising, relatively new technique for the study of biological molecules, *cryoelectron microscopy* (cryo-EM), was perfected in the 1990s and is at this point widely used in many research laboratories throughout the world. The three principal developers of the innovations that made this new technique possible, Jacque Dubochet (University of Lausanne), Joachim Frank (Columbia University), and Richard Henderson (Cambridge University), were awarded, for this work, the 2017 Nobel Prize in Chemistry.

Until recently (around the year 2000), information about the structure of biomolecules such as proteins, DNA, and smaller organisms, among them viruses and bacteria, was obtained mainly using two techniques: electron microscopy and X-ray diffraction (see Sections 16.4 and 16.5). In some applications, nuclear magnetic resonance techniques (see Section 17.2) have also been used to obtain information about the structure and some functional aspects of smaller biological molecules. These techniques have been highly successful in providing information about the basic building blocks of life and created the fields of modern biochemistry, chemical biology, and genetics.

However, each of these older techniques, successful as they have been, has drawbacks that limit its utility. One of the key limitations of X-ray crystallography is carried in its name. The molecules to be studied have to be formed into a crystal. For large biological molecules, this is in all cases a challenging task. Formation of crystals with some of the large flexible biomolecules has turned out to be impossible. Further, crystalized biomolecules are dry, whereas the natural environment of biomolecules is an aqueous solution. Another problematic issue associated with crystallization is the close proximity of molecules to one another that may produce structural distortions that are difficult to take into account. Thus, X-ray crystallography in many cases does not provide data corresponding to the natural state of the biomolecules.

Conventional electron microscopy also presents severe limitations. The high intensity of the electron beam required to obtain good image resolution breaks bonds and damages the biomolecule being studied. Further, electron microscopy has to be done in a low-pressure chamber to avoid scattering of the electron beam by the background molecules. In conventional microscopy, this is accomplished by placing the biomolecules in an evacuated environment. This again necessitates working with dry samples that do not faithfully represent the natural state of the biomolecules being studied.

Both conventional older techniques yield only static structures of the biomolecules. Cryo-EM provides the possibility of capturing molecules in various stages of their functions, thus providing information about the structure as well as function of the biomolecules.

Over several decades of work, the three Nobel Prize winners and their colleagues have perfected and incorporated new methodologies into electron microscopy that substantially removed the limitations of older electron microscopy and X-ray crystallography discussed earlier.

Joachim Frank and his colleagues realized that the damaging effect of the electron beam could be greatly reduced, and even eliminated, if the power of the electron beam were distributed over a large number of identical molecules. The image produced by each separate molecule would of course be too weak and noisy to provide useful information. However, if the orientation and position of every molecule in the assembly were known, then the weak images of each of the molecules could be appropriately combined to produce a high-resolution single image. Frank and his research group, utilizing newly evolving computer technologies, developed algorithms that could achieve this task and eventually yield high, 2 or 3 Å, resolution, three-dimensional images of biological molecules.

The problem of having to work with dry biomolecules outside of their natural environment was solved by the research group of Jacques Dubochet. As was mentioned, electron microscopy requires that the electron beam chamber be at a low pressure. The required low pressures can be attained with aqueous solutions at a low temperature, typically about 100 K. However, under normal conditions, aqueous solutions at such low temperatures freeze into regular crystalline structures, producing electron diffraction patterns that make it impossible to obtain electron microscopy images of the biomolecules. After considerable experimentation, Dubochet and his research group developed a technique for rapid cooling of biological molecules in thin layers of aqueous solutions in a way that avoids crystallization and maintains the liquid structure of the solution in a form called vitrified water. In this vitrified (glass-like) state, water does not produce a diffraction pattern that interferes with the electron microscopic images of the biological molecules.

Using the best electron microscopes available and incorporating ongoing advances in cryo-EM, in 1990, Henderson and his group perfected cryo-EM techniques to a point where they were able to demonstrate that cryo-EM can produce images with as high a resolution as those obtained with X-ray diffraction techniques, with the additional advantages of studying the biomolecules in their near-natural environment, and in some cases capturing the details of their functional aspects. This work paved the way for cryo-EM to replace X-ray diffraction techniques as the prime tool in biological studies. The wide applicability of the technique is expected to yield new insights into the functioning of drugs and lead to the development of new pharmaceutical interventions for a variety of diseases.

## 16.10 Laws of physics and life

We have discussed in this book many phenomena in the life sciences that are well explained by the theories of physics. Now we come to the most fundamental question: Can physics explain life itself? In other words, if we put together the necessary combination of atoms, at each step following the known laws of physics, do we inevitably end up with a living organism, or must we invoke some new principles outside the realm of current physics in order to explain the occurrence of life? This is a very old question that still cannot be answered with certainty. But it can be clarified.

Quantum mechanics, which is the fundamental theory of modern atomic physics, has been very successful in describing the properties of atoms and the interaction of atoms with each other. Starting with a single proton and one electron, the theory shows that their interaction leads to the hydrogen atom with its unique configuration and properties. The quantum mechanical calculations for larger atoms are more complicated. In fact, so far, a complete calculation has been performed only for the hydrogen atom. The properties of heavier atoms have to be computed using various approximation and computer techniques. Yet, there is no doubt that quantum mechanics does describe all the properties of atoms from the lightest to the heaviest. The experimental evidence gathered over the past 100 years fully confirms this view.

The interactions between atoms, which result in the formation of molecules, are likewise in the domain of quantum mechanics. Here again, exact solutions of the quantum mechanical equations have been obtained only for the simplest molecule, $H_2$. Still, it is evident that all the rules for both organic and inorganic chemistry follow from the principles of quantum mechanics. Even though our present numerical techniques cannot cope with the enormous calculations required to predict the exact configuration of a complex molecule, the concepts developed in physics and chemistry are applicable. The strengths of the interatomic bonds and the orientations of the atoms within the molecules are all in accord with the theory. This is true even for the largest organic molecules such as proteins and DNA.

Past this point, however, we encounter a new level of organization: the cell. The organic molecules, which are in themselves highly complex, combine to form cells, which in turn are combined to form larger living organisms, which possess all the amazing properties of life. These organisms take nourishment from the environment, grow, reproduce, and at some level begin to govern their own actions. Here it is no longer obvious that the theories governing the interaction of atoms lead directly to the functions that characterize life. We are now in the realm of speculations.

The phenomena associated with life show such remarkable organization and planning that we may be tempted to suggest that perhaps some new undiscovered law governs the behavior of organic molecules that come together to form life. Yet there is no evidence for any special laws operating within living systems. So far on all levels of examination, the observed phenomena associated with life obey the well-known laws of physics. This does not mean that the existence of life follows from the basic principles of physics, but it may. In fact, the large organic molecules inside cells are sufficiently complex to contain within their structures the information necessary to guide in a predetermined way the activities associated with life. Some of these codes contained in the specific groupings of atoms within the molecules have now been unraveled. Due to these specific structures, a given molecule always participates in a well-defined activity within the cell. It is very likely that all the complex functions of cells and of cell aggregates are simply the collective result of the enormously large number of predetermined but basically well-understood chemical reactions.

This still leaves the most important question unanswered: What are the forces and the principles that initially cause the atoms to assemble into coded molecules, which then ultimately lead to life? The answer here is probably again within the scope of our existing theories of matter.

In 1951, S. L. Miller simulated in his laboratory the type of conditions that may have existed perhaps 3.5 billion years ago in the atmosphere of the primordial earth. He circulated a mixture of water, methane, ammonia, and hydrogen through an electric discharge. The discharge simulated the energy sources that were then available from the sun, lightning, and radioactivity. After about 1 week, Miller found that the chemical activities in the mixture produced organic molecules including some of the simple amino acids, which are the building blocks of proteins. Since then, hundreds of other organic molecules have been synthesized under similar conditions. Many of them resemble the components of the important large molecules found in cells. It is thus plausible that in the primordial oceans, rich in organic molecules produced by the prevailing chemical reactions, life began spontaneously. A number of smaller organic molecules combined accidentally to form a large self-replicating molecule such as DNA. These in turn combined into organized aggregates and finally into living cells.

Although the probability of the spontaneous occurrence of such events is small, the time span of evolution is probably long enough to make this scenario plausible. If that is indeed the case, the current laws of physics can explain all of life. At the present state of knowledge about life processes, the completeness of the descriptions provided by physics cannot be proved. The principles of physics have certainly explained many phenomena, but mysteries remain. At present, however, there seems to be no need to invoke any new laws.

## Exercises

**16.1.** Explain the operation of a spectrometer and describe two possible uses for this device.

**16.2.** Describe the process of X-ray computerized tomography. What information does this process provide that ordinary X-ray images do not?

**16.3.** Describe the operation of a helium−neon laser. Include a description of the method for obtaining the inverted population distribution.

**16.4.** Two lasers commonly used in laser surgery are the $CO_2$ laser and the argon-ion laser. Describe the method for obtaining the inverted population distribution in these two lasers.

**16.5.** The probe cantilever assembly in Fig. 16.13 can be considered a spring with a spring constant K. Typically $K = 1$ N/m. Assume that the AFM tip is 1 nm above the surface of a sample and that the attraction between the surface and the tip displaces the tip by 0.5 nm.

(a) Compute the force between the tip and the surface. (See Chapter 5.)

(b) What is the magnitude of the local charge on the sampling tip (and the equal and opposite charge on the facing sample) that produces this force? (See Appendix B.)

**16.6.** Discuss some of the most notable attributes of living systems that distinguish them from inanimate ones.

# Chapter 17

# Nuclear physics

## 17.1 The nucleus

Although all the atoms of a given element have the same number of protons in their nucleus, the number of neutrons may vary. Atoms with the same number of protons but different number of neutrons are called *isotopes*. All the nuclei of the oxygen atom, for example, contain 8 protons but the number of neutrons in the nucleus may be 8, 9, or 10. These are the isotopes of oxygen. They are designated as $^{16}_{8}O$, $^{17}_{8}O$, and $^{18}_{8}O$. This is a general type of nuclear symbolism in which the subscript to the chemical symbol of the element is the number of protons in the nucleus and the superscript is the sum of the number of protons and neutrons. The number of neutrons often determines the stability of the nucleus.

The nuclei of most naturally occurring atoms are stable. They do not change when left alone. There are, however, many unstable nuclei that undergo transformations accompanied by the emission of energetic radiation. It has been found that the emanations from these radioactive nuclei fall into three categories: (1) alpha ($\alpha$) particles, which are high-speed helium nuclei consisting of two protons and two neutrons; (2) beta ($\beta$) particles, which are very high-speed electrons; and (3) gamma ($\gamma$) rays, which are highly energetic photons.

The radioactive nucleus of a given element does not emit all three radiations simultaneously. Some nuclei emit alpha particles, others emit beta particles, and the emission of gamma rays may accompany either event.

Radioactivity is associated with the transmutation of the nucleus from one element to another. Thus, for example, when radium emits an alpha particle, the nucleus is transformed into the element radon. The details of the process are discussed in most physics texts (see Burns and MacDonald, 1970).

The decay or transmutation of a given radioactive nucleus is a random event. Some nuclei decay sooner; others decay later. If, however, we deal with a large number of radioactive nuclei, it is possible, by using the laws of probability, to predict accurately the decay rate for the aggregate. This decay rate is characterized by the half-life, which is the time interval for half the original nuclei to undergo transmutation.

There is a great variation in the half-life of radioactive elements. Some decay very quickly and have a half-life of only a few microseconds or less.

Physics in Biology and Medicine. https://doi.org/10.1016/B978-0-443-21558-2.00017-1

Others decay slowly with a half-life of many thousands of years. Only the very long-lived radioactive elements occur naturally in the Earth's crust. One of these, for example, is the uranium isotope $^{238}_{92}U$, which has a half-life of $4.51 \times 10^9$ years. The short-lived radioactive isotopes can be produced in accelerators by bombarding certain stable elements with high-energy particles. Naturally occurring phosphorus, for example, has 15 protons and 16 neutrons in its nucleus $\left(^{31}_{15}P\right)$. The radioactive phosphorus isotope $^{32}_{15}P$ with 17 neutrons can be produced by bombarding sulfur with neutrons. The reaction is

$$^{32}_{16}S + \text{neutron} \ \rightarrow \ ^{32}_{15}P + \text{proton}$$

This radioactive phosphorus has a half-life of 14.3 days. Radioactive isotopes of other elements can be produced in a similar way. Many of these isotopes have been very useful in biological and clinical work.

## 17.2 Magnetic resonance imaging

Images of the shapes of internal organs obtained with computerized X-ray tomography are excellent. However, X-rays do not provide information about the internal structure of tissue. CT scans may, therefore, miss changes in tissue structure and pathological alteration inside internal organs. *Magnetic resonance imaging (MRI)*, introduced in the early 1980s, is the most recent addition to medical imaging techniques. This technique utilizes the magnetic properties of the nucleus to provide images of internal body organs with detailed information about soft-tissue structure.

The imaging techniques we have discussed so far (X-ray and ultrasound) are in principle relatively simple. They utilize reflected or transmitted energy to visualize internal structures. MRI is more complex. It utilizes the principles of *nuclear magnetic resonance (NMR)* developed in the 1940s. A detailed description of MRI is beyond the scope of this text, but the principles are relatively simple to explain. A discussion of MRI begins with an introduction to the principles of NMR.

### 17.2.1 Nuclear magnetic resonance

Protons and neutrons which are the constituents of atomic nuclei possess the quantum mechanical property of spin, which has magnitude and direction. We can imagine these particles as if they were small spinning tops. As a result of spin, the nuclear particles act as small bar magnets. Inside the nucleus, these small magnets associated with the nucleons (protons and neutrons) line up so as to cancel each other's magnetic fields. However, if the number of nucleons is odd, the cancellation is not complete, and the nucleus possesses a net magnetic moment. Therefore, nuclei with an odd number of nucleons behave as tiny magnets. Hydrogen, which has a nucleus consisting of a single proton, does, of course, have a nuclear magnetic moment. The human body is made of mostly water and other hydrogen-containing molecules. Therefore, MRI images of structures within the body can be most effectively produced using the magnetic properties of the hydrogen nucleus. Our discussion will be restricted to the nuclear magnetic properties of hydrogen.

Normally, the little nuclear magnets in bulk material are randomized in space, as is shown in Fig. 17.1a, and the material does not possess a net magnetic moment $(M = 0)$. The nuclear magnets are represented as small arrows. However, the situation is altered in the presence of an external magnetic field. When an external magnetic field is applied to a material possessing nuclear magnetic moments, the tiny nuclear magnets line up either parallel or antiparallel with the magnetic field, as shown in Fig. 17.1b. The direction of the external magnetic field is usually designated as the $z$-axis. As shown in the figure, the $x$-$y$ plane is orthogonal to the $z$-axis. Because the nuclear magnets parallel to the field $(+z)$ have a somewhat lower energy than those that are

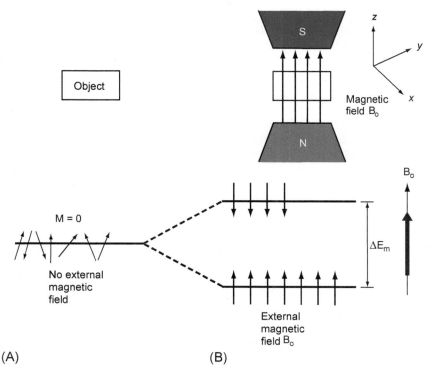

(A)　　　　　　　　　　　　(B)

**FIGURE 17.1** (A) In the absence of an external magnetic field, nuclear spins are randomized. (B) When an external magnetic field is applied to a material possessing nuclear magnetic moments, the tiny nuclear magnets line up either parallel or antiparallel with the magnetic field. The parallel configuration is at a lower energy.

antiparallel $(-z)$, more of the nuclei are in the parallel state than in the antiparallel state. In an external magnetic field, the assembly of parallel/antiparallel nuclear spins as a whole has a net magnetic moment M that behaves as a magnet pointing in the direction of the magnetic field.

The energy spacing $\Delta E_m$ between the parallel and antiparallel alignments is

$$\Delta E_m = \gamma h B / 2\pi \tag{17.1}$$

Here B is the externally applied magnetic field, h is the Planck's constant as defined earlier, and $\gamma$ is called *gyromagnetic ratio,* which is a property of a given nucleus. Typically the strength of magnetic fields used in MRI is about 2 tesla (T). (By comparison, the strength of the magnetic field of the Earth is on the order of $10^{-4}$ T.)

The discrete energy spacing, $\Delta E_m$, between the two states shown in Fig. 17.1b makes this a resonant system. The frequency corresponding to the energy difference between the two states is called the *Larmor frequency* and in accordance with Eq. 16.1 is given by

$$f_L = \Delta E_m / h = \gamma B / 2\pi \tag{17.2}$$

The gyromagnetic ratio $\gamma$ for a proton is $2.68 \times 10^8$ T$^{-1}$ sec$^{-1}$. Magnetic fields used in MRI are typically in the range of 1 to 4 T. The corresponding Larmor frequencies are about 43 to 170 MHz. These frequencies are in the radio frequency (RF) range, which are much lower than X-rays and do not disrupt living tissue.

If by some means the magnetic moment is displaced from the field, as shown in Fig. 17.2, it will *precess* (rotate) around the field as a spinning top precesses in the gravitational field of the Earth. The frequency of precession is the Larmor frequency given by Eq. 17.2. The displacement of the magnetic moment is due to a reversal of alignment for some of the individual nuclear magnetic moments from parallel to antiparallel alignment as shown in Fig. 17.2. A displacement of 90° corresponds to equalizing the population of the spin up and spin down states. To reverse the alignment of antiparallel spins requires energy which must be supplied by an external source.

The energy required to displace the magnetic moment from the direction of the external field is supplied by a short RF driving pulse at the Larmor frequency, which is the natural (resonant) frequency of precession. (This is analogous to setting a pendulum swinging by applying to it a force at the frequency of the pendulum resonance.) The driving pulse is applied by a coil surrounding the sample, as shown in Fig. 17.2. At the end of the spin-flipping driving pulse, the magnetic moment is displaced from the external magnetic field by an angle determined by the strength and duration of the driving pulse.

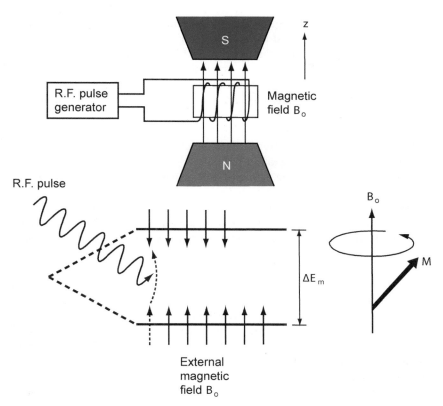

**FIGURE 17.2**   A short radio frequency driving pulse at the Larmor frequency displaces the magnetic moment from the external magnetic field by an angle determined by the strength and duration of the driving pulse.

The displaced magnetic moment produced by the RF driving pulse precesses around the external magnetic field and itself generates a RF signal at the Larmor frequency of rotation. This emitted NMR signal can be detected by a separate coil or by the driving coil itself. The detected NMR signal decreases exponentially with time due to two distinct processes: (1) the return of the nuclear spin orientations to the equilibrium distribution and (2) variations in the local magnetic field.

As was stated earlier, in the presence of an external field more of the nuclei are lined up parallel to the field than antiparallel. The RF pulse flips some of the parallel spins into the antiparallel configuration. As soon as the driving pulse is over, the nuclear spins and the associated magnetic moment begin to return back to the original equilibrium alignment. The equilibration is brought about by the exchange of energy between the nuclear spins and the surrounding molecules. With the return of the magnetic moment to the original alignment with the external magnetic field, the precession angle decreases, as does the associated NMR signal. The decay of the NMR signal is exponential with time constant $T_1$, called the spin lattice relaxation time.

The local magnetic field throughout the object under examination is not perfectly uniform. Variations in the magnetic field are produced by the magnetic properties of molecules adjacent to the nuclear spins. Such variations in the local magnetic field cause the Larmor frequency of the individual nuclear magnetic moments to differ slightly from each other. As a result, the precessions of the nuclei get out of phase with each other, and the total NMR signal decreases. This dephasing is likewise exponential, with a time constant $T_2$, called the spin-spin relaxation time.

The driving pulse and the emitted NMR signal are shown schematically in Fig. 17.3. The NMR signal detected after the driving pulse contains information about the material being studied. For a given initial driving pulse, the magnitude of the emitted NMR signal is a function of the number of hydrogen nuclei in the material. Bone, for example, which contains relatively few water or other hydrogen-containing molecules, produces a relatively low NMR signal. The postpulse radiation emitted by fatty tissue is much higher.

The time constants $T_1$ and $T_2$ characterizing the rate of decay of the emitted NMR signal provide information about the nature of the material within which the precessing nuclei are located. The spinning top provides a useful analogy. A well-designed top in vacuum will spin for a long time. In air, the duration of the spin will be somewhat shorter because collisions with air molecules will dissipate its rotational energy. In water, where the frictional losses are yet greater, the top will spin hardly at all. The decay rate of the spinning top provides information about the nature of the medium surrounding the top. Similarly, the characteristic time constants $T_1$ and $T_2$ provide information about the

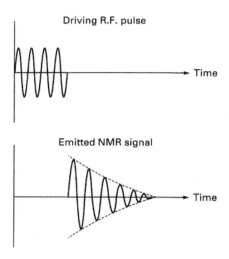

FIGURE 17.3 **The driving pulse and the emitted NMR signal.**

matter surrounding the precessing nuclei (see Dowsett et al., 1998). Using rather complex techniques, the two time constants $T_1$ and $T_2$ can be separately determined. For example, with an external magnetic field of 1 T, for fat $T_1 = 240$ milliseconds and $T_2 = 80$ milliseconds; for heart tissue $T_1 = 570$ milliseconds and $T_2 = 57$ milliseconds (see [109]). Malignant tissue is often characterized by higher values of $T_1$.

The NMR principles described have been used since the 1940s to identify molecules in various physics, chemistry, and biological applications. In this application the detected NMR signal is derived from the entire volume exposed to the magnetic field. The technique as discussed so far cannot provide information about the location of the signal within the volume studied.

## 17.2.2 Imaging with NMR

In order to obtain a three-dimensional image using NMR, we must isolate and identify the location of signals from small sections of the body and then build the image from these individual signals. In CT scans, such tomographic spatial images are obtained by extracting the information from intersection points of narrowly focused X-ray beams. This cannot be done with NMR because the wavelengths of the RF driving signals are long, in the range of meters, which cannot be collimated into the narrow beams required to examine small regions of interest.

In the 1970s, several new techniques were developed to utilize NMR signals for the construction of two-dimensional tomographic images similar to those provided by CT scans. One of the first of these was described by P. C. Lauterbur in 1973. He demonstrated the principle using two tubes of water, $A$ and $B$, as shown in Fig. 17.4. In a uniform magnetic field ($B_0$) the Larmor frequency of the two tubes is the same. Therefore, the postpulse NMR signals from tubes $A$ and $B$ cannot be distinguished. The NMR signals from the two tubes can be made distinguishable by superimposing on the uniform field $B_0$ a magnetic field gradient $B(x)$ as shown in Fig. 17.4b. The total magnetic field now changes with position along the $x$-axis, and the associated Larmor frequencies at the location of tubes $A$ and $B$ are now different.

As is evident, each point (actually small region $\Delta x$) on the $x$-axis is now characterized by its unique Larmor frequency. Therefore, the NMR signal observed after excitation with a pulse of a given frequency can be uniquely associated with a specific region in the $x$-space. A field gradient in one direction yields projection of the object onto that axis. To obtain a tomographic image in the $x$-$y$ plane, a field gradient in both the $x$- and $y$-directions must be introduced (see Fig. 17.5). A magnetic field gradient is also applied in the $z$-direction to select within the body the slice to be examined. A very large number of NMR signals have to be collected and synthesized to construct an MRI image. For this purpose, the intensity as well as the time constants $T_1$ and $T_2$ of the NMR signal are needed. The process is more complex than for a CT scan and requires highly sophisticated computer programs.

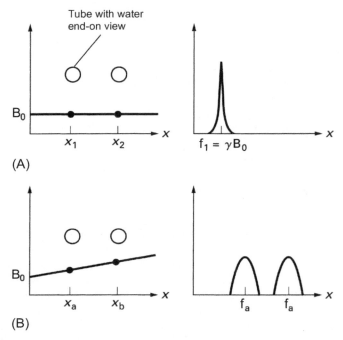

**FIGURE 17.4**    (A) In a uniform magnetic field ($B_0$) the Larmor frequency of two locations in space $A$ and $B$ is the same. (B) When a magnetic field gradient is superimposed on the uniform field, the Larmor frequencies at locations $A$ and $B$ are different.

A sketch of a whole-body MRI apparatus is shown in Fig. 17.5. Most such devices use liquid helium-cooled superconducting magnets to produce the high magnetic fields required for the production of high-resolution images. An MRI image of the brain is shown in Fig. 17.6.

The MRI technique yields detailed visualization of soft tissue structures with a resolution of about 0.5 mm. Such visualizations have been particularly useful in neurology. All parts of the brain structure including arteries as thin as a hair can be clearly seen deep inside the brain. However, conventional MRI does not provide information about the functions performed by the brain. Such a display *in vivo* of neural activity in the brain as it performs various tasks and functions can be obtained with a modified MRI technique called functional magnetic resonance imaging (fMRI).

### 17.2.3 Functional magnetic resonance imaging

Prior to the development of modern imaging techniques, information about specific localized functions of the brain was obtained primarily from studies (usually postmortem) of brain tumors and injuries. For example, in 1861

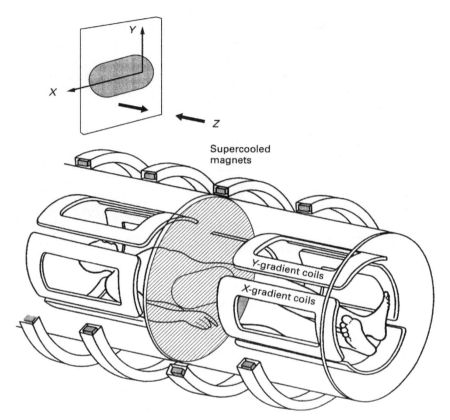

**FIGURE 17.5   Sketch of a whole-body MRI imaging system.**

through a postmortem examination of an aphasic patient, a French physician, Paul Pierre Broca, determined that the patient had a lesion in the left cerebral hemisphere and confirmed earlier studies suggesting that this part of the brain controlled speech production. The development of fMRI has made it possible to observe noninvasively a wide range of neural functions of interest in psychology and clinical medicine.

When a specific region of the brain is activated, the energy requirement of that region rises. Oxygenated blood flow to that part of the brain increases to ensure an adequate supply of oxygen required to meet the increased energy requirements. The fMRI technique makes use of the fact that oxygenated hemoglobin does not have a magnetic moment while deoxygenated hemoglobin does. In the presence of deoxygenated hemoglobin with its magnetic moment, the dephasing of the hydrogen NMR signal is more rapid and the signal intensity is weaker than with the hemoglobin oxygenated. Therefore, regions of greater brain activity, infused with more oxygenated blood, will

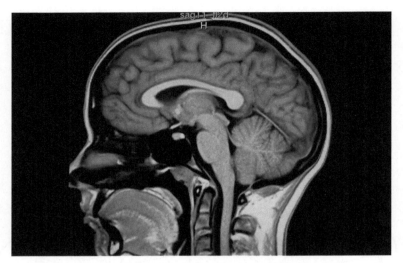

**FIGURE 17.6    MRI image of the brain.** *(Source: SMK4pix/Shutterstock.com.)*

produce more intense $T_2$-weighted NMR signals. In this way regions of increased brain activity can be clearly identified.

fMRI has been applied to identify regions of brain activity in the performance of a wide range of cognitive, motor, and sensory activities. Most of the fMRI applications to date have been research related. For example, in a 2003 study, Eisenberg and colleagues set up a situation where the subjects experienced social exclusion. Their fMRI studies showed that in such a situation the same part of the brain is activated as is during physical pain.

The clinical applications of fMRI are still in their infancy, but there is little doubt that within a few years, fMRI will have a major role in medicine. Research is being conducted to use fMRI as an early diagnostic tool to detect neurological diseases such as Alzheimer's, Parkinson's, and Huntington's diseases. Emerging results indicate that fMRI can provide important information to make neurosurgical procedures such as removal of tumors more accurate. Pain management and more accurate functional testing of psychoactive drugs are some of the other areas where fMRI is likely to become a useful tool.

## 17.3 Radiation therapy

The photons of X-rays and gamma-rays and the particles emitted by radioactive nuclei all have energies far greater than the energies that bind electrons to atoms and molecules. As a result, when such radiation penetrates into biological materials, it can rip off electrons from the biological molecules and produce substantial alterations in their structure. The ionized molecule may break up, or it may chemically combine with another molecule to form an undesirable complex. If the damaged molecule is an important component of a cell, the whole cell may die. Water molecules in the tissue are also broken up by the radiation into reactive fragments (H + OH). These fragments combine with the biological molecules and alter them in a detrimental way. In addition, radiation passing through tissue may simply give up its energy and heat the tissue to a dangerously high temperature. A large dose of radiation may damage so many cells that the whole organism dies. Smaller but still dangerous doses may produce irreversible changes such as mutations, sterility, and cancer.

In controlled doses, however, radiation can be used therapeutically. In the treatment of certain types of cancer, an ampul containing radioactive material such as radium or cobalt 60 is implanted near the cancerous growth. By careful placement of the radioactive material and by controlling the dose, the hope is to destroy the cancer without greatly damaging the healthy tissue. Unfortunately some damage to healthy tissue is unavoidable. As a result, this treatment is often accompanied by the symptoms of radiation sickness (diarrhea, nausea, loss of hair, loss of appetite, and so on). If long-lived isotopes are used in the therapy, the material must be removed after a prescribed period. Short-lived isotopes, such as gold 198 with a half-life of about 3 days, decay quickly enough so that they do not need to be removed after treatment.

Certain elements introduced into the body by injection or by mouth tend to concentrate in specific organs. This phenomenon is used to advantage in radiation therapy. The radioactive isotope phosphorus 32 (half-life, 14.3 days) mentioned earlier accumulates in the bone marrow. Iodine 131 (half-life, 8 days) accumulates in the thyroid and is given for the treatment of hyperthyroidism.

An externally applied beam of gamma rays or X-rays can also be used to destroy cancerous tumors. The advantage here is that the treatment is administered without surgery. The effect of radiation on healthy tissue can be reduced by frequently altering the direction of the beam passing through the body. The tumor is always in the path of the beam, but the dosage received by a given section of healthy tissue is reduced.

## 17.4 Food preservation by radiation

Without some attempt at preservation, all foods decay rather quickly. Within days and often within hours, many foods spoil to a point where they become inedible. The decay is usually caused by microorganisms and enzymes that decompose the organic molecules of the food.

Over the years, a number of techniques have been developed to retard spoilage. Keeping the food in a cold environment reduces the rate of activity for both the enzymes and the microorganisms. Dehydration of food achieves the same goal. Heating the food for a certain period of time destroys many microorganisms and again retards decay. This is the principle of *pasteurization*. These methods of retarding spoilage are all at least 100 years old. There is now a new technique for preserving food by irradiation.

High-energy radiation passing through the food destroys microorganisms that cause decay. Radiation is also effective in destroying small insects that attack stored foods. This is especially important for wheat and other grains, which at present are often fumigated before shipping or storage. Chemical fumigation kills the insects but not their eggs. When the eggs hatch, the new insects may destroy a considerable fraction of the grain. Radiation kills both the insects and the eggs.

Gamma rays are used most frequently in food preservation. They have a great penetrating power and are produced by relatively inexpensive isotopes such as cobalt 60 and cesium 137. High-speed electrons produced by accelerators have also been used to sterilize food. Electrons do not have the penetrating power of gamma rays, but they can be aimed better and can be turned off when not in use.

In the United States and in many other countries, there are now a number of facilities for irradiating food. In the usual arrangement, the food on a conveyor passes by the radioactive source, where it receives a controlled dose of radiation. The source must be carefully shielded to protect the operator. This problem is relatively simple to solve, and at present, the technical problems seem to be well in hand. One plant for irradiating food, in Gloucester, Massachusetts, initially built by the Atomic Energy Commission, has been operating successfully since 1964. It can process 1000 lb of fish per hour.

There is no doubt that irradiation retards spoilage of food. Irradiated strawberries, for example, remain fresh for about 15 days after they have been picked, whereas strawberries that have not been treated begin to decay after about 10 days. Irradiated unfrozen fish also lasts a week or two longer. Tests have shown that the taste, nutritional value, and appearance of the food remain acceptable. The important question is the safety of the procedure. Irradiation at the levels used in the treatment does not make the food radioactive. There is, however, the possibility that the changes induced by radiation may make the food harmful. Over the past three decades, there have been many test programs both with animals and with human volunteers to ascertain the safety of food irradiation. At this point, the technique has been judged safe and is in commercial use (see Exercise 17.2).

## 17.5 Isotopic tracers

Most elements have isotopes differing from each other by the number of neutrons in their nuclei. The isotopes of a given element are chemically identical—that is, they participate in the same chemical reactions—but they can be distinguished from each other because their nuclei are different. One difference is, of course, in their mass. This property alone can be used to separate one isotope from another. A mass spectrometer is one of the devices that can be used to perform this task. Another way to distinguish isotopes is by their radioactivity. Many elements have isotopes that are radioactive. These isotopes are easily identified by their activity. In either case, isotopes can be used to trace the various steps in chemical reactions and in metabolic processes. Tracer techniques have been useful also in the clinical diagnoses of certain disorders.

Basically the technique consists of introducing a rare isotope into the process and then following the course of the isotope with appropriate detection techniques. We will illustrate this technique with a few examples. Nitrogen is one of the atoms in the amino acids that compose the protein molecules. In nature, nitrogen is composed primarily of the isotope $^{14}$N. Only 0.36% of natural nitrogen is in the form of the nonradioactive isotope $^{15}$N. Ordinarily the amino acids reflect the natural composition of nitrogen.

It is possible to synthesize amino acids in a laboratory. If the synthesis is done with pure $^{15}$N, the amino acids are distinctly marked. The amino acid glycine produced in this way is introduced into the body of a subject, where it is incorporated into the hemoglobin of the blood. Periodic sampling of the blood measures the number of blood cells containing the originally introduced glycine. Such experiments have shown that the average lifetime of a red blood cell is about four months.

Radioactive isotopes can be traced more easily and in smaller quantities than the isotopes that are not radioactive. Therefore, in reactions with elements that have radioactive isotopes, radioactive tracer techniques are preferred. Since the 1950s, when radioactive isotopes first became widely available, hundreds of important experiments have been conducted in this field.

An example of this technique is the use of radioactive phosphorus in the study of nucleic acids. The element phosphorus is an important component of the nucleic acids DNA and RNA. Naturally occurring phosphorus is all in the form $^{31}$P, and, of course, this is the isotope normally found in the nucleic acids. However, as discussed earlier, by bombarding sulfur 32 with neutrons, it is possible to produce the radioactive phosphorus $^{32}$P, which has a half-life of 14.3 days. If the $^{32}$P isotope is introduced into the cell, the nucleic acids synthesized in the cell incorporate this isotope into their structure. The nucleic acids are then removed from the cell and their radioactivity is measured. From these measurements, it is possible to calculate the rate at which nucleic acids are manufactured by the cell. These measurements, among others, provided evidence for the roles of DNA and RNA in cell functions.

Radioactive tracers have been useful also in clinical measurements. In one technique, the radioactive isotope of chromium is used to detect internal hemorrhage. This isotope is taken up by the blood cells, which then become radioactive. The radioactivity is, of course, kept well below the danger level. If the circulation is normal, the radioactivity is distributed uniformly throughout the body. A pronounced increase in radioactivity in some region indicates a hemorrhage at that point.

## Exercises

**17.1.** Describe the basic principles of magnetic resonance imaging.

**17.2.** What is your (considered) opinion of food preservation by radiation?

**17.3.** Through a literature search, describe the most recent use of fMRI.

# Chapter 18

# Nanotechnology in biology and medicine

## 18.1 Nanostructures

*Nanostructures* are material configurations in the size range of 1–100 nano-meters (1 nanometer, abbreviated nm, is equal to $10^{-9}$ m). Nanometer-size structures are indeed very small. The size of 1 nm is about that of 10 hydrogen atoms in a row. The smallest object our eyes can resolve (say, a small pencil dot) is about 100,000 nm. Following are two frequently given illustrations of the nanometer size scale: Typically our fingernails grow 1 nm in about 1 s. Further, if the diameter of a marble were reduced to 1 nm, the diameter of the Earth on the same scale would be 1 m (see Exercise 18.1).

Nanosized structures are ubiquitous in nature. Central components in living matter such as proteins, DNA, and many other constituents of cells are in the nanosize range. Airborne clay dust, salt, and soot particles are typically *nanoparticles* (NPs) in the range of 30–200 nm. Such airborne particles called aerosol serve as nuclei in the formation of cloud droplets, thereby affecting climate. The quantitative effect of aerosol on climate is not yet fully understood and remains a subject of intense studies. The deleterious health effects of nanosized aerosol are well documented but they're not yet fully understood.

## 18.2 Nanotechnology

*Nanotechnology* is the construction, manipulation, and utilization of nanosized materials of controlled composition and shape. In limited ways nanoparticles have been utilized in a range of applications before their attributes were fully understood. In the middle ages artisans learned that by mixing gold chloride into melted glass, they could obtain a richly colored stained glass. It is now known that they were producing nanosized gold particles within the glass. Carbon black nanoparticles have been used for decades as reinforcing material in rubber compounds, an application central to the manufacturing of tires.

The era of modern nanotechnology was initiated by the development in the 1980s of scanning probe microscopic imaging techniques of the type described in Section 16.8. These instruments made it possible not only to visualize but

Physics in Biology and Medicine. https://doi.org/10.1016/B978-0-443-21558-2.00018-3

also to manipulate nanostructures. Techniques have been developed to produce nanoparticles of controlled shape and composition. As an example, it is now possible to manufacture nanoparticles with a thin metal shell surrounding another material with properties designed for specific applications.

The field of nanotechnology has proceeded in two main directions. One branch of the effort is centered on the development of new nanomaterials with desired properties such as more efficient utilization of solar energy, enhanced catalytic activity, and superhydrophobic self-cleaning. The other area of intense activity is the applications of nanostructures in biology and medicine. The latter is the focus of our discussion.

To date, in the area of biology and medicine, gold and silver nanoparticles have been most widely studied and utilized, silver because of its strong anti-bacterial properties. A key factor in the utilization of gold nanoparticles has been the development of techniques for attaching biologically important molecules to gold particles. Although in most cases gold is chemically inert, with proper preparation, it does react with sulfur-containing molecules, forming a stable, relatively strong bond. Via the sulfur–gold bond, a wide range of sulfur-containing molecules such as DNA and proteins including enzymes and antibodies can be attached to gold nanoparticles.

## 18.3 Some properties of nanostructures

The properties of materials on a nanosize scale are often very different from the properties of the bulk material. Here we will discuss briefly nanoparticle properties with implications in biology and medicine.

### 18.3.1 Optical properties of metal nanoparticles

In metals (bulk or nanosized) such as silver, gold, or copper, positively charged metal atoms (ions) are in fixed positions surrounded by delocalized electrons. These electrons are free to move within the metal and specifically can move in response to an electric field including the electric field of a light wave. The interaction of the delocalized electrons with light can be described as a forced collective oscillation of the electron cloud at the frequency of the light. (The relationship between the frequency of light f and its wavelength $\lambda$ is $f = c/\lambda$, where c is the speed of light.) The collective oscillation of electrons results in the absorption and subsequent emission of light. The electron cloud in bulk silver is easily induced to oscillate at frequencies corresponding to the full visible spectrum; that is, the wavelength range from blue (390 nm) to red (750 nm). At optical energies, the efficiency of absorption and reemission of light is about equal. Therefore, when silver is illuminated with white light, the full spectrum is reemitted, yielding the white silver color of the metal. The structure of gold is such that its electron cloud is not readily induced to oscillate at frequencies corresponding to the full visible spectrum. The response of the electrons diminishes at frequencies corresponding to the blue end of the spectrum. Therefore, in bulk gold (and also copper), the efficiency of light absorption and the corresponding reemission decrease toward the blue end of the spectrum. As a result, when gold is illuminated with white light, it assumes the characteristic yellow color (white light minus the blue end of the spectrum). Light that is not reflected enters the bulk and is dissipated within the bulk material. If the gold is in the form of a thin foil, less than about 100 nm thick, most of the bluish-green light that enters is transmitted and observed.

The color properties of metals as described above change dramatically when particle size is decreased down to about 100 nm. In this nanometer size range, the motion of electrons is more tightly coupled and their oscillating motion is constrained by the size of the particle. The absorption—emission properties of the particle are dominated by the resonance frequencies of the electron cloud. This collective resonance motion of the electron cloud is called the *plasmon resonance*. The smaller the particle, the higher the plasmon resonance frequency (i.e., shorter the wavelength) at which the electron cloud can absorb and subsequently reemit electromagnetic radiation. Gold particles in the 10 nm size range absorb, and subsequently reemit, light in the blue range of the spectrum. Therefore, suspended particles in this size range appear blue on reflection and red in transmitted light. As particle size increases, light absorption and subsequent reemission shift toward longer wavelengths and transmission toward shorter wavelengths. In this way, with increasing size, the color of suspended metal nanoparticles changes from red to blue on transmission, and from blue to red on reflection. The optical properties of nanoparticles, both the intensity and spectral response, depend on the shape as well as the size of the particles.

## 18.3.2 Surface properties of metal nanoparticles

In particles larger than about 1000 nm (1 μm), the ratio of the number of surface atoms to atoms in the bulk volume is negligible. For a 1000-nm silver particle, it is about $3 \times 10^{-4}$ (see Exercise 18.2). However, as the particle size decreases, that ratio becomes significant. For a 10-nm particle, the ratio of the number of surface atoms to atoms in the bulk volume is about $0.03 = 3\%$. Because atoms on the surface of a small particle have to conform to a sharper curvature, the bonds between surface atoms on a nanoparticle are more strained. That is, surface atoms of a nanoparticle are less tightly bound to each other than those on the surface of a larger particle. The surface bond strain affects the bonds in the interior of the nanoparticle. Nanoparticle melting point manifests this strain in the interatomic bonds. When surface atoms represent a significant fraction of the total number of atoms, the transition from solid to liquid occurs at a lower temperature. For example, 10-nm gold particles melt at ~700°C compared to the 1064°C melting point of bulk gold. More importantly, from the perspective of this chapter, because surface bonds are strained, the reactivity of surface atoms of nanoparticles is enhanced, with consequences for some medical applications.

### 18.3.3 Superhydrophilicity of nanostructured surfaces

Nanostructured surfaces can be highly water repellant (hydrophobic). That is, a surface composed of nanorods that are spaced relatively far apart can be so hydrophobic that a liquid droplet will not stick to the surface. The droplet will bead up and will be shed off the surface. This so-called *superhydrophobic property* can be quantitatively analyzed in terms of the contact angle a liquid droplet makes with the surface (see Fig. 18.1).

The contact angle depends on the type of liquid and the nature of the surface in contact with the liquid. The contact angle for hydrophilic surfaces is about 30° or less. A surface is considered hydrophobic if the contact angle is greater than 90°. The contact angle for superhydrophobic surfaces is typically greater than 150°. A surface composed of widely spaced nanorods is mostly air that exhibits no attraction to water or other liquids. Molecules of the liquid are attracted much more strongly to each other than to such a surface. The surface therefore exhibits superhydrophobic behavior. Such surfaces are self-cleaning because dirt particles on the low contact area surface are picked up by water droplets and are rolled off.

Superhydrophobic behavior was noted thousands of years ago on the leaves of several plants, particularly the lotus flower. Superhydrophobicity is now

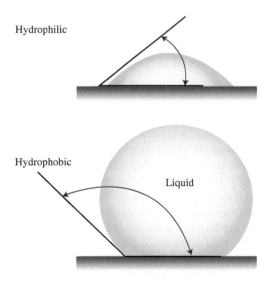

FIGURE 18.1   Contact angle on hydrophilic and hydrophobic surface.

**FIGURE 18.2   Leaf of a lady's mantle plant exhibiting superhydrophobicity.** *(Courtesy of Krzysztof Belczyński, Flickr, CC BY-SA 2.0 https://www.flickr.com/photos/x-oph/3494583806/.)*

often referred to as the lotus effect. Electron microscopic images of the lotus plant leaf show a surface composed of waxy nanometer-sized crystals. Even though the lotus plant grows in wet, muddy environments, its leaves remain clean and dry. Lady's mantle, shown in Fig. 18.2, is another plant exhibiting superhydrophobicity.

Superhydrophobic surfaces can now be manufactured using well-developed techniques. Particles with superhydrophobic properties are being incorporated into an increasing number of products such as paints, roof tiles, and a variety of coatings such as used for self-cleaning windows. Super-hydrophobic self-cleaning textile fabrics coated with nanofilaments of silicone and other suitable materials are now in active commercial development (see Exercise 18.3).

## 18.4  Medical applications of nanotechnology

### 18.4.1  Nanoparticles as biosensors

An important application of nanotechnology is the detection and identification of biological agents such as DNA, a range of proteins, and viruses. The most frequently used *biosensor* techniques are based on the optical properties of gold nanoparticles discussed in Section 18.3.1.

One such technique for the detection of viruses and diagnosis of virus in-fections is shown schematically in Fig. 18.3 (Driskell et al., 2011). Antibodies specific to the target virus are attached to gold nanoparticles. The particle—antibody complexes are introduced into a vial containing the suspected virus (Fig. 18.3a). A laser illuminates the vial and the scattered laser light is detected. Because the virus does not scatter light, the detected light is characteristic of the

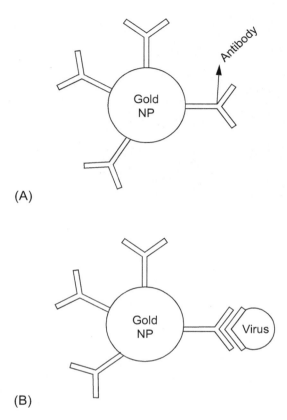

(A)

(B)

**FIGURE 18.3 Detection of viruses.** (A) Antibodies specific to the target virus are attached to gold nanoparticles and the particle–antibody complexes are introduced into a vial containing the suspected virus. Laser light (not shown) is scattered from the complex and is detected. (B) Virus locked onto the virus-specific antibody alters the scattered laser light. The change in the scattered light signals the presence and identity of the virus.

particle–antibody complex. A virus encountered by the antibody is attached, creating a new complex with an altered plasmon frequency (Fig. 18.3b). The presence of the new complex is manifested by a change in the scattered light. In addition to confirming the presence of the target virus, an analysis of the scattered light also provides information about the viral load. Modification of this technique can be used to detect DNA and a variety of proteins.

## 18.4.2 Nanotechnology in cancer therapy

The development of techniques to manipulate materials on the nanometer scale has made it possible to envision a range of medical applications that had been previously in the realm of science fiction. In this section we will present some of the proposed applications of nanotechnology in cancer therapy. Most of these applications are still in the development stage but several are in clinical trials. This section is not a complete survey of the field. The intent here is to illustrate the range of possibilities offered by the technology (see Exercise 18.4).

The most promising *nanoparticle cancer therapy* techniques involve specific targeting of tumors for destruction. Such targeted therapies are made possible by the difference between blood vessels that perfuse tumors and vessels that perfuse normal cells. Nanoparticles in a size range smaller than about 200 nm, typically used in targeted therapy, cannot penetrate normal blood vessels and therefore they circulate with the blood. However, tumor blood vessels, because of their rapid growth, have large pores that allow such nanoparticles to pass out of the blood stream into the tumor tissue. Further, because lymphatic drainage from tumors is slow, nanoparticles accumulate in the tumor. Specific properties of the concentrated nanoparticles can be used to destroy the tumor without excessively damaging healthy tissue.

### 18.4.3 Passive targeted heating of tumors

Passive heating of tumors is the most straightforward technique of targeted tumor destruction. Nanoparticles designed to absorb light at a specific wavelength are introduced into the blood stream. Nanoparticles leak through the porous blood vessels into the tumor tissue. The tumor is then illuminated with laser light of the suitable wavelength, heating the particles and the surrounding tissue cells. Heated proteins within cancer cells begin to unfold and cease functioning. The goal is to destroy the cancer cells without excessive damage to surrounding tissue.

If the tumor is close to the surface, the intensity of near-infrared light transmitted through the skin may be adequate to heat-destroy the tumor. Optical fibers can be used to conduct laser light to tumors located deeper inside the body.

Several nanoparticle designs are being tested for tumor heating. Gold is the most commonly used material. Solid gold nanoparticles can be heated by infrared laser light but other configurations are more effective. In one study hollow gold nanoshells encapsulating a nonconducting heat-absorbing material are used. The near-infrared laser light penetrates through the shell and heats the absorbing material. The heat is then conducted into the surrounding tissue. Another research group has demonstrated efficient laser-heating of nanoparticles in the form of a gold mesh.

Tumors can also be heated magnetically. In this technique, nanoparticles made of materials with magnetic properties such as the iron oxide maghemite ($\gamma-Fe_2O_3$) are introduced into the tumor. An alternating magnetic field changing directions typically at a frequency of a few hundred kilohertz is applied to the body. The nanomagnets flip direction at the frequency of the applied field. The kinetic energy of this motion is dissipated, heating the surrounding tissue. Healthy tissue does not contain magnetic material and therefore is not heated by the applied field.

### 18.4.4 Targeted drug delivery

Chemotherapy is one of the main methods of cancer treatment. However, in most cases the toxic chemicals attack both the healthy and the cancerous tissues. The efficacy of the technique is due to higher uptake of the toxic chemical by the fast-dividing cancer cells. Still the healthy tissue is also affected, often causing serious side effects. The amount of tumor-toxic chemical that can be administered is limited by the toxic side effects. Targeted drug delivery is being developed to maximize the infusion of the drug into the tumor while minimizing the side effects of chemotherapy.

A wide variety of techniques are being developed to address this important task. We will describe one such technique. For chemotherapy, molecules are sought that disrupt the functioning of the cell and cause its death. Through decades of research, molecules have been discovered that are effective cytotoxins for specific types of cancer. The tumor-toxic chemical is attached to the gold nanoparticle via a sulfur bond. One such cell toxin is the protein tumor necrosis factor-alpha (TNF-$\alpha$) that causes cell inflammation and death. The nanoparticles with the attached toxin are injected into the blood flow and accumulate in the tumor, as described in Section 18.4.2. However, for the therapy to be effective, another step is needed. Gold particles by themselves are not attacked by the immune system. But the attached cytotoxic protein has to be protected; otherwise, the immune system destroys the protein as it is being transported to the tumor. Techniques have been developed to mask the protein from the immune system such as coating the protein—gold particle combination with a form of polyethylene glycol (PEG-Thiol).

### 18.4.5 Silver nanoparticles in medicine

Through the ages, silver has been utilized mostly as an ornamental and coinage metal. But its medicinal properties had already been known in ancient Greece, Rome, and other parts of the world at that time. Experience showed that spoilage of wine, milk, water, and other foods was significantly retarded when stored in silver containers. In the 4th century BCE, the Greek physician Hippocrates wrote about the use of silver in wound care. Gastrointestinal infections were thought to be reduced by utilization of silver eating utensils.

In the early 1880s, Carl Credé introduced silver, in the form of silver nitrate, into modern medical practice. Till that time, septic neonatal conjunctivitis caused by infective agents, acquired as the neonate passed through the birth canal, was a major cause of permanent eye damage including blindness. In a series of convincing studies published in 1884, Credé demonstrated that wiping the eyes of newborn babies with 2% silver nitrate solution eliminated this infection. Prior to 1939, more than 90 silver-containing medications were in use. The silver nitrate eye treatment was used well into the 1950s when it was replaced by antibiotic ointments.

In these applications, the antibacterial activity of silver is due to silver ions ($Ag^+$) that interact with and disrupt the functioning of the bacterial cell walls as well as the interior components of the cell. Silver ions exhibit a high level of human toxicity. Therefore, with the advent of antibiotics, the medicinal use of silver has been largely discontinued, with the exception of silver sulfadiazine ointment introduced in the early 1970s that remains the topical treatment of choice for burns.

The development of techniques for manufacturing *silver nanoparticles* and the emergence of a wide range of antibiotic-resistant bacteria has brought resurgence in the use of silver as a bactericidal agent. Silver nanoparticles have been shown to have a higher antibacterial activity than silver ions alone. It appears that in addition to the antibacterial activity of silver ions, simply contact with the surface of the silver nanoparticle destroys the cell walls of the bacterium with high effectiveness. The bactericidal effect is size and shape dependent. Smaller nanoparticles as well as multifaceted shapes such as octahedral and decahedral are more effective. This higher bactericidal activity is likely due to the higher reactivity of strained surface atoms, as discussed in Section 18.3.2. Further, initial experiments indicate that in addition to its antibacterial activity, silver nanoparticles also have significant antiin-flammatory properties and may also exhibit antiviral activity.

Because of the highly effective bactericidal activity, utilization of silver nanoparticles in medicine has been increasing at a rapid pace (see Exercise 18.5). An important and promising application is in the manufacture of catheters. Catheters are tubes inserted into arteries to deliver drugs or into body cavities to drain excess fluids. They are commonly used in many hospital procedures. Because part of the catheter is exposed to open air, catheters are a frequent cause of hospital infections. Nanosilver coatings on these devices appear to significantly reduce bacterial infections.

The benefits of nanosilver-coated wound dressings have now been demonstrated. They reduce infection and inflammation and speed up healing when compared to other available treatments.

Bone cement is used to secure attachments in procedures such as hip and knee replacement. With conventional cement as the gluing agent, infection rates are high, 1%−4%. Incorporation of nanosilver into the cement has been shown to reduce infections by about a factor of 2.

## 18.5 Concerns over use of nanoparticles in consumer products

The website of the organization "Project on Emerging Nanotechnologies" currently lists over 1300 products that contain nanoparticles and the list is growing rapidly. The environmental, health, and safety risks associated with these products are highly uncertain. At the request of the Environmental Protection Agency (EPA), the National Research Council convened a committee to address the risks presented by nanotechnology. Their 154-page report, posted on the Web in January 2012, stresses the need for research on the safety of products containing nanoparticles. Two areas where the effect of nanoparticles is almost completely unknown are the ingestion of nanoparticles and the effect of disseminating these particles into the environment. In this connection, products containing silver nanoparticles are likely to present the greatest hazard.

About 20% of the consumer products containing nanoparticles incorporate, in one way or another, silver nanoparticles. Among the products containing nanosilver are: Tableware, zip-top food storage bags, towels, sheets, pillowcases, blankets, slippers, socks, mouthwash, and a variety of children's toys including stuffed bears and dogs. According to the manufacturers, the silver nanoparticles provide health benefits due to antibacterial and antifungal activity of silver.

The uses of nanosilver in medical applications are demonstrably beneficial. The use and final disposition of the medical products are likely to be done with care, under controlled conditions. However, the benefits of silver in many other consumer products are questionable. The health benefits claimed for these products have not been demonstrated, nor have their risks been assessed. The nanoparticles in these products have been shown to dislodge, making them likely to be swallowed, particularly if they are components in children's clothing and toys. Silver nanoparticles ingested or in contact with the body are likely, over time, to disrupt the essential bacterial colonies that are indispensable to the proper functioning of the body. Further, it has now been demonstrated that a significant fraction of the nanoparticles in clothing is removed with each washing. These particles, in one way or another, end up in the water system where they are likely to harm aquatic life including bacterial colonies necessary for the fertility of soil.

Health concerns are not limited to silver nanoparticles. Because of their small size, all nanoparticles have the potential to enter cells and cause damage. The hazards of nanoparticles in an ever-increasing number of products need to be examined and consumer awareness about the possible dangers of such products needs to be heightened (see Exercise 18.6).

## 18.6 Health impact of nanoparticles in polluted air

The deleterious effects of *air pollution*, specifically pollution consisting of suspended particles in air, called haze or smog, have been known for thousands of years. The Roman philosopher Seneca (4 BCE to CE 65) wrote to a friend: "No sooner had I left behind the oppressive atmosphere of the city (Rome) and that reek of smoking cookers which pour out, along with clouds of ashes, all the poisonous fumes .... I noticed the change in my condition ...."

Pollution became worse as cities and industry grew. It was recognized that smog is formed via the interaction of fog with suspended particulates called aerosol generated by the combustion of coal and wood. Because of special weather patterns and high population density, smog was particularly thick and its effects severe in London. However, other large cities have been affected as well. In an attempt to control smog pollution, King Edward I, in 1306, prohibited the burning in London of sea coal, an especially smoky fuel found washed up on beaches. However, people ignored the order and pollution continued to get worse. In the 1830s, a London writer commented: "The fog was so thick that the shops in Bond Street had lights at noon. I could not see people in the street from my windows." During the Victorian period, health problems particularly associated with lungs began to be noted.

In London, the problem came to a head on December 5, 1952. Cold weather, with its increased coal burning and windless conditions, resulted in an unusually thick smog that lasted till Tuesday, December 9. Four thousand people died as a direct result of pulmonary complications caused by the inhalation of the foul air. Subsequent estimates indicate that the total death toll was about 12,000 and that 100,000 people were sickened by respiratory complications. This very direct evidence of the deadly effect of air pollution led to the enactment in 1956, in Great Britain, of the very influential Clear Air Act. Unfortunately, with increased industrialization and growing urban congestion, some cities in Asia and South America remain inflicted with deadly urban air pollution.

Since the 1970s, attention has been focused on the more detailed aspects of the health effects associated with particulate air pollution. Animal studies as well as public health statistical surveys have made it evident that particles 2.5 μm in diameter and smaller present the greatest health hazard. This particle size range is designated $PM_{2.5}$. Actually, the greatest damage is caused by particles in the lower range of $PM_{2.5}$, that is, particles smaller than 0.1 μm (0.1 μm $= 10^{-7}$ m $= 100$ nm). The particle size effect is easily explained. Larger particles do not penetrate deep into the lungs. They are intercepted by the membranes in the nose, throat, or wider lung passages and are then expectorated by coughing or spitting. On the other hand, the smaller particles follow the streamlines of the breathed-in air and are transported deeper into the lungs. Nanometer-sized particles (100 nm or smaller) penetrate into the alveoli where they lodge. Particles in the size range of 30 nm can penetrate through the lung membrane from the alveoli into the lymphatic system, and those in the 10 nm range can enter the blood stream.

The alveoli do not possess an efficient mechanism to eliminate lodged particles. The particles that remain in the lungs are thought to be responsible for aggravated asthma, chronic bronchitis, and decreased lung function. Nanoparticles that enter the blood stream can affect every organ of the body, but most notably the arteries. Here, depending on their chemical composition, they may cause vascular inflammation and atherosclerosis. Rheumatoid arthritis and other autoimmune disorder have also been associated with exposure to nanosized pollutants. Here animal experiments are beginning to provide clues about biochemical changes induced by nanoparticles in certain amino acids. Nanoparticles are thought to be involved in several malfunctions of the autoimmune system.

The EPA's limit on $PM_{2.5}$ pollution is 12 $\mu g/m^3$ (see Exercise 18.7). However, exposures even to significantly lower $PM_{2.5}$ concentrations can be harmful. A report published in June 2016 by the International Energy Agency estimates that annually 6.5 million deaths worldwide are linked to anthropogenic air pollution, mostly due to incomplete fuel combustion. According to the report, this death toll is "much greater than the number from HIV/AIDS, tuberculosis and road accidents combined." Most of these deaths occur in China, India, and other industrially developing Asian countries.

Over the past decade, animal studies have been performed indicating that particles smaller than 200 nm can pass through nasal cavities, travel along neurons, and enter the brain. These particles then trigger damaging inflammations in parts of the brain. (See Underwood, 2017) Several epidemiological studies indicate that such particles may produce severe brain damage. For example, USC researchers report that people living in regions where the concentration of $PM_{2.5}$ is higher than the EPA standard have nearly double the dementia risk. A University of Toronto study compared the records of 6.6 million inhabitants of the province of Ontario of those living 150 m away from a major road to records of those living within 50 m of a major road, where $PM_{2.5}$ is 10 times higher. The study showed that those living within 50 m were 12% more likely to develop dementia. Several animal studies show pathological changes in the brain associated with elevated exposure to $PM_{2.5}$ air pollution.

## Exercises

**18.1.** Confirm the statement in Section 18.1 related to illustrations of the nanometer size scale in terms of (a) Earth—marble comparison and (b) the rate of nail growth.

**18.2.** (a) Consider a 10-nm-diameter silver particle composed of 0.3-nm-diameter atoms. Calculate the number ratio of surface atoms to atoms in the interior of the particle.
(b) Repeat the calculation for a particle 1 mm in diameter particle.

**18.3.** What self-cleaning materials using superhydrophobic properties of nanocoatings are now commercially available? (The Web is a good source of information.)

**18.4.** Survey the status of clinical trials for the nanotechnology techniques described in Section 18.4.4.

**18.5.** Describe one recent medical application of silver nanoparticles not discussed in the text.

**18.6.** Examine the current state of research on the safety of products containing nanoparticles.

**18.7.** Calculate the number of particles entering the lung in a single breath at the EPA-set limit on $PM_{2.5}$ pollution of 12 $\mu g/m^3$. Assume an average particle mass density of 1 $g/cm^3$ and an average particle diameter of 100 nm. Typically, volume of a breath intake is 500 $cm^3$.

**18.8.** The effects of indoor air pollution may be more deleterious than outdoor air pollution.

(a) Discuss the sources of indoor air pollution.

(b) Why might you expect indoor air pollution to be more harmful than outdoor pollution?

Chapter 19

# Metamaterials

## 19.1 Background

The prefix *"meta"* is a Greek word meaning beyond or transcending. Thus the word *"metamaterial"* means a material with properties beyond ordinary materials. Metamaterials are usually constructed of components of specific shape and size, designed to yield properties beyond those displayed by matter ordinarily found in nature. The size of the components of a metamaterial, often called *metaatoms,* are usually etched onto a wafer-like surface or embedded in an appropriate material, and they must be smaller than the wavelength of the phenomenon of interaction (e.g., microwaves or visible light). The metaatoms are arrayed in an ordered configuration determined by the intended function of the metamaterial device. The functional operation of a metamaterial device depends primarily on the structure and ordering of the metaatom components rather than on the chemistry of the components. The development of specific metamaterial devices is at the interface of physics, chemistry, and engineering and is therefore usually a cooperative effort of workers in these disciplines.

The study of metamaterials gives us an opportunity to explore a field of science that is still evolving and expanding. Metamaterials are likely to have a profound impact on many aspects of our lives, among them wireless communication, medicine, and material science. This study will also lead us to encounter interesting concepts in optics, electromagnetics, and medicine, not usually met in introductory physics. Many applications for metamaterials have been suggested and predictions for their future impact have been made. This chapter will provide us with the opportunity to participate in these prognostications and evaluate their outcome.

An early theoretical model for this novel matter we now call metamaterial was first published in 1967 by the Russian physicist *V.G. Veselago (1968).* The work of Veselago that predicted the possibility of negative refraction (to be discussed in a subsequent section) lay dormant for nearly 30 years till it was discovered by the San Diego research group, who demonstrated the reality of negative refraction. However, it took nearly another 30 years for the technology to reach the stage where Veselago's model could be studied experimentally and tested.

Physics in Biology and Medicine. https://doi.org/10.1016/B978-0-443-21558-2.00019-5

In the 1990s John Pendry, a physicist from the Imperial College in London, began to study the type of processes predicted by Veselago. Pendry observed that the radiation absorption properties of carbon fibers depended not on the chemistry of the carbon fibers but on the physical fine structure of the fibers on a scale significantly smaller than the applied electromagnetic radiation (Ball, 2018). This observation was in accord with Veselago's prediction.

The field of metamaterial studies is relatively new, with the start of intensive studies dating to about the year 2010. While impressive progress has been made, yielding some metamaterial devices already in commercial development, so far, significant applications of metamaterial devices have been few. However, it is widely recognized that metamaterials have the potential to transform the fields of optics, wireless communication, audio engineering, material science, and medicine. Nearly all research universities, as well as industrial and national laboratories, have well-funded research groups working on the development and application of metamaterials. Most people in this field predict that within 5 years, metamaterials will attain wide-ranging applications. The study of this field will also give us an opportunity to observe the transformation, often tortuous, of ideas into viable technologies.

Devices constructed of metamaterials can be designed to display complex manipulation of electromagnetic waves that result in highly unusual properties such as altering the visibility of objects (called cloaking, see Exercise 19.1), production of flat lenses, design of lenses with higher resolution than the diffraction limit that governs conventional lenses, control of material properties with external signals, and other phenomena that a few years ago were in the realm science fiction. An attractive feature of metamaterial devices is the potential for inexpensive production. In most cases the metaatoms can be produced by adapting the photolithographic processes developed for the mass production of semiconductor devices.

We will begin the discussion of the basic construction, principles, and use of metamaterials with two devices, *metalenses* and *metamaterial antennas,* that have so far received the most attention and are closest to widespread application. We will then discuss other potential applications of metamaterials under consideration with a focus on medicine and biology. In the course of our discussion we will describe one type of commonly used metaatom, the split ring resonator (SRR), and we will explain two concepts; negative refraction and evanescent waves not discussed in previous editions of this text.

## 19.2 Construction of a metalens: a simplified description

We begin the description of the basic principles of metamaterial design and function by examining the operation of a convex lens. Two concepts previously discussed in Appendix C are reviewed here for completeness of presentation. A simplified representation of light emanating from a source is shown in Fig. 19.1. The line with an arrowhead shows the propagation

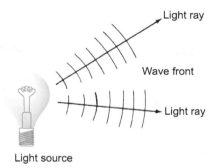

Light ray

Wave front

Light ray

Light source

FIGURE 19.1 **Simplified representation of light.** (For the convenience of readers, this figure is a repeat of Figure C.1 shown in Appendix C.)

direction of the light ray. The lines perpendicular to the light ray represent a snapshot of the wave, with the distance between the lines indicating the wavelength ($\lambda$) of the wave, that is, the distance between adjacent peaks (or other equivalent intensities) of the wave.

As is stated in the Appendix, the speed of light in vacuum is, within our relevant accuracy, $3 \times 10^8$ m/sec, and it is only slightly less in air. However, the speed of light in all material media is significantly less. In glass it is about $2 \times 10^8$ m/sec. The index of refraction (n), defined as n = c/v, where c is the speed of light in vacuum and v is the speed of light in a given material, is about 1.5 in glass. (See Eq. C.1.) As the light beam enters the lens, it is slowed down and therefore delayed. Light entering different parts of the lens (in this case a convex lens) is slowed down and delayed more during transit through the middle of the lens where the lens is thickest than away from the middle where the glass is thinned out. As a consequence of the delay time variability through the lens, and the specific shape of the lens, a converging wave front is created that comes to a focused spot at a specific point and then diverges, as shown in Fig. 19.2.

Conventional glass lenses have served us well for nearly 1000 years, but they have several limitations. Their performance is set by the intrinsic bulk properties of the glass and the carefully controlled shape of the glass. To change the location of the focal point requires the construction of a new lens or the addition of another lens to the configuration. Further, the spatial resolution of a conventional lens, that is, the fine detail that can be discerned in an image produced by the lens, is limited by diffraction to about $\lambda/2$. Further, because the index of refraction is a function of the wavelength of light, light passing through the lens is subject to color aberrations. Conventional lenses are relatively bulky and expensive to produce. Using metamaterials to construct devices usually leads to greater design flexibility, including more facile

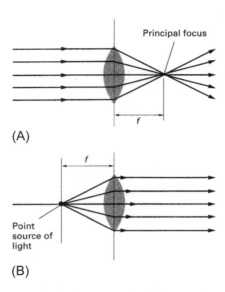

(A)

(B)

**FIGURE 19.2** (A) A convex lens focusing a parallel beam of light to a point. (B) A point source of light at the focal point of a convex lens producing a parallel beam of light. (For the convenience of readers, this figure is a repeat of Figure C.3 shown in Appendix C.)

miniaturization, improved performance, and lower price. This is certainly the case for lenses and most other metamaterial devices.

In designing metamaterial devices, the function of the device has to be clearly separated from its conventional physical form. As an example, when designing a metamaterial lens (usually referred to as *metalens*), we are freed from the constraint of using glass or from employing a specific shape for the lens. In our example of a convex lens, all we have to achieve is a configuration that focuses the electromagnetic wave (light wave in our example) into a specific point, as a conventional lens does. In the case of a convex lens discussed here, the metalens must slow the light down most as it passes along the axes of the device, that is, through its middle, and slow it appropriately less toward the edge of the device. That is, the slowdown has to follow the same dependence exhibited by the conventional lens.

To start the construction of a metalens, we have to create an array of components that are small compared to the wavelength ($\lambda$) of the electromagnetic radiation we plan to control (in this case to focus). These components have been named *metaatoms*. The size of the components is typically a fraction of the light wavelength to be controlled (about a tenth to a quarter $\lambda$). Because the components, as well as their spacing, are small compared to $\lambda$, the radiation entering the structure is subjected to a medium that appears relatively smooth. Whether the metaatoms are formed from metal or a dielectric material does in some cases significantly affect the properties of the component. It has been found that dielectrics have lower losses and are more suitable for some metamaterial applications than metals. On the other hand, several metamaterial properties can only be obtained with metaatoms made of nonmagnetic metals, such as copper. However, within these categories, the primary determinant of the metamaterial properties is the structure of the metaatoms comprising the metamaterial rather than the chemistry of the constituents. Most (but not all) metamaterials are built on a wafer-like surface, giving such a configuration the name *metasurface*.

Fig. 19.3 shows an electron microscope image of a section of a lens metasurface. Specifically the figure shows a side view at an angle of the metalens surface comprised of an array of nanometer-sized cylinders (Khorasaninejad et al., 2016). These cylinders are the metaatoms of the metamaterial lens. The shape of the metaatoms does not have to be cylindrical. Various other shapes have been used such as triangles, pyramids, and wedges. The light beam to be focused impinges on the lens in a direction perpendicular to the metasurface.

The metaatoms shown in the figure are made of a dielectric material, titanium dioxide, that has an index of refraction of about 2.9 for a wavelength of light $\lambda = 600$ nm. The nanostructures for metalenses can be formed from metals or dielectric materials using well-developed techniques including photolithographic processes developed for the mass production of solid-state devices. The metaatoms are usually etched into a wafer-thin optically transparent surface.

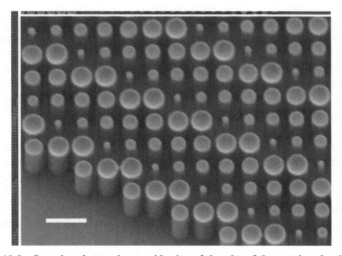

**FIGURE 19.3** **Scanning electron image, side view of the edge of the metalens for the visible range of the spectrum showing the vertical profile of the nanopillars.** Scale bar: 600 nm. *(Reproduced (adapted) with permissions from Khorasaninejad MA, Zhu Y, Roques-Carmes C, Chen WT, Oh J, Mishra I, Devlin RC, Capasso F. Polarization-insensitive metalenses at visible wavelengths. Nano Lett 2016;16(11):7229–34, https://doi.org/10.1021/acs.nanolett.6b03626. Copyright © 2016 American Chemical Society.)*

The response of the metalens is controlled by the structure and configuration of the metaatoms (here, the nanocylinders) rather than the glass and the shape of the lens, which are the key determinants of the optical response for a conventional lens. When the light enters the metalens, each metaatom (here, the nanocylinder) responds to the electromagnetic wave individually. That is, it delays, transmits, and scatters light in accordance with its individual design specifications, reemitting the light as a separate wave front. The individual wave fronts add to produce the desired new wave front that emerges from the metalens. Of course, the shape, size, and location of the metaatoms must be carefully designed to yield the desired wave front emerging from the lens.

The arrangement of components in the grid duplicates the transmission properties of a conventional convex lens. That is, as in a conventional convex lens, the transmission of light in an optical metalens is slowest in the middle and fastest toward the edges of the configuration. In between those two boundaries the light transmission of the metaatoms duplicates light transmission of the convex glass lens. For a visual representation of this process, see the YouTube presentation by Prof. Steve Cummer, Duke University. Other optical focusing functions would require correspondingly different metaatom configurations.

The metalens with the dielectric metaatoms discussed in this section functions as an ordinary glass lens with diffraction-limited resolution. That is a resolution limited to half the wavelength of the illuminating light. Still there are three features that distinguish the metalens from a conventional lens: (i) The lens is flat; (ii) the lens is thin with an attainable thickness of less than 1 mm; and (iii) using techniques developed for solid-state mass production, the lens is estimated to be inexpensive to fabricate and amenable to easy miniaturization. An important point to stress is that the shape of the emerging wave front is not limited to that of a lens. It can be tailored by the appropriate shaping and positioning of the metaatoms to any desired shape not easily attained by conventional glass-lens shaping methods. Recently metalenses in the infrared region of the spectrum were constructed with focal lengths that could be changed without moving the lens. This was accomplished by altering the phase of the metasurface material with applied heat.

The properties of the metalens and other metamaterials are significantly altered if the dielectric metaatom components of the metamaterial are replaced by specially configured metaatoms made of a nonmagnetic metal such as copper. The SRR is one such commonly used metaatom configuration. A properly designed metamaterial using such metaatoms displays optical properties that cannot be obtained with any conventional material. Two such properties stand out: (i) Such metamaterials can be used to construct lenses that have a resolution significantly below the diffraction limit, and (ii) such materials can display negative diffraction phenomena. An accurate quantitative treatment of these two and several other phenomena displayed by metamaterials is beyond the scope of this presentation. Here we will present a qualitative explanation only, with the aim of providing a basic phenomenological understanding of the processes.

We begin with a description of the split ring resonator (SRR), a frequently used metal-based nonmagnetic metaatom component.

## 19.3 Split ring resonator

The SRR was invented in the 1990s by the previously mentioned metamaterial pioneer, physicist John Pendry (Pendry et al., 1999). As shown in Fig. 19.4, the structure consists of two concentric nonmagnetic metal rings (e.g., copper), each with a slit in the ring with the slits opposite one another. Depending on the application, the two slitted loops can also be shaped in the form of squares. The structure is easily miniaturized and etched on a dielectric substrate, transparent if required.

Electrically the configuration can be represented by a lumped parameter-tuned resonant circuit. (For a brief discussion on resonant circuits, see Appendix B.8.) The loops act as the effective inductance $L_{eff}$ and the gaps in the rings act as the effective capacitor $C_{eff}$. The capacitive impedance provides continuity of alternating current flow through the loops.

The resonant frequency (see Appendix B.8) $f_{res}$ of the configuration is:

$$f_{res} = \frac{1}{2\pi\sqrt{L_{eff}\,C_{eff}}} \tag{19.1}$$

By changing the width of the gaps in the rings, the capacitance $C_{eff}$ can be conveniently changed up or down as needed. Consequently, the resonance

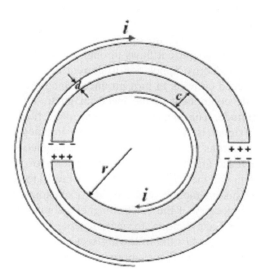

**FIGURE 19.4** **Split ring resonator.** (吴艺, *CC BY-SA 3.0. https://creativecommons.org/licenses/ by-sa/3.0, via Wikimedia Commons.*)

frequency $f_{res}$ of the electromagnetic signal of the SRR is likewise adjustable. The effective inductance and capacitance are functions of the physical construction of the split rings, that is, the radius, the thickness of the rings, and the size of the gaps in the rings.

To continue our discussion of metalenses, we need to examine briefly (and from a highly oversimplified perspective) the interaction of electromagnetic radiation with matter. Electromagnetic radiation (including light) consists of two interrelated components: the electric field and the magnetic field vibrating at right angles to each other at a frequency f that characterizes the specific electromagnetic radiation. The frequency f and wavelength $\lambda$ are related to the speed v of propagation of the electromagnetic radiation in the medium by the relationship $v = \lambda f$. (See Chapter 12.)

Electromagnetic radiation propagates through conventional material primarily via the interaction of negatively charged electrons in the material with the electric field component of the electromagnetic wave. The electrons in the material are set into vibration by the electric field at the frequency of the electromagnetic radiation and reemit that wave in the direction of the propagating wave. While electrons, because of their spinning charge, do possess a magnetic dipole moment, in ordinary materials interactions of electrons with the magnetic component of the electromagnetic field are nearly zero. In conventional materials interactions with the electric part of the field determine the direction of propagation through the material as well as the other familiar phenomena displayed by such materials.

An electromagnetic signal applied to the SRR metaatoms composing the metamaterial induces circulating currents in the loops. When the frequency of the electromagnetic radiation impinging on the SRR metaatom is at the resonant frequency of the SRR, the opposition to the current flow in the loops is nearly zero and a large current flows through the loops. (That is so because the capacitive and inductive opposition to current flow cancel.)

The large induced currents in turn produce a magnetic field that interacts with the magnetic component of the electromagnetic signal and greatly increases its magnitude. The intensity of the magnetic field component of the electromagnetic signal is now large enough to interact with the magnetic moment of the electrons in the path of the electromagnetic wave. As a result, now both the electric and the magnetic components of the electromagnetic wave interact with the metamaterial and affect the trajectory of the electromagnetic wave. This interaction with the SRR metaatoms provides additional control over the trajectory of the electromagnetic signal through the material and yields other effects unique to metamaterials, among them negative refraction, to be discussed in the following section.

While the size of each SRR metaatom is physically much smaller than the wavelength of the impinging electromagnetic radiation, at resonance, the SRR metaatom behaves as if it were much larger than its physical size. It sustains and stores energy of the longer wavelength electromagnetic signal. That is, the wavelength of the resonant signal is significantly larger than the physical size of the SRR, as is required for the operation of the metamaterial. However, because the spacing between metaatoms is small compared to the wavelength of the radiation, the effect of the individual metaatoms is smoothed out. Further, by appropriate design, each SRR can be tailored to produce its own response, yielding additional control over the magnetic component of the electromagnetic signal. A metasurface array of SRRs is shown in Fig. 19.5 (Wilson and Schwartz, 2005).

## 19.4 Refraction

When dealing with conventional materials, with a light beam that enters from a medium of lower index of refraction ($n_1$), where it travels faster, into a conventional material, where it travels slower (higher $n_2$), the light ray is bent toward the normal to the interface. See Fig. 19.6. This is referred to as positive refraction. Such refraction is observed with all conventional materials, for example, a light ray entering glass or water from air.

A light beam entering a metamaterial will be bent toward or away from the normal to the interface depending on the design of the metamaterial. The phenomenon of the light bending away from the normal is referred to as

FIGURE 19.5 **A metasurface array of split ring resonators.** *(Jeffrey D. Wilson (NASA), via Wikipedia Commons.)*

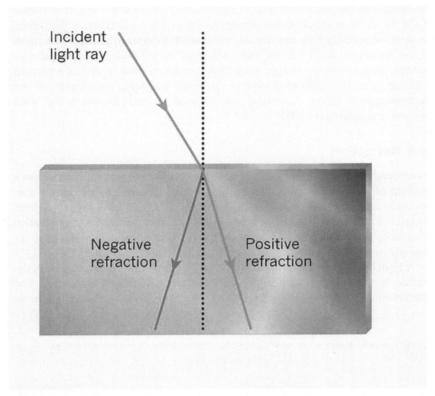

**FIGURE 19.6   Ordinary (positive) and negative refraction.** In ordinary refraction observed in conventional materials, light is bent toward the normal axis. In metamaterials exhibiting negative refraction, light is bent away from the normal axis.

negative refraction. Such behavior is not observed with conventional materials. As was mentioned, the expanded control of the electromagnetic beam trajectory is due to the designed interaction between the electrons in the metaatom and the enhanced magnetic component of the electromagnetic beam. With an appropriately designed metamaterial, the incident beam can be bent in any desired direction. That is, the refractive index can be made negative, positive, or zero. (In a more rigorous treatment refraction is formulated in terms of electric permittivity $\varepsilon$ and magnetic permeability $\mu$. Depending on requirements, the metamaterial can be designed so that these two parameters may be independently varied and controlled.)

Most of the studies with SRRs and other metaatoms have been performed at microwave frequencies where $f_{res}$ is on the order and $10 \times 10^9$ Hz and the corresponding wavelength $\lambda$ is 3 cm. The size of the SRR in the metamaterial array would have to be about 3 mm. In the visible region of about 600 nm, the diameter of the rings of the SRR metaatoms composing the effective meta-materials would have to be about 60 nm. This size range, and even that an order of magnitude smaller, is attainable with current nanofabrication techniques.

## 19.5 Superlens

Resolution is a key parameter in evaluating the quality of an optical imaging system. In simplest terms, the resolution of an optical system is the smallest detail that can be resolved in an image produced by the system. The optimum resolution of a conventional lens is subject to the so-called diffraction limit given by $\lambda/2$, where $\lambda$ is the wavelength of the light used to form the image. A lens made of a metamaterial composed of metaatoms such as SRRs can attain a resolution significantly below the diffraction limit. This type of lens is named a superlens.

Our elementary explanation of superlens performance requires a brief description of electromagnetic wave generation. Electromagnetic waves are generated by motion of electrons in a wire or on the surface of a metal. Two components of such an electromagnetic wave are identified. One component, called the far field component, propagates in the forward direction and can form an image. The other component remains close to the source of the electromagnetic signal, that is, near the surface of the metal, and does not propagate in a forward direction. That component is called the evanescent wave or the near-field component, and it contains the subdiffraction infor-mation in the electromagnetic signal. In a conventional lens with a positive index of refraction, the evanescent signal decays with a decay length of about the wavelength of the electromagnetic radiation. Therefore, the subdiffraction information does not extend to the plane where the image is formed. How-ever, when the electromagnetic wave travels in negative refraction material and the transverse magnetic component of the signal is reenforced via its interaction with a metaatom such as an SRR, surface electron oscillations are set up that significantly increase the amplitude of the evanescent waves. The superlens proposed in 2000 by John Pendry is based on the concept of enhancing the evanescent fields to restore the near-field components of the electromagnetic signal (Pendry, 2000). The amplitude of the evanescent wave becomes adequate for the subdiffraction information to reach the image formation plane and become incorporated into the image, breaking the diffraction limit. In recent experiments with red light, $\lambda = 650$ nm, a reso-lution of 50 nm was observed. That is about a sixfold improvement in res-olution over the diffraction limit of 325 nm.

Other methods for probing near-field evanescent waves have been devel-oped, among them the near-field scanning optical microscope and a variety of higher-resolution scanning probe techniques. (See Chapter 16.) But these instruments are expensive and not portable. It is likely that once the superlens is perfected, it will make possible very useful, widely available high-resolution microscopes.

## 19.6 Metamaterial antennas

Antennas are built to receive and/or transmit electromagnetic signals. In the process of reception the signal being received impinges on the antenna and creates in the antenna a current that is detected, and the information contained in the current is extracted by electronic circuitry designed for that purpose. In the process of transmission the signal to be transmitted is coupled to the antenna, where it produces a current that in turn produces an electromagnetic signal that is transmitted by cable or launched into space to be detected and decoded.

A wire of any length will receive and transmit electromagnetic radiation, but for optimum operation (i.e., maximum reception or transmission of the electromagnetic radiation), the length of a conventional antenna wire should be about half the wavelength ($\lambda$) of the electromagnetic signal. At that length, the wire is in resonance with the electromagnetic signal. That is, maximum energy is stored in the wire of that length, in much the same way that maximum sound energy is stored in an organ pipe with its length set at half the wavelength of the given sound wave. Conventional antennas that are short, that is, that do not meet the $\lambda/2$ requirement, will reflect a significant fraction of the electromagnetic signal back to the source.

As an illustration, consider the following example. At a frequency of 2.4 GHz $= 2.4 \times 10^9$ Hz (a common frequency band for cell phones), $\lambda = c/f = 3 \times 10^8/2.4 \times 10^9 = 0.125$ m $= 12.5$ cm. The optimum length of the antenna would therefore be about $\lambda/2 = 6.25$ cm. By an appropriate design, the antenna can be configured into an area of about 2 cm$^2$ or a square of about 1.4 cm to the side. This is the size of a typical antenna in currently built cell phones. The efficiency of such antennas is rather low, less than about 50%. That is, only about 50% of the power delivered to the antenna is radiated from the antenna. The remainder of the energy is dissipated, in one way or another, into heat.

New smartphones have four or more antennas to cover the range of frequency bands used in cell phone operation and to respond optimally to the directionality of the signals to be transmitted and received. These antennas have to be separated as far as possible from each other to avoid interactions and interference. The antennas currently designed and manufactured perform adequately the required tasks. However, as the technology progresses, more communication channels and hence more antennas will have to be added to service the system. A significant engineering challenge is to fit the increasing number of antennas into the small space available in smartphones and other miniature devices and to improve the efficiency of the antennas. Smaller antennas and greater radiation and reception efficiencies would significantly enhance design and production of improved electronic and medical instrumentation in fields such as wireless communication, remote equipment control, and a variety of wirelessly monitored microsensors. Such improvements can be implemented with antennas constructed using specifically designed metamaterials. As in the design of metamaterial lenses, here also, SRRs are often the choice for the metaatom components.

As was stated, for efficient operation, a standard antenna has to be half a wave long (or some multiple of that length) corresponding to the frequency of the electromagnetic signal to be transmitted or received. For example, at f = 300 MHz, $\lambda = c/f = 3 \times 10^8/3 \times 10^8 = 1$ m and $\lambda/2 = 50$ cm. At that length, the antenna is in resonance with the corresponding frequency. That is, at that frequency (300 MHz), electromagnetic energy builds up in the antenna for efficient radiation or detection.

Metamaterial antennas function as if they were much larger than their physical size. They are not constrained by the wavelength of the electromagnetic radiation. For the SRR antenna metaatom, the resonant frequency at which electromagnetic energy builds up in the antenna for efficient reception and/or transmission is set by the values of $L_{eff}$ and $C_{eff}$ in Eq. 19.1, rather than the physical size of the device. SRR metaatoms as small as $6 \times 10^{-6}$ cm have been fabricated. The minimum size is usually limited by the power requirement of the device.

For efficient operation, electromagnetic radiation has to be coupled into the metamaterial antenna when it acts as a transmitter and coupled out of the antenna when it serves as a receiver. These functions are performed very effectively by metamaterial negative refractive index superlenses. Recently developed metamaterial antenna systems have reported transmission and reception efficiencies as high as 95%.

In 2009 LG Electronics, a South Korean company, was the first to introduce a mobile handset with a metamaterial antenna. Technically the instrument met the expectations of the designers. The phone had elegant slim lines, an improved antenna efficiency of up to 70%, and high-quality transmission and reception. However, after years of trying, LG Electronics was not able to profitably compete against smartphones made by Apple and Samsung, and in 2021, it discontinued that part of its business.

Several start-up companies are now actively developing metamaterial antennas and are introducing them into the market. An increasing number of satellite antennas are using metamaterial construction. The use of metamaterial antennas has significantly reduced the size of the antennas and has also improved other antenna parameters such as transmission and reception efficiency, operating bandwidth, and multiband frequency of operation. As reported by MarketsandMarkets Research Company, in 2021 the metamaterial market was valued at $305 million and is expected to grow to $1.457 billion by 2026.

## 19.7 Metamaterials in medicine

In this section we will briefly discuss several of the large number of potential metamaterial applications in medicine that are likely to become widely used in the near future.

(a) **Metamaterials in wireless endoscopy.** An endoscope is a medical instrument for examining and/or monitoring of an internal cavity or organ that is accessible through an external opening. Colonoscopy (examination of the colon and large intestine) and gastroscopy (examination of the stomach and duodenum) are among the most common endoscopic procedures. Most colonoscopes and gastroscopes are connected to the operator via flexible bundle of fiber optic cables that carry a strong light to illuminate the organ and a miniature video camera to return images. A miniature surgical instrument is often included to excise polyps found in the wall of the intestine. The procedure with such an endoscope is also called "push endoscopy." Push endoscopy can be at times painful and does not provide full visualization of the small intestine.

In 2001 the wireless capsule endoscope, designed to be swallowed, was introduced. The pill-sized device consists of a miniature camera with an antenna and a miniature battery power supply that transmits images to a recorder strapped to the patient. The location of the pill, as it makes its way from mouth to anus, is tracked. The transit time varies from person to person, but it usually takes about 10 hours. However, during this period, the patient can continue normal activities, and the recording provides images for examination of the entire gastrointestinal system at a later time. On the other hand, the capsule cannot provide surgical intervention simultaneously such as removal of a polyps, should that be required.

The high resolution of superlenses and the high efficiency of metamaterial antennas combined with the small size of both these devices make them ideal components for the wireless capsule endoscope. A publication describing the construction of a metamaterial antenna for a wireless endoscope states that the antenna provided a 74% size reduction compared to a conventional antenna (Cheng et al., 2011). The additional available space could be used for another antenna, a larger battery, or additional sensors, or as will be discussed, to facilitate other future uses for the capsule.

(b) **Metamaterials and MRI.** Magnetic resonance imaging (MRI) has become one of the most important diagnostic tools in medicine. It provides accurate and detailed noninvasive images of organs and other structures inside patients' bodies. The basic theory and procedures of MRI are described in Chapter 17. Here we will provide only the salient points of the process needed to understand the contributions metamaterials might make to MRI procedures.

To obtain an MRI image, a strong magnetic field of 1.5 tesla (T) or more is applied to the sample (i.e., the part of the body that is being examined). By comparison, the Earth's magnetic field at its strongest is about $6.5 \times 10^{-5}$ T. Atoms with nuclear magnetic moments such as hydrogen are lined up by the applied magnetic field, some parallel, others antiparallel. In equilibrium, more magnetic moments are parallel to the field than antiparallel, creating two energy levels with a separation $\Delta E_m$. The energy separation is linearly proportional to the strength of the magnetic field, as given by Eq. 17.1. To form an MRI image, a radio frequency (RF) pulse at frequency $\Delta E_m/h$ is applied to the sample (h is Plank's constant). This RF pulse, called the driving pulse, produces a transition of atoms from the lower to the higher energy state. At the end of the driving pulse, atoms (actually the nuclei) return to their equilibrium distribution, emitting the excess energy in the form of an RF pulse of the same frequency as the driving pulse. The postexcitation pulse, called the nuclear magnetic resonance (NMR) RF signal, contains information about the source tissue and is used to form the magnetic resonance image.

The MRI technique is not without its problems. The main difficulty is the inherently low signal-to-noise ratio (SNR) of the procedure. The initiating RF driving pulse is a high-amplitude signal at frequency $f = \Delta E_m/h$, and as was stated, the actual information-carrying signal emitted after the driving pulse is a weak signal at the same frequency. Inevitably, the large amplitude driving signal interferes with the weak signal that follows it, contributing to the low SNR. To form a clear image with a low SNR signal requires long signal integration times, that is, long averaging times. To obtain the full complement of MRI images, the procedure often takes an hour or longer. This is an uneconomical use of very expensive apparatus and it is also very uncomfortable for the patient confined inside the instrument. The conventional ways of improving the signal-to-noise ratio in MRI such as increasing the driving pulse intensity and/or the strength of the main magnetic field, have by and large been optimized.

With the advent of metamaterials, several groups have been experimenting with ways to utilize metamaterials to improve the signal-to-noise ratio of MRI. Here we will describe briefly a technique proposed and used by scientists from Boston University that shows promising early results (Duan et al., 2019). Our explanation of the system is, out of necessity, oversimplified and is somewhat analogous to our description of the superlens.

To increase the signal-to-noise ratio, the postexcitation emitted RF pulse carrying the MRI information has to be increased without increasing the interference due to the driving RF pulse. The metamaterial configuration in Fig. 19.7 is an array of 5-cm-high polymer cylinders helically wound with copper wire. This is a resonant system with each cylinder having a capacitance due to spacing of the windings and an inductance due to the

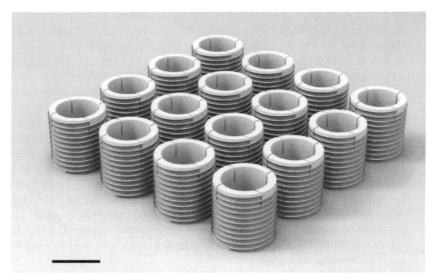

**FIGURE 19.7    Schematic of a metamaterial for enhancing the signal-to-noise ratio in MRI.**
Scale bar is 3 cm. *(From Duan G, et al. Boosting magnetic resonance imaging signal-to-noise ratio using magnetic metamaterials. Commun Phys 2019;2:35. https://doi.org/10.1038/s42005-019-0135-7. May need permission to reproduce.)*

helical wound wires. To calculate the effective capacitance and inductance of the array as a whole, the interactions between the cylinders have to be also taken into account. It is envisioned that to improve MRI instrument sensitivity, a metamaterial array such as shown in Fig. 19.7 would be made flat and flexible so that it could be placed on the part of the body being imaged.

When the frequency of the RF driving pulse and the following NMR RF pulse are at the resonance frequency of the metamaterial ensemble, the RF current in the wires of the cylinders (metaatoms) greatly increases. The increased current produces an increased magnetic component of the RF signals and hence the magnitude of both the driving RF pulse and the emitted RF pulse carrying the MRI information increase. The increased driving pulse does not contain MRI information but does contribute to the noise. Therefore, one would like to amplify the NMR-emitted RF pulse without amplifying the driving pulse. A method for doing this has been implemented. An SRR containing a varactor diode has been incorporated into the metamaterial array shown in Fig. 19.7. A varactor diode is a capacitor designed such that its capacitance changes with voltage. When the high amplitude driving voltage impinges on the varactor, its capacitance shifts and the metamaterial is pulled off resonance. As a result, the driving pulse is not amplified. The weak postexcitation emitted NMR pulse that follows allows the capacitance of the varactor to return to its original value, restoring the system to resonance. The emitted NMR pulse is now amplified without the interfering effect of the driving pulse. Reported initial experiments indicate that this system improves the signal-to-noise ratio by a factor of 10.

Several other uses of metamaterials in medicine have been proposed utilizing the unique properties of metamaterial antennas and superlenses. Here we will simply list a few such proposed applications.

(c) **Metamaterials for early breast tumor detection.** Preliminary studies have shown that with a superlens, tumors as small as 1 mm could be detected at a depth of 12.5 mm to 24 mm.

(d) **Metamaterial antennas and lenses in hyperthermia cancer treatment.** High temperatures (about 42°C, 107°F) will damage cancer cells without excessive damage to the neighboring healthy tissue. The high temperature may at times destroy the tumor but more often will make it considerably more susceptible to other therapies such as radiation or chemotherapy. The tumor is usually heated by microwave radiation that can be more accurately positioned and focused with metamaterial antennas and lenses than with their conventional counterparts.

(e) **Implantable metamaterial monitors.** Methods now exist to monitor certain internal bodily processes continuously. Among them are gastric pH, broken bone healing, and wound healing. The wireless capsule endoscope described earlier can be easily modified to be suitable for implanting into various parts of the body as needed. As pointed out, the basic components of the capsule, namely the conventional camera lenses and the antenna, can be replaced with smaller metamaterial components, providing space for additional monitoring instruments.

## 19.8 Other types of metamaterials

To date (2022), most of the metamaterial research (as well as our discussions) have focused on electromagnetic phenomena. However, metamaterials related to other phenomena are also being developed and are likely to become useful in several fields, among them acoustics, thermal conduction, and mechanical structures. One could list other types of metamaterials depending on classifications. We will close this chapter with a brief discussion of the application of mechanical metamaterial to prostheses.

Prostheses for below-knee amputations are the ones that most often cause problems for wearers (Garland et al., 2020). The interfacial load between the residual limb and the prosthetic socket is unevenly distributed, as is the hardness of the tissue. Some of the residual limb is bone and some is soft tissue of variable hardness. Studies show that the average person wearing prostheses has them attached about 12 hours a day. The uneven friction and load bearing often cause wounds and infections to develop at the interface of the prostheses and the residual limb. Currently, the main approach to solving the problem has been to provide interfacial liners of uniform composition to protect the residual limb. This simple approach is inadequate to deal with a situation where the interfacial pressure is variable from location to location. In some locations the interface is hard bone, and in other locations it is soft tissue.

The strategy of current research is to create a material with individual cells (metaatoms) of composition shape and distribution to meet the specific requirements of each region of the interface to minimize pressure, stiffness, and friction. Such a specialized interface can be created for each individual's use with current three-dimensional computer printing technology.

## EXERCISES

**19.1.** In 2006 the engineering group at Duke University, with Prof. David R. Smith as one of the leaders of this group, demonstrated the first rudimentary cloaking at a microwave frequency.
  (a) What was the experimental demonstration?
  (b) Explain briefly the basic principle of cloaking.
  (c) Suggest a practical application for cloaking.

  (Relevant articles are found on the web.)

**19.2.** Explain the operating principles for
  (a) Metamaterial antennas
  (b) Superlenses
  (c) MRI enhancement using metamaterial technology

**19.3.** What are acoustic metamaterials. How do they work and what are some of the effects they produce?

**19.4.** Of all the applications of metamaterials you have encountered in this chapter or elsewhere, which do you think will have the greatest impact and why?

# Appendix A

# Basic concepts in mechanics

In this section, we will define some of the fundamental concepts in mechanics. We assume that the reader is familiar with these concepts and that here a simple summary will be sufficient. A detailed discussion can be found in basic physics texts, some of which are listed in the Bibliography.

## A.1 Speed and velocity

Velocity is defined as the rate of change of position with respect to time. Both magnitude and direction are necessary to specify velocity. Velocity is, therefore, a vector quantity. The magnitude of the velocity is called *speed*. In the special case when the velocity of an object is constant, the distance s traversed in time t is given by

$$s = vt \qquad \text{(A.1)}$$

In this case, velocity can be expressed as

$$v = \frac{s}{t} \qquad \text{(A.2)}$$

If the velocity changes along the path, the expression $s/t$ yields the average velocity.

## A.2 Acceleration

If the velocity of an object along its path changes from point to point, its motion is said to be *accelerated* (or decelerated). Acceleration is defined as the rate of change in velocity with respect to time. In the special case of uniform acceleration, the final velocity v of an object that has been accelerated for a time t is

$$v = v_0 + at \qquad \text{(A.3)}$$

Here $v_0$ is the initial velocity of the object, and a is the acceleration.[1] Acceleration can, therefore, be expressed as

$$a = \frac{v - v_0}{t} \tag{A.4}$$

In the case of uniform acceleration, a number of useful relations can be simply derived. The average velocity during the interval t is

$$v_{av} = \frac{v + v_0}{2} \tag{A.5}$$

The distance traversed during this time is

$$s = v_{av}t \tag{A.6}$$

Using Eqs. A.4 and A.5, we obtain

$$s = v_0 t + \frac{at^2}{2} \tag{A.7}$$

By substituting $t = (v - v_0)/a$ (from Eq. A.4) into Eq. A.7, we obtain

$$v^2 = v_0^2 + 2as \tag{A.8}$$

## A.3 Force

Force is a push or a pull exerted on a body that tends to change the state of motion of the body.

## A.4 Pressure

Pressure is the force applied to a unit area.

---

1. Both velocity and acceleration may vary along the path. In general, velocity is defined as the time derivative of the distance along the path of the object; that is,

$$v = \lim_{\Delta t \to a0} \frac{\Delta s}{\Delta t} = \frac{ds}{dt}$$

Acceleration is defined as the time derivative of the velocity along the path; that is,

$$a = \frac{dv}{dt} = \frac{d}{dt}\left(\frac{ds}{dt}\right) = \frac{d^2s}{dt^2}$$

## A.5  Mass (m)

We have stated that a force applied to a body tends to change its state of motion. All bodies have the property of resisting change in their motion. Mass is a quantitative measure of inertia or the resistance to a change in motion.

## A.6  Weight (w)

Every mass exerts an attractive force on every other mass; this attraction is called the *gravitational force*. The weight of a body is the force exerted on the body by the mass of the Earth. The weight of a body is directly proportional to its mass (W = mg). Weight being a force is a vector, and it points vertically down in the direction of a suspended plumb line.

Mass and weight are related but distinct properties of an object. If a body were isolated from all other bodies, it would have no weight, but it would still have mass.

## A.7  Linear momentum

Linear momentum of a body is the product of its mass and velocity; that is,

$$\text{Linear momentum} = mv \tag{A.9}$$

## A.8  Newton's laws of motion

The foundations of mechanics are Newton's three *laws of motion*. The laws are based on observation, and they cannot be derived from more basic principles. These laws can be stated as follows.

1. *A body remains at rest or in a state of uniform motion in a straight line unless it is acted on by an applied force.*
2. *The time rate of change of the linear momentum of a body is equal to the force F applied to it.*

Except at very high velocities, where relativistic effects must be considered, the second law can be expressed mathematically in terms of the mass m and acceleration a of the object as[2]

---

2. The second law can be expressed mathematically in terms of the time derivative of momentum: that is,

$$\text{Force} = \big|_{\Delta t \to 0} \frac{mv(t+\Delta t) - mv(t)}{\Delta t} = \frac{d}{dt}(mv) = m\frac{dv}{dt} = ma$$

$$F = ma \tag{A.10}$$

This is one of the most commonly used equations in mechanics. It shows that if the applied force and the mass of the object are known, the acceleration can be calculated. When the acceleration is known, the velocity of the object and the distance traveled can be computed from the previously given equations.

The Earth's gravitational force, like all other forces, causes an acceleration. By observing the motion of freely falling bodies, this acceleration has been measured. Near the surface of the Earth, it is approximately $9.8 \, \text{m/sec}^2$. Because gravitational acceleration is frequently used in computations, it has been given a special symbol g. Therefore, the gravitational force on an object with mass m is

$$F_{\text{gravity}} = mg \tag{A.11}$$

This is, of course, also the weight of the object.

**3.** *For every action, there is an equal and opposite reaction.*

This third law implies that when two bodies $A$ and $B$ interact so that $A$ exerts a force on $B$, a force of the same magnitude but opposite in direction is exerted by $B$ on $A$. A number of illustrations of the third law are given in the text.

## A.9 Conservation of linear momentum

It follows from Newton's laws that the total linear momentum of a system of objects remains unchanged unless acted on by an outside force.

## A.10 Radian

In the analysis of rotational motion, it is convenient to measure angles in a unit called a *radian*. With reference to Fig. A.1, the angle in radian units is defined as

$$\theta = \frac{s}{r} \tag{A.12}$$

where s is the length of the circular arc and r is the radius of rotation. In a full circle, the arc length is the circumference $2\pi r$. Therefore in radian units the angle in a full circle is

$$\theta = \frac{2\pi r}{r} = 2\pi \, \text{rad}$$

Hence,

$$1 \, \text{rad} = \frac{360°}{2\pi} = 57.3°$$

**FIGURE A.1**    **The radian.**

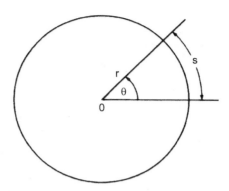

## A.11 Angular velocity

The angular velocity ω is the angular displacement per unit time; that is, if a body rotates through an angle θ (in radians) in a time t, the angular velocity is

$$\omega = \frac{\theta}{t} \ (\text{rad/sec}) \tag{A.13}$$

## A.12 Angular acceleration

Angular acceleration α is the time rate of change of angular velocity. If the initial angular velocity is $\omega_0$ and the final angular velocity after a time t is $\omega_f$, the angular acceleration is[3]

$$\alpha = \frac{\omega_f - \omega_0}{t} \tag{A.14}$$

## A.13 Relations between angular and linear motion

As an object rotates about an axis, each point in the object travels along the circumference of a circle; therefore, each point is also in linear motion. The linear distance s traversed in angular motion is

$$s = r\theta$$

---

3. Both angular velocity and angular acceleration may vary along the path. In general, the instantaneous angular velocity and acceleration are defined as

$$\omega = \frac{d\theta}{dt}; \ \alpha = \frac{d\omega}{dt} = \frac{d^2\theta}{dt^2}$$

**TABLE A.1** Equations for rotational motion (angular acceleration, $\alpha = $ constant).

| |
|---|
| $\omega = \omega_0 + \alpha t$ |
| $\theta = \omega_0 t + \frac{1}{2}\alpha t^2$ |
| $\omega^2 = \omega_0^2 + 2\alpha\theta$ |
| $\omega_{av} = \frac{(\omega_0 + \omega)}{2}$ |

The linear velocity $v$ of a point that is rotating at an angular velocity $\omega$ at a distance r from the center of rotation is

$$v = r\omega \tag{A.15}$$

The direction of the vector v is at all points tangential to the path s. The linear acceleration along the path s is

$$a = r\alpha \tag{A.16}$$

## A.14 Equations for angular momentum

The equations for angular motion are analogous to the equations for translational motion. For a body moving with a constant angular acceleration $\alpha$ and initial angular velocity $\omega_0$, the relationships are shown in Table A.1.

## A.15 Centripetal acceleration

As an object rotates uniformly around an axis, the magnitude of the linear velocity remains constant, but the direction of the linear velocity is continuously changing. The change in velocity always points toward the center of rotation. Therefore, a rotating body is accelerated toward the center of rotation. This acceleration is called *centripetal* (center-seeking) *acceleration*. The magnitude of the centripetal acceleration is given by

$$a_c = \frac{v^2}{r} = \omega^2 r \tag{A.17}$$

where r is the radius of rotation and v is the speed tangential to the path of rotation. Because the body is accelerated toward its center of rotation, we conclude from Newton's second law that a force pointing toward the center of rotation must act on the body. This force, called the *centripetal force* $F_c$, is given by

$$F_c = ma_c = \frac{mv^2}{r} = m\omega^2 r \tag{A.18}$$

where m is the mass of the rotating body.

For a body to move along a curved path, a centripetal force must be applied to it. In the absence of such a force, the body moves in a straight line, as required by Newton's first law. Consider, for example, an object twirled at the end of a rope. The centripetal force is applied by the rope on the object. From Newton's third law, an equal but opposite reaction force is applied on the rope by the object. The reaction to the centripetal force is called the *centrifugal force*. This force is in the direction away from the center of rotation. The centripetal force, which is required to keep the body in rotation, always acts perpendicular to the direction of motion and, therefore, does no work (see Eq. A.28). In the absence of friction, energy is not required to keep a body rotating at a constant angular velocity.

## A.16 Moment of inertia

The moment of inertia in angular motion is analogous to mass in translational motion. The moment of inertia I of an element of mass m located a distance r from the center of rotation is

$$I = mr^2 \tag{A.19}$$

In general, when an object is in angular motion, the mass elements in the body are located at different distances from the center of rotation. The total moment of inertia is the sum of the moments of inertia of the mass elements in the body.

Unlike mass, which is a constant for a given body, the moment of inertia depends on the location of the center of rotation. In general, the moment of

**TABLE A.2** Moments of inertia of some simple bodies.

| Body | Location of axis | Moment of inertia |
| --- | --- | --- |
| A thin rod of length l | Through the center | $ml^2/12$ |
| A thin rod of length l | Through one end | $ml^2/3$ |
| Sphere of radius r | Along a diameter | $2mr^2/5$ |
| Cylinder of radius r | Along axis of symmetry | $mr^2/2$ |

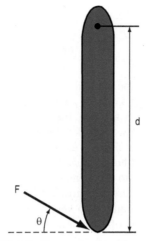

**FIGURE A.2    Torque produced by a force.**

inertia is calculated by using integral calculus. The moments of inertia for a few objects useful for our calculations are shown in Table A.2.

## A.17 Torque

Torque is defined as the tendency of a force to produce rotation about an axis. Torque, which is usually designated by the letter L, is given by the product of the perpendicular force and the distance d from the point of application to the axis of rotation; that is (see Fig. A.2),

$$L = F\cos\theta \times d \tag{A.20}$$

The distance d is called the *lever arm* or *moment arm.*

## A.18 Newton's laws of angular motion

The laws governing angular motion are analogous to the laws of translational motion. Torque is analogous to force, and the moment of inertia is analogous to mass.

A body in rotation will continue its rotation with a constant angular velocity unless acted upon by an external torque.

The mathematical expression of the second law in angular motion is analogous to Eq. A.10. It states that the torque is equal to the product of the moment of inertia and the angular acceleration; that is,

$$L = I\alpha \tag{A.21}$$

For every torque, there is an equal and opposite reaction torque.

**FIGURE A.3**   **The resolution of a force into its vertical and horizontal components.**

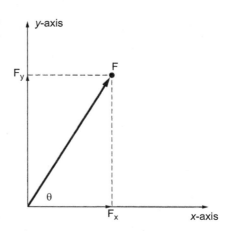

## A.19 Angular momentum

Angular momentum is defined as

$$\text{Angular momentum} = I\omega \qquad (A.22)$$

From Newton's laws, it can be shown that angular momentum of a body is conserved if there is no unbalanced external torque acting on the body.

## A.20 Addition of forces and torques

Any number of forces and torques can be applied simultaneously to a given object. Because forces and torques are vectors, characterized by both a magnitude and a direction, their net effect on a body is obtained by vectorial addition. When it is required to obtain the total force acting on a body, it is often convenient to break up each force into mutually perpendicular components. This is illustrated for the two-dimensional case in Fig. A.3. Here we have chosen the horizontal $x$- and the vertical $y$-directions as the mutually perpendicular axes. In a more general three-dimensional case, a third axis is required for the analysis.

The two perpendicular components of the force F are

$$F_x = F\cos\theta$$
$$F_y = F\sin\theta \qquad (A.23)$$

The magnitude of the force F is given by

$$F = \sqrt{F_x^2 + F_y^2} \qquad (A.24)$$

When adding a number of forces $(F_1, F_2, F_3, ...)$ the mutually perpendicular components of the total force $F_T$ are obtained by adding the corresponding components of each force; that is,

$$(F_T)_x = (F_1)_x + (F_2)_x + (F_3)_x + \cdots$$
$$(F_T)_y = (F_1)_y + (F_2)_y + (F_3)_y + \cdots \tag{A.25}$$

The magnitude of the total force is

$$F_T = \sqrt{(F_T)_x^2 + (F_T)_y^2} \tag{A.26}$$

The torque produced by a force acts to produce a rotation in either a clockwise or a counterclockwise direction. If we designate one direction of rotation as positive and the other as negative, the total torque acting on a body is obtained by the addition of the individual torques, each with the appropriate sign.

## A.21 Static equilibrium

A body is in static equilibrium if both its linear and angular accelerations are zero. To satisfy this condition, the sum of the forces F acting on the body, as well as the sum of the torques L produced by these forces, must be zero; that is,

$$\sum F = 0 \text{ and} \sum L = 0 \tag{A.27}$$

## A.22 Work

In our everyday language, the word *work* denotes any type of effort, whether physical or mental. In physics, a more rigorous definition is required. Here work is defined as the product of force and the distance through which the force acts. Only the force parallel to the direction of motion does work on the object. This is illustrated in Fig. A.4. A force F applied at an angle $\theta$ pulls the object along the surface through a distance D. The work done by the force is

$$\text{Work} = F \cos\theta \times D \tag{A.28}$$

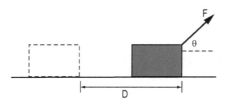

**FIGURE A.4 Work done by a force.**

## A.23 Energy

Energy is an important concept. We find reference to energy in connection with widely different phenomena. We speak of atomic energy, heat energy, potential energy, solar energy, chemical energy, and kinetic energy; we even speak of people as being full of energy. The common factor that ties together these manifestations is the possibility of obtaining work from these sources. The connection between energy and work is simple: Energy is required to do work. Energy is measured in the same units as work; in fact, there is a one-to-one correspondence between them. It takes 2 J of energy to do 2 J of work. In all physical processes, energy is conserved. Through work, one form of energy can be converted into another, but the total amount of energy remains unchanged.

## A.24 Forms of energy

### A.24.1 Kinetic energy

Objects in motion can do work by virtue of their motion. For example, when a moving object hits a stationary object, the stationary object is accelerated. This implies that the moving object applied a force on the stationary object and performed work on it. The kinetic energy (KE) of a body with mass m moving with a velocity v is

$$KE = \frac{1}{2}\,mv^2 \tag{A.29}$$

In rotational motion, the kinetic energy is

$$KE = \frac{1}{2}I\omega^2 \tag{A.30}$$

### A.24.2 Potential energy

Potential energy of a body is the ability of the body to do work because of its position or configuration. A body of weight W raised to a height H with respect to a surface has potential energy (PE)

$$PE = WH \tag{A.31}$$

This is the amount of work that had to be performed to raise the body to height H. The same amount of energy can be retrieved by lowering the body back to the surface.

A stretched or compressed spring possesses potential energy. The force required to stretch or compress a spring is directly proportional to the length of the stretch or compression (s); that is,

$$F = ks \tag{A.32}$$

Here k is the spring constant. The potential energy stored in the stretched or compressed spring is

$$PE = \frac{1}{2} ks^2 \tag{A.33}$$

### A.24.3 Heat

Heat is a form of energy, and as such, it can be converted to work and other forms of energy. Heat, however, is not equal in rank with other forms of energy. While work and other forms of energy can be completely converted to heat, heat energy can only be converted partially to other forms of energy. This property of heat has far-reaching consequences, which are discussed in Chapter 10.

Heat is often measured in calorie units. One calorie (cal) is the amount of heat required to raise the temperature of 1 g of water by 1 C°. The heat energy required to raise the temperature of a unit mass of a substance by 1°C is called *specific heat*. One calorie is equal to 4.184 J.

A heat unit frequently used in chemistry and in food technology is the *kilocalorie* or Cal, which is equal to 1000 cal.

### A.25 Power

The amount of work done—or energy expended—per unit time is called *power*. The algebraic expression for power is

$$P = \frac{\Delta E}{\Delta t} \tag{A.34}$$

where $\Delta E$ is the energy expended in a time interval $\Delta t$.

## A.26 Units and conversions

In our calculations we will mostly use SI units in which the basic units for length, mass, and time are meter, kilogram, and second, respectively. However, other units are also encountered in the text. Units and conversion factors for the most commonly encountered quantities are listed here with their abbreviations.

### A.26.1 Length

$$SI\ unit: meter\ (m)$$
$$Conversions: 1\ m = 100\ cm\ (centimeter) = 1000\ mm\ (milimeter)$$
$$1000\ m = 1\ km$$
$$1\ m = 3.28\ feet = 39.37\ in$$
$$1\ km = 0.621\ miles$$
$$1\ in = 2.54\ cm$$

In addition, the micron and the angstrom are used frequently in physics and biology.

$$1\ micron\ (\mu m) = 10^{-6}\ m = 10^{-4}\ cm$$
$$1\ angstrom\ \left(\overset{\circ}{A}\right) = 10^{-8} cm$$

### A.26.2 Mass

$$SI\ unit: kilogram\ (kg)$$
$$Conversions: 1\ kg = 1000\ g$$
$$The\ weight\ of\ a\ 1-kg\ mass\ is\ 9.8\ newton\ (N).$$

### A.26.3 Force

$$SI\ unit: kg\ m/s^2, name\ of\ unit: newton\ (N)$$
$$Conversions: 1\ N = 10^5\ dynes\ (dyn) = 0.225\ lb$$

## A.26.4  Pressure

$$\textbf{\textit{SI unit}}: kg/m\ s^2, name\ of\ unit: pascal\ (Pa)$$

$$\textbf{\textit{Conversions}}: 1\ Pa\ =\ 10\frac{dynes}{cm^2}\ =\ 9.87 \times 10^{-6} atmospheres\ (atm)$$

$$=\ 1.45 \times 10^{-4}\ lb/in^2$$

$$1\ atm\ =\ 1.01 \times 10^5\ Pa\ =\ 760\ mm\ Hg\ (torr)$$

## A.26.5  Energy

$$\textbf{\textit{SI unit}}: kg\ m^2/s^2, name\ of\ unit: joule\ (J)$$

$$\textbf{\textit{Conversions}}: 1\ J\ =\ 1\ Nm\ =\ 10^7\ ergs\ =\ 0.239\ cal\ =\ 0.738\ ft\ lb.$$

## A.26.6  Power

$$\textbf{\textit{SI unit}}: J/s, name\ of\ unit: watt\ (W)$$

$$\textbf{\textit{Conversions}}: 1\ W\ =\ \frac{10^7\ ergs}{sec}\ =\ 1.34 \times 10^{-3}\ horsepower\ (hp)$$

# Appendix B

# Review of electricity

## B.1 Electric charge

Matter is composed of atoms. An atom consists of a nucleus surrounded by electrons. The nucleus itself is composed of protons and neutrons. Electric charge is a property of protons and electrons. There are two types of electric charge: Positive and negative. The proton is positively charged, and the electron is negatively charged. All electrical phenomena are due to these electric charges.

Charges exert forces on each other. Unlike charges attract and like charges repel each other. The electrons are held around the nucleus by the electrical attraction of the protons. Although the proton is about 2000 times heavier than the electron, the magnitude of the charge on the two is the same. There are as many positively charged protons in an atom as negatively charged electrons. The atom as a whole is, therefore, electrically neutral. The identity of an atom is determined by the number of protons in the nucleus. Thus, for example, hydrogen has 1 proton; nitrogen has 7 protons; and gold has 79 protons.

It is possible to remove electrons from an atom, making it positively charged. Such an atom with missing electrons is called a *positive ion*. It is also possible to add an electron to an atom which makes it a *negative ion*.

Electric charge is measured in coulombs (C). The magnitude of the charge on the proton and the electron is $1.60 \times 10^{-19}$ C. The force F between two charged bodies is proportional to the product of their charges $Q_1$ and $Q_2$ and is inversely proportional to the square of the distance R between them; that is,

$$F = \frac{-KQ_1Q_2}{R^2} \tag{B.1}$$

The negative sign takes into account that like charges repel and unlike charges attract.

This equation is known as *Coulomb's law*. If R is measured in meters, the constant K is $9 \times 10^9$, and F is obtained in newtons.

## B.2 Electric field

An electric charge exerts a force on another electric charge; a mass exerts a force on another mass; and a magnet exerts a force on another magnet. All these forces have an important common characteristic: Exertion of the force does not require physical contact between the interacting bodies. The forces act at a distance. The concept of *lines of force* or *field lines* is useful in visualizing these forces which act at a distance.

Any object that exerts a force on another object without contact can be thought of as having lines of force emanating from it. The complete line configuration is called a *force field*. The lines point in the direction of the force, and their density at any point in space is proportional to the magnitude of the force at that point.

The lines of force emanate from an electric charge uniformly in all directions. By convention, the lines point in the direction of the force that the source charge exerts on a positive charge. Thus, the lines of force point away from a positive source charge and into a negative source charge (see Fig. B.1). The number of lines emanating from the charge is proportional to the magnitude of the electric charge. If the size of the source charge is doubled, the number of force lines is also doubled.

Lines of force need not be straight lines; as we mentioned, they point in the direction in which the force is exerted. As an example, we can consider the net field due to two charges separated by a distance d. To determine this field, we must compute the direction and size of the net force on a positive charge at all points in space. This is done by adding vectorially the force lines due to each charge. The force field due to a positive and negative charge of equal

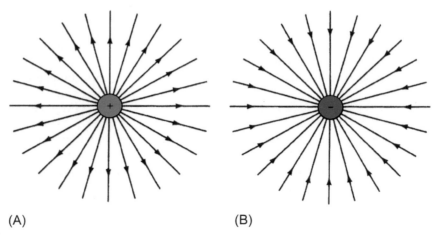

(A)                                      (B)

**FIGURE B.1** Two-dimensional representation of the electric field produced by **(A)** a positive point charge and **(B)** a negative point charge.

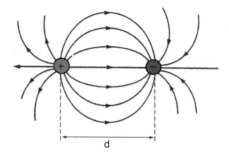

**FIGURE B.2**   Lines of force produced by a positive and a negative charge separated by a distance d.

magnitude separated by a distance d from each other is shown in Fig. B.2. Here the lines of force are curved. This is, of course, the direction of the net force on a positive charge in the region surrounding the two fixed charges. The field shown in Fig. B.2 is called a *dipole field*, and it is similar to the field produced by a bar magnet.

## B.3 Potential difference or voltage

The electric field is measured in units of volts per meter (or volts per centimeter). The product of the electric field and the distance over which the field extends is an important parameter which is called *potential difference* or *voltage*. The voltage (V) between two points is a measure of energy transfer as the charge moves between the two points. Potential difference is measured in volts. If there is a potential difference between two points, a force is exerted on a charge placed in the region between these points. If the charge is positive, the force tends to move it away from the positive point and toward the negative point.

## B.4 Electric current

An electric current is produced by a motion of charges. The magnitude of the current depends on the amount of charge flowing past a given point in a given period of time. Current is measured in amperes (A). One ampere is 1 coulomb (C) of charge flowing past a point in 1 second (sec).

## B.5 Electric circuits

The amount of current flowing between two points in a material is proportional to the potential difference between the two points and to the electrical properties of the material. The electrical properties are usually represented by three parameters: Resistance, capacitance, and inductance. Resistance measures the opposition to current flow. This parameter depends on the property of the material called *resistivity*, and it is analogous to friction in mechanical motion. Capacitance measures the ability of the material to store electric charges. Inductance measures the opposition in the material to changes in current flow. All materials exhibit to some extent all three of these properties; often, however, one of these properties is predominant. It is possible to manufacture components with specific values of resistance, capacitance, or inductance. These are called, respectively, *resistors, capacitors*, and *inductors*.

The schematic symbols for these three electrical components are shown in Fig. B.3. Electrical components can be connected together to form an electric circuit. Currents can be controlled by the appropriate choice of components and interconnections in the circuit. An example of an electric circuit is shown in Fig. B.4. Various techniques have been developed to analyze such circuits and to calculate voltages and currents at all the points in the circuit.

### B.5.1 Resistor

The resistor is a circuit component that opposes current flow. Resistance (R) is measured in units of ohm ($\Omega$). The relation between current (I) and voltage (V) is given by Ohm's law, which is

$$V = IR \tag{B.2}$$

Materials that present a very small resistance to current flow are called *conductors*. Materials with a very large resistance are called *insulators*. A flow of current through a resistor is always accompanied by power dissipation as electrical energy is converted to heat. The power (P) dissipated in a resistor is given by

**FIGURE B.3   Circuit components.**

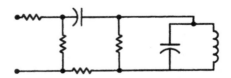

**FIGURE B.4   Example of an electric circuit.**

$$P = I^2R \qquad (B.3)$$

The inverse of resistance is called *conductance*, which is usually designated by the symbol G. Conductance is measured in units of *mho*, also called *Siemens*. The relationship between conductance and resistance is

$$G = \frac{1}{R} \qquad (B.4)$$

## B.5.2 Capacitor

The capacitor is a circuit element that stores electric charges. In its simplest form it consists of two conducting plates separated by an insulator (see Fig. B.5). Capacitance (C) is measured in *farads*. The relation between the stored charge (Q) and the voltage across the capacitor is given by

$$Q = CV \qquad (B.5)$$

In a charged capacitor, positive charges are on one side of the plate, and negative charges are on the other. The amount of energy (E) stored in such a configuration is given by

$$E = \frac{1}{2}CV^2 \qquad (B.6)$$

## B.5.3 Inductor

The *inductor* is a device that opposes a change in the current flowing through it. Inductance is measured in units called *henry*.

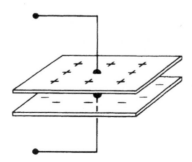

FIGURE B.5    **A simple capacitor.**

## B.6 Voltage and current sources

Voltages and currents can be produced by various batteries and generators. Batteries are based on chemical reactions that result in a separation of positive and negative charges within a material. Generators produce a voltage by the motion of conductors in magnetic fields. The circuit symbols for these sources are shown in Fig. B.6.

## B.7 Electricity and magnetism

Electricity and magnetism are related phenomena. A changing electric field always produces a magnetic field, and a changing magnetic field always produces an electric field. All electromagnetic phenomena can be traced to this basic interrelationship. A few consequences of this interaction are given as follows:

1. An electric current always produces a magnetic field in a direction perpendicular to the current flow.
2. A current is induced in a conductor that moves perpendicular to a magnetic field.
3. An oscillating electric charge emits electromagnetic waves at the frequency of oscillation. This radiation propagates away from the source at the speed of light. Radio waves, light, and X-rays are examples of electromagnetic radiation.

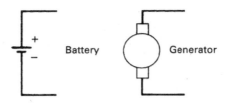

**FIGURE B.6** Circuit symbols for a battery and a generator.

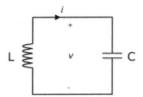

**FIGURE B.7** A parallel inductor-capacitor circuit. *(First Harmonic, CC BY-SA 3.0 https://creativecommons.org/licenses/by-sa/3.0, via Wikimedia Commons.)*

## B.8 Parallel L-C resonant circuit

A parallel inductor-capacitor circuit is characterized by a resonance frequency $f_{res}$ given by

$$f_{res} = \frac{1}{2\pi\sqrt{L_{eff}\,C_{eff}}}$$

$L_{eff}$ and $C_{eff}$ are the effective inductor and capacitor in the electrical circuit, respectively. An electromagnetic signal at the resonance frequency produces a large response in the circuit.

# Appendix C

# Review of optics

## C.1 Geometric optics

The characteristics of optical components, such as mirrors and lenses, can be completely derived from the wave properties of light. Such detailed calculations, however, are usually rather complex because one has to keep track of the wave front along every point on the optical component. It is possible to simplify the problem if the optical components are much larger than the wavelength of light. The simplification entails neglecting some of the wave properties of light and considering light as a ray traveling perpendicular to the wave front (Fig. C.1). In a homogeneous medium, the ray of light travels in a straight line; it alters direction only at the interface between two media. This simplified approach is called *geometric optics*.

The speed of light depends on the medium in which it propagates. In vacuum, light travels at a speed of $3 \times 10^8$ m/sec. In a material medium, the speed of light is always less. The speed of light in a material is characterized by the index of refraction (n), defined as

$$n = \frac{c}{v} \tag{C.1}$$

where c is the speed of light in vacuum and v is the speed in the material. When light enters from one medium into another, its direction of propagation

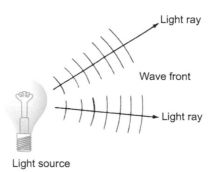

Light ray

Wave front

Light ray

Light source

FIGURE C.1 **Light rays perpendicular to the wave front.**

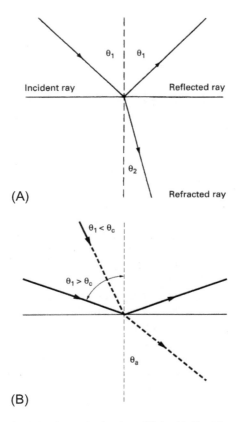

(A)

(B)

**FIGURE C.2** **(A)** Reflection and refraction of light. **(B)** Total internal reflection.

is changed (see Fig. C.2). This phenomenon is called *refraction*. The relationship between the angle of incidence ($\theta_1$) and the angle of refraction ($\theta_2$) is given by

$$\frac{\sin\theta_1}{\sin\theta_2} = \frac{n_2}{n_1} \tag{C.2}$$

The relationship in Eq. C.2 is called *Snell's law*. As shown in Fig. C.2, some of the light is also reflected. The angle of reflection is always equal to the angle of incidence.

In Fig. C.2a, the angle of incidence $\theta_1$ for the entering light is shown to be greater than the angle of refraction $\theta_2$. This implies that $n_2$ is greater than $n_1$, as would be the case for light entering from air into glass, for example (see Eq. C.2). If, on the other hand, the light originates in the medium of higher refractive index, as shown in Fig. C.2b, then the angle of incidence $\theta_1$ is smaller than the angle of refraction $\theta_2$. At a specific value of angle $\theta_1$ called the *critical angle* (designated by the symbol $\theta_c$), the light emerges tangent to the surface, that is, $\theta_2 = 90°$. At this point, $\sin\theta_2 = 1$ and, therefore, $\sin\theta_1 = \sin\theta_c = n_2/n_1$. Beyond this angle, that is, for $\theta_1 > \theta_c$, light originating in the medium of higher refractive index does not emerge from the medium. At the interface, all the light is reflected back into the medium. This phenomenon is called *total internal reflection*. For glass, $n_2$ is typically 1.5, and the critical angle at the glass–air interface is $\sin\theta_c = 1/1.5$ or $\theta_c = 42°$.

Transparent materials such as glass can be shaped into lenses to alter the direction of light in a specific way. Lenses fall into two general categories: Converging lenses and diverging lenses. A converging lens alters the direction of light so that the rays are brought together. A diverging lens has the opposite effect; it spreads the light rays apart.

Using geometric optics, we can calculate the size and shape of images formed by optical components, but we cannot predict the inevitable blurring of images that occurs as a result of the wave nature of light.

## C.2 Converging lenses

A simple converging lens is shown in Fig. C.3. This type of lens is called a convex lens.

Parallel rays of light passing through a convex lens converge at a point called the *principal focus of the lens*. The distance of this point from the lens is called the *focal length f*. Conversely, light from a point source at the focal point emerges from the lens as a parallel beam. The focal length of the lens is

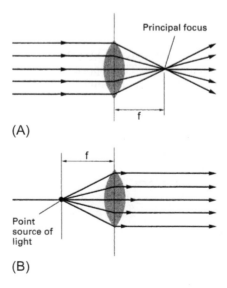

**FIGURE C.3**  The convex lens illuminated (**A**) by parallel light and (**B**) by point source at the focus.

determined by the index of refraction of the lens material and the curvature of the lens surfaces. We adopt the following convention in discussing lenses.

1. Light travels from left to right.
2. The radius of curvature is positive if the curved surface encountered by the light ray is convex; it is negative if the surface is concave.

It can be shown that for a thin lens, the focal length is given by

$$\frac{1}{f} = (n-1)\left(\frac{1}{R_1} - \frac{1}{R_2}\right) \qquad (C.3)$$

where $R_1$ and $R_2$ are the curvatures of the first and second surfaces, respectively (Fig. C.4). In Fig. C.4, $R_2$ is a negative number.

Focal length is a measure of the converging power of the lens. The shorter the focal length, the more powerful the lens. The focusing power of a lens is often expressed in diopters, defined as

$$\text{Focusing power} = \frac{1}{f\,(\text{meters})} (\text{diopters}) \qquad (C.4)$$

If two thin lenses with focal lengths $f_1$ and $f_2$, respectively, are placed close together, the focal length $f_T$ of the combination is

$$\frac{1}{f_T} = \frac{1}{f_1} + \frac{1}{f_2} \qquad (C.5)$$

Light from a point source located beyond the focal length of the lens is converged to a point image on the other side of the lens (Fig. C.5a). This type of image is called a *real image* because it can be seen on a screen placed at the point of convergence.

If the distance between the source of light and the lens is less than the focal length, the rays do not converge. They appear to emanate from a point on the source side of the lens. This apparent point of convergence is called a *virtual image* (Fig. C.5b).

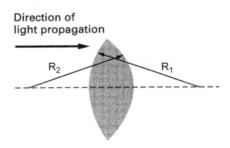

**Direction of light propagation**

$R_2$    $R_1$

**FIGURE C.4    Radius of curvature defined for a lens.**

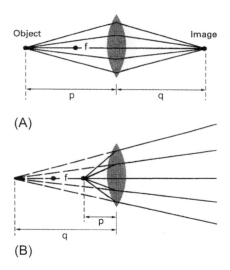

**FIGURE C.5    Image formation by a convex lens.** (A) Real image, (B) virtual image.

For a thin lens, the relationship between the source and the image distances from the lens is given by

$$\frac{1}{p}+\frac{1}{q} = \frac{1}{f} \tag{C.6}$$

Here p and q, respectively, are the source and the image distances from the lens. By convention, q in this equation is taken as positive if the image is formed on the side of the lens opposite to the source and negative if the image is formed on the source side.

Light rays from a source very far from the lens are nearly parallel; therefore, by definition, we would expect them to be focused at the principal focal point of the lens. This is confirmed by Eq. C.6, which shows that as p becomes very large (approaches infinity), q is equal to f.

If the source is displaced a distance x from the axis, the image is formed at a distance y from the axis such that

$$\frac{y}{x} = \frac{q}{p} \tag{C.7}$$

This is illustrated for a real image in Fig. C.6. The relationship between p and q is still given by Eq. C.6.

## C.3 Images of extended objects

So far, we have discussed only the formation of images from point sources. The treatment, however, is easily applied to objects of finite size.

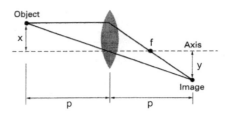

**FIGURE C.6**  **Image formation off-axis.**

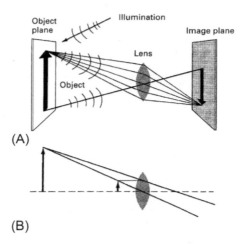

(A)

(B)

**FIGURE C.7**  **Image of an object.** (A) Real image, (B) virtual image.

When an object is illuminated, light rays emanate from every point on the object (Fig. C.7a). Each point on the object plane a distance p from the lens is imaged at the corresponding point on the image plane a distance q from the lens. The relationship between the object and the image distances is given by Eq. C.6. As shown in Fig. C.7, real images are inverted and virtual images are upright. The ratio of image to object height is given by

$$\frac{\text{Image height}}{\text{Object height}} = -\frac{q}{p} \qquad \text{(C.8)}$$

## C.4 Diverging lenses

An example of a diverging lens is the concave lens shown in Fig. C.8. Parallel light diverges after passing through a concave lens. The apparent source of origin for the diverging rays is the focal point of the concave lens. All the equations we have presented for the converging lens apply in this case also, provided the sign conventions are obeyed. From Eq. C.3, it follows that the

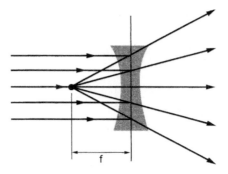

FIGURE C.8 **A diverging lens.**

focal length for a diverging lens is always negative and the lens produces only virtual images (Fig. C.8).

## C.5 Lens immersed in a material medium

The lens equations that we have presented so far apply in the case when the lens is surrounded by air that has a refraction index of approximately 1. Let us now consider the more general situation shown in Fig. C.9, which we will need in our discussion of the eye. The lens here is embedded in a medium that has a different index of refraction ($n_1$ and $n_2$) on each side of the lens. It can be shown (see for e.g., Marshall, 1967) that under these conditions, the relationship between the object and the image distances is

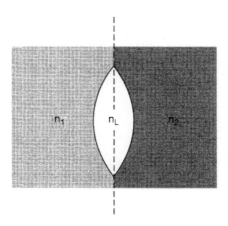

FIGURE C.9 **Lens immersed in a material medium.**

$$\frac{n_1}{p} + \frac{n_2}{q} = \frac{n_L - n_1}{R_1} - \frac{n_L - n_2}{R_2} \tag{C.9}$$

Here $n_L$ is the refraction index of the lens material. The effective focal length in this case is

$$\frac{1}{f} = \frac{n_L - n_1}{R_1} - \frac{n_L - n_2}{R_2} \tag{C.10}$$

Note that in air $n_1 = n_2 = 1$ and Eq. C.10 reduces to Eq. C.3.

The lens equations we have presented in this appendix assume that the lenses are thin. This is not a fully valid assumption for the lenses in the eye. Nevertheless, these equations are adequate for our purposes.

# Bibliography

## CHAPTERS 1 TO 6

Alexander RMN. Animal mechanics. London: Sidgwick and Jackson; 1968.

Baez AV. The new college physics: a spiral approach. San Francisco, CA: W. H. Freeman and Co.; 1967.

Blesser WB. A systems approach to biomedicine. New York, NY: McGraw-Hill Book Co.; 1969.

Bootzin D, Muffley HC. Biomechanics. New York, NY: Plenum Press; 1969.

Cameron JR, Skofronick JG, Grant RM. Physics of the body. Madison, WI: Medical Physics Publishing; 1992.

Chapman RF. The insects. New York, NY: American Elsevier Publishing Co.; 1969.

Conaghan PG. Update on osteoarthritis part 1: current concepts and the relation to exercise. Br J Sports Med 2002;36:330−3.

Cooper JM, Glassow RB. Kinesiology. 3rd ed. St. Louis, MO: The C. V. Mosby Co.; 1972.

Cromer AH. Physics for the life sciences. New York, NY: McGraw-Hill Book Co.; 1974.

Frankel VH, Burstein AH. Orthopaedic biomechanics. Philadelphia, PA: Lea and Febiger; 1970.

French AP. Newtonian mechanics. New York, NY: W. W. Norton & Co., Inc.; 1971.

Frost HM. An introduction to biomechanics. Springfield, IL: Charles C Thomas, Publisher; 1967.

Grandfield K. Bone, implants, and their interfaces. Phys Today 2015;68:40−5.

Gray J. How animals move. Cambridge: Cambridge University Press; 1953.

Heglund NC, Willems PA, Penta M, Cavagna GA. Energy-saving gait mechanics with head-supported loads. Nature 1995;375:52−4.

Hobbie R, Intermediate K. Physics for medicine and biology. New York, NY: Springer; 1997.

Ingber DE. The architecture of life. Sci Am 1998;278(1):47.

Jensen CR, Schultz GW. Applied kinesiology. New York, NY: McGraw-Hill Book Co.; 1970.

Kenedi RM, editor. Symposium on biomechanics and related bioengineering topics. New York, NY: Pergamon Press; 1965.

Latchaw M, Egstrom G. Human movement. Englewood Cliffs, NJ: Prentice-Hall; 1969.

Lauk M, Chow CC, Pavlik AE, Collins JJ. Human balance out of equilibrium: nonequilibrium statistical mechanics in posture control. Phys Rev Lett 1998;80:413.

Mathews DK, Fox EL. The physiological basis of physical education and athletics. Philadelphia, PA: W. B. Saunders and Co.; 1971.

McCormick EJ. Human factors engineering. New York, NY: McGraw-Hill Book Co.; 1970.

Morgan J. Introduction to university physics. 2nd ed. Vol. 1. Boston, MA: Allyn and Bacon; 1969.

Novacheck TF. The biomechanics of running. Gait Posture 1998;7:77−95.

Offenbacher EL. Physics and the vertical jump. Am J Phys 1970;38:829−36.

Richardson IW, Neergaard EB. Physics for biology and medicine. New York, NY: John Wiley & Sons; 1972.

Roddy E, et al. Evidence-based recommendations for the role of exercise in the management of osteoarthritis. Rheumatology 2005;44:67−73.

Rome LC. Testing a muscle's design. Am Sci 1997;85:356.

Strait LA, Inman VT, Ralston HJ. Sample illustrations of physical principles selected from physiology and medicine. Am J Phys 1947;15:375.

Sutton RM. Two notes on the physics of walking. Am J Phys 1955;23:490.

Wells KF. Kinesiology: the scientific basis of human motion. Philadelphia, PA: W. B. Saunders and Co.; 1971.

Williams M, Lissner HR. Biomechanics of human motion. Philadelphia, PA: W. B. Saunders Co.; 1962.

Winter DA. Human balance and posture control during standing and walking. Gait Posture 1995;3:193−214.

Wolff HS. Biomedical engineering. New York, NY: McGraw-Hill Book Co.; 1970.

Zhou ZR, Jin ZM. Biotribology: recent progresses and future perspectives. Biosurf Biotribol 2015;1:3−24.

## CHAPTER 7

Alexander RMN. Animal mechanics. London: Sidgwick and Jackson; 1968.

Bush JWM, Hu DL. Walking on water: biolocomotion at the interface. Annu Rev Fluid Mech 2006;38:339−69.

Chapman RF. The insects. New York, NY: American Elsevier Publishing Co.; 1969.

Foth HD, Turk LM. Fundamentals of soil science. New York, NY: John Wiley & Sons; 1972.

Gamow G, Ycas M. Mr. Tomkins inside himself. New York, NY: The Viking Press; 1967.

Gao X, Jiang L. Water-repellent legs of water striders. Nature 2004;432:36.

Hobbie R, Intermediate K. Physics for medicine and biology. New York, NY: Springer; 1997.

Morgan J. Introduction to university physics. 2nd ed. Boston, MA: Allyn and Bacon; 1969.

Murray JM, Weber A. The cooperative action of muscle proteins. Sci Am 1974;230(2):59.

Rome LC. Testing a muscle's design. Am Sci 1997;85:356.

## CHAPTER 8

Ackerman E. Biophysical sciences. Englewood Cliffs, NJ: Prentice Hall; 1962.

Hademenos GJ. The biophysics of stroke. Am Sci 1997;85:226.

Mahesh K, Vaidya S. Microfluidics: a boon for biological research. Curr Sci 2017;112:2021−8.

Morgan J. Introduction to university physics. 2nd ed. Boston, MA: Allyn and Bacon; 1969.

Myers GH, Parsonnet V. Engineering in the heart and blood vessels. New York, NY: John Wiley & Sons; 1969.

Richardson IW, Neergaard EB. Physics for biology and medicine. New York, NY: John Wiley & Sons; 1972.

Ruch TC, Patton HD, editors. Physiology and biophysics. Philadelphia, PA: W. B. Saunders Co.; 1965.

Strait LA, Inman VT, Ralston HJ. Sample illustrations of physical principles selected from physiology and medicine. Am J Phys 1947;15:375.

## CHAPTERS 9 TO 11

Ackerman E. Biophysical science. Englewood Cliffs, NJ: Prentice-Hall; 1962.

Akaishi T, et al. Chaos theory for clinical manifestations in multiple sclerosis. Med Hypotheses 2018 Jun;115:87−93. https://doi.org/10.1016/j.mehy.2018.04.004. Epub 2018 Apr 11. PMID, 29685206.

Angrist SW. "Perpetual motion machines." Scientific American; January 1968. p. 114.

Atkins PW. The 2nd law. New York, NY: W. H. Freeman and Co.; 1994.

Brown JHU, Gann DS, editors. Engineering principles in physiology. Vols. 1 and 2. New York, NY: Academic Press; 1973.

Carbone C, et al. Energetic constraints on the diet of terrestrial carnivores. Nature 1999 Nov 18;402:286—8. https://doi.org/10.1038/46266.

Casey E. J. Biophysics. New York, NY: Reinhold Publishing Corp.; 1962.

Church T, Martin CK. The obesity epidemic: a consequence of reduced energy expenditure and the uncoupling of energy Intake? Obesity 2018;26:14—6. https://doi.org/10.1002/oby.22072.

Cox CE. Role of physical activity for weight loss and weight maintenance. Diabetes Spectr 2017 Aug;30(3):157—60. https://doi.org/10.2337/ds17-0013.

Dash S, et al. Hybrid chaotic firefly decision making model for Parkinson's disease diagnosis. Int J Distributed Sens Netw 2020;16(1). https://doi.org/10.1177/1550147719895210.

Flack KD, et al. Energy compensation in response to aerobic exercise training in overweight adults. Am J Physiol Regul Integr Comp Physiol 2018;315:R619—26.

Franz MJ, et al. Weight-loss outcomes: a systematic review and meta-analysis of weight-loss clinical trials with a minimum 1-year follow-up. J Am Diet Assoc 2007 Oct;107(10):1755—67. https://doi.org/10.1016/j.jada.2007.07.017.

Gupta V, Mittal M, Mittal V. Chaos theory and ARTFA: emerging tools for interpreting ECG signals to diagnose cardiac arrhythmias. Wireless Pers Commun 2021;118:3615—46. https://doi.org/10.1007/s11277-021-08411-5.

Hall KD. Energy compensation and metabolic adaptation: "The Biggest Loser" study reinterpreted. Obesity 2022;30(1):11—3. https://doi.org/10.1002/oby.23308/.

Klassen M. Deterministic chaos: applications in cardiac electrophysiology. Occam's Razor 2016;6:7. Available at: https://cedar.wwu.edu/orwwu/vol6/iss1/7.

Loewenstein WR. The touchstone of life: molecular information, cell communication, and the foundations of life. New York, NY: Oxford University Press; 1999.

Mibaile J, et al. Chaos in human brain phase transition. Research advances in chaos theory. Intech Open 2020. https://doi.org/10.5772/intechopen.86667.

Miguel AF. Dendritic design as an archetype for growth patterns in nature: fractal and constructal views. Front Phys 2014. https://doi.org/10.3389/fphy.2014.00009.

Morgan J. Introduction to university physics. 2nd ed. Boston, MA: Allyn and Bacon; 1969.

Morowitz HJ. Energy flow in biology. New York, NY: Academic Press; 1968.

Peters RH. The ecological implications of body size. Cambridge, England: Cambridge University Press; 1983.

Pontzer H. Exercise paradox. Sci Am 2017 Feb;316:28—31.

Ripple WJ, et al. Status and ecological effects of the world's largest carnivores. Science 2014;343:1241384. https://doi.org/10.1126/science.124184.

Rose AH, editor. Thermobiology. London: Academic Press; 1967.

Ruch TC, Patton HD, editors. Physiology and biophysics. Philadelphia, PA: W. B. Saunders Co.; 1965.

Schmidt-Nielsen K. Scaling, why is animal size so important? Cambridge, UK: Cambridge University Press; 1984.

Schuldberg D, Richards R, Shan G, editors. "Chaos and nonlinear psychology: keys to creativity in mind and life" New York, 2022; online ed, Oxford Academic, 2022 Apr 21. https://doi.org/10.1093/oso/9780190465025.001.0001.

Schurch S, Lee M, Gehr P. Pulmonary surfactant: surface properties and function of alveolar and airway surfactant. Pure Appl Chem 1992;64(11):1745—50.

Stacy RW, Williams DT, Worden RE, McMorris RW. Biological and medical physics. New York, NY: McGraw-Hill Book Co.; 1955.

Swift DL, et al. The role of exercise and physical activity in weight loss and maintenance. Prog Cardiovasc Dis 2014 Jan—Feb;56(4):441—7. https://doi.org/10.1016/j.pcad.2013.09.012.

Szczesna A. Quaternion entropy for analysis of gait data. Entropy 2019;21:79. https://doi.org/10.3390/e21010079.

## CHAPTER 12

Alexander RMN. Animal mechanics. Seattle, WA: University of Washington Press; 1968.

Brown JHU, Gann DS, editors. Engineering principles in physiology. Vols. 1 and 2. New York, NY: Academic Press; 1973.

Burns DM, MacDonald SGG. Physics for biology and pre-medical students. Reading, MA: Addison-Wesley Publishing Co.; 1970.

Casey EJ. Biophysics. New York, NY: Reinhold Publishing Corp; 1962.

Cromwell L, Weibell FJ, Pfeiffer EA, Usselman LB. Biomedical instrumentation and measurements. Englewood Cliffs, NJ: Prentice-Hall; 1973.

Marshall JS, Pounder ER, Stewart RW. Physics. 2nd ed. New York, NY: St. Martin's Press; 1967.

Mizrach A, et al. Acoustic trap for female Mediterranean fruit flies. Trans ASAE 2005;48:2017−22.

Morgan J. Introduction to university physics. 2nd ed. Boston, MA: Allyn and Bacon; 1969.

Richardson IW, Neergaard EB. Physics for biology and medicine. New York, NY: John Wiley & Sons; 1972.

Stacy RW, Williams DT, Worden RE, McMorris RW. Biological and medical physics. New York, NY: McGraw-Hill Book Co.; 1955.

ter Haar G. Acoustic surgery. Phys Today 2001;54(12):29.

## CHAPTER 13

Ackerman E. Biophysical science. Englewood Cliffs, NJ: Prentice-Hall, Inc.; 1962.

Bassett CAL. Electrical effects in bone. Sci Am 1965;213:18.

Bullock TH. Seeing the world through a new sense: electroreception in fish. Am Sci 1973;61:316.

Delchar TA. Physics in medical diagnosis. New York, NY: Chapman and Hall; 1997.

Hobbie RK. Intermediate physics for medicine and biology. New York, NY: Springer; 1997.

Hobbie RK. Nerve conduction in the pre-medical physics course. Am J Phys 1973;41:1176.

Katz B. How cells communicate. Sci Am 1961;205:208.

Katz B. Nerve muscle and synapse. New York, NY: McGraw-Hill, Inc.; 1966.

Mckenzie I, et al. Motor skill learning requires active central myelination. Science 2014;346(6207):318−22. https://doi.org/10.1126/science.1254960.

Miller WH, Ratcliff F, Hartline HK. How cells receive stimuli. Sci Am 1961;205:223.

Scott BIH. Electricity in plants. Sci Am 1962;207:107.

## CHAPTER 14

Ackerman E. Biophysical science. Englewood Cliffs, NJ: Prentice-Hall, Inc.; 1962.

Blesser WB. A systems approach to biomedicine. New York, NY: McGraw-Hill Book Co.; 1969.

Cromwell L, Weibell FJ, Pfeiffer EA, Usselman LB. Bio-medical instrumentation and measurements. Englewood Cliffs, NJ: Prentice-Hall, Inc.; 1973.

Davidovits P. Communication. New York, NY: Holt, Rinehart and Winston; 1972.

Loizou PC. Mimicking the human ear. IEEE Signal Process Mag 1998;15(5):101−30.

Scher AM. The electrocardiogram. Sci Am 1961;205:132.

## CHAPTER 15

Ackerman E. Biophysical science. Englewood Cliffs, NJ: Prentice-Hall, Inc.; 1962.

Davidovits P, Egger MD. Microscopic observation of endothelial cells in the cornea of an intact eye. Nature 1973;244:366.

Katzir A. Optical fibers in medicine. Sci Am 1989;260:120.

Marshall JS, Pounder ER, Stewart RW. Physics. 2nd ed. New York, NY: St. Martin's Press; 1967.

Muntz WRA. Vision in frogs. Sci Am 1964;210:110.

Oberst J, Sanders S. Changing perspectives on daylight, science, technology and culture. Washington, DC: Science/AAAS; 2017.

Ruch TC, Patton HD. Physiology and biophysics. Philadelphia, PA: W. B. Saunders and Co.; 1965.

Wald G. Eye and the camera. Sci Am 1950;183:32.

## CHAPTERS 16 AND 17

Ackerman E. Biophysical sciences. Englewood Cliffs, NJ: Prentice-Hall, Inc.; 1962.

Burns DM, MacDonald SGG. Physics for biology and premedical students. Reading, MA: Addison-Wesley Publishing Co.; 1970.

Delchar TA. Physics in medical diagnosis. New York, NY: Chapman and Hall; 1997.

Dowsett DJ, Kenny PA, Johnston RE. The physics of diagnostic imaging. New York, NY: Chapman and Hall Medical; 1998.

Hobbie RK. Intermediate physics for medicine and biology. New York, NY: Springer; 1997.

Miller J. Cryo-electron microscopy pioneers win chemistry Nobel. Phys Today 2017;70:22−4.

Pizer V. Preserving food with atomic energy. Oak Ridge, TN: United States Atomic Energy Commission Division of Technical Information; 1970.

Pykett IL. NMR imaging in medicine. Sci Am 1982;246:78.

Schrödinger E. "What is life?" and other scientific essays. Garden City, NY: Anchor Books, Doubleday and Co.; 1956.

## CHAPTER 18

Chaloupka K, Malam Y, Seifalian AM. Nanosilver as a new generation of nanoproduct in biomedical applications. Trends Biotechnol 2010;28:580−8.

Driskell JD, Jones CA, Tompkins SM, Tripp RA. One-step assay for detecting influenza virus using dynamic light scattering and gold nanoparticles. Analyst 2011;136:3083−90.

Heath JR, Davis ME, Hood L. Nanomedicine targets cancer. Sci Am 2009;300:44−51.

Peer D, Karp JM, Hong S, Farokhzad OC, Margalit R, Langer R. Nanocarriers as an emerging platform for cancer therapy. Nat Nanotechnol 2007;2:753−60.

Shirtcliffe NJ, McHale G, Newton MI. Learning from superhydrophobic plants: the use of hydrophilic areas on superhydrophobic surface for droplet control. Langmuir 2009;25:14121−8.

Underwood E. The polluted brain. Science 2017;123:342.

William JR, et al. Status and ecological effects of the world's largest carnivores. Science 2014;343:1241384. https://doi.org/10.1126/science.124184.

## CHAPTER 19

Ball P. Bending the laws of optics with metamaterials: an interview with John B. Pendry National Sci Rev 2018 Mar;5(2):200−2.

Cheng X, Senior DE, Kim C, Yoon YK. A compact omnidirectional self-packaged patch antenna with complementary split-ring resonator loading for wireless endoscope applications. IEEE Antenn Wireless Propag Lett 2011;10:1532−5. https://doi.org/10.1109/LAWP.2011.2181315.

Duan G, et al. Boosting magnetic resonance imaging signal-to-noise ratio using magnetic metamaterials. Commun Phys 2019;2:35. https://doi.org/10.1038/s42005-019-0135-7.

Garland A, et al. Metamaterial design for targeted limb-socket interface pressure offloading in transtibial amputees. Paper No: DETC2020-22044, V11AT11A011ASME Digital Collection. https://doi.org/10.1115/DETC2020-22044. Published Online: November 3, 2020.

Khorasaninejad MA, et al. Polarization-insensitive metalenses at visible wavelengths. Nano Lett 2016;16(11):7229−34.

Pendry JB, Holden AJ, Robbins DJ, Stewart WJ. Magnetism from conductors and enhanced nonlinear phenomena. IEEE Trans Microw Theory Tech 1999 Nov;47:2075. https://doi.org/10.1109/22.798002.

Pendry JB. Negative refraction makes a perfect lens. Phys Rev Lett 2000 Oct 30;85:3966.

Veselago VG. The electrodynamics of substances with simultaneously negative values of $\varepsilon$ and $\mu$. *Sov Phys Usp* 1968, 10 509

Wilson JD, Schwartz ZD. Multifocal flat lens with left-handed metamaterial. Appl Phys Lett 2005;86(2).

# Answers to Numerical Exercises

Answers to numerical exercises that are provided in the text are not listed here.

## Chapter 1

**1.1** (b). $F = 254$ N (57.8 lb)
**1.3** $\theta = 72.6°$
**1.4** Maximum weight $= 335$ N (75 lb)
**1.5** (a). $F_m = 2253$ N (508 lb), $F_r = 2386$ N (536 lb)
**1.6** $F_m = 720$ N, $F_r = 590$ N
**1.7** (a). $F_m = 2160$ N, $F_r = 1900$ N
**1.8** $\Delta F_m = 103$ N, $\Delta F_r = 84$ N
**1.10** $\Delta x = 19.6$ cm, v of tendon $= 4$ cm/sec, v of weight $= 38$ cm/sec
**1.11** $F_m = 0.47$ W, $F_r = 1.28$ W
**1.12** (a). $F_m = 2000$ N, $F_r = 2200$ N; (b). $F_m = 3220$ N, $F_r = 3490$ N
**1.13** $F_A = 2.5$ W, $F_T = 3.5$ W

## Chapter 2

**2.1** (a). Distance $= 354$ m; (b). Independent of mass
**2.2** $\mu = 0.067$
**2.3** (a). $\mu = 1.95$; (b). with $\mu = 1.0$, $\theta = 39.4°$, with $\mu = 0.01$, $\theta = 0.6°$
**2.4** Thickness of saliva layer $= 0.0775$ mm

## Chapter 3

**3.1** $P = 4120$ watt
**3.2** $H'_L = 60$ cm
**3.3** $F_r = 1.16$ W, $\theta = 65.8°$
**3.4** $T = 0.534$ seconds
**3.5** (a). $R = 13.5$ m; (b). $H = 3.39$ m; (c). 4.08 seconds
**3.6** $v = 8.6$ m/sec
**3.7** $r = 1.13$ m
**3.8** (a). $v = 8.3$ m/sec; (b). 16.6 cm/sec
**3.9** Energy expended/sec $= 1350$ J/sec
**3.10** $P = 371$ watt

## Chapter 4

**4.2** $F = 10.1$ N

**4.3** $\omega = \frac{1.25 \text{ rad}}{\text{sec}}$; linear velocity $= \frac{6.25 \text{ m}}{\text{sec}}$

**4.4** (b). $\omega = 1.25$ rad/sec $= 33.9$ rpm

**4.5** $v = 31.4$ m/sec

**4.6** Speed $= 1.13$ m/sec $= 4.07$ km/h $= 2.53$ mph

**4.7** $T = 1.6$ sec

**4.8** $E = 1.64$ mv$^2$

**4.9** Fall time $= 1$ second

**4.10** Distance $= 0.11$ m

## Chapter 5

**5.1** $v = \frac{2.9 \text{ m}}{\text{sec}} = 5.3$ mph

**5.2** $v = 8$ m/sec; with 1 cm$^2$ area $v = 2$ m/sec

**5.3** $h = 5.1$ m

**5.4** $t = 3 \times 10^{-2}$ sec

**5.5** $v = 17$ m/sec (37 mph)

**5.6** Force/cm$^2$ $= 4.6 \times 10^6$ dyn/cm$^2$, yes

**5.7** $v = 0.7$ m/sec, no

## Chapter 6

**6.1** $F = 2$ W

**6.2** $\ell = 0.052$ mm

**6.3** $h = 18.4$ cm

**6.4** $\ell = 10.3$ cm

## Chapter 7

**7.2** $P = 7.8$ W

**7.3** $v = [gV(\rho_w - \rho)/(A\rho_w)]^{1/2}$;

$$P = \left(1/2\left[W\{(\rho_w/\rho) - 1\}^{(3/2)}\right]\right)\Big/(A\rho_w)^{1/2}$$

**7.5** $P = 1.51 \times 10^7$ dyn/cm$^2$ $= 15$ atm

**7.6** Volume of swim bladder $= 3.8\%$

**7.7** $\rho_2 = \rho_1 (W_1/W_1 - W_2)$

**7.8** $p = 1.46 \times (10^5 \text{ dyn})/\text{cm}^2$

**7.11** Perimeter $= 9.42$ km

**7.12** (b). Speed $= 115$ cm/sec

## Chapter 8

**8.1** $\Delta P = 3.19 \times 10^{-2}$ torr
**8.2** $\Delta P = 4.8$ torr
**8.3** h $= 129$ cm
**8.4** (a). p $= 61$ torr; (b). p $= 200$ torr
**8.5** (b). $R_1/R_2 = 0.56$
**8.6** v $= 26.5$ cm/sec
**8.7** N $= 5.03 \times 10^9$
**8.8** $\Delta p = 79$ torr
**8.9** P $= 10.1$ W
**8.10** (a). P $= 0.25$ W; (b). P $= 4.5$ W
**8.12** (a). L $= D/2$; (b). R $= 5 \times 10^{-2}$

## Chapter 9

**9.2** V $= 29.3$ L
**9.3** (a). t $= 10^{-2}$ seconds; (b). t $= 10^{-5}$ seconds
**9.5** N $= 1.08 \times 10^{20}$ molecules/sec
**9.6** No. of breaths/min $= 10.4$
**9.7** (a). Rate $= 1.71 \times 10^{-5}$ L/h-cm$^2$; (b). diameter $= 0.5$ cm
**9.8** $\Delta P = 0.0287$ atm

## Chapter 11

**11.4** t $= 373$ hours
**11.5** v $= 4.05$ m$^3$
**11.6** t $= 105$ days
**11.7** Weight loss $= 0.892$ kg
**11.8** H $= 18.7$ Cal/h
**11.10** (b). Change $= 22\%$; (c). $K_r = 6.0$ Cal/m$^2$-h-C$^\circ$
**11.11** Heat removed $= 8.07$ Cal/h
**11.12** Heat loss $= 660$ Cal/m$^2$-h
**11.13** H $= 14.4$ Cal/h
**11.18** (a). The basal metabolism of 190 kg lion is 24,820 Cal/wk; (b). Number of antelopes required by one lion per week is 0.827

## Chapter 12

**12.1** R $= 31.6$ km
**12.2** 1.75 times
**12.3** p $= 2.9 \times 10^{-4}$ dyn/cm$^2$
**12.6** D $= 11.5$ m
**12.8** Minimum detectable power of the sound entering the tube $= 2.49 \times 10^{-16}$ W

**12.9** Smallest size of detectable objects in soft tissue is 0.308 mm
**12.11** Increase in temperature of the heated body $= 19.1°C$

## Chapter 13

**13.1** (a). No. of ions $= 1.88 \times 10^{11}$;
(b). No. of $Na^+$ions $= 7.09 \times 10^{14}/m$;
No. of $K^+$ions $= 7.09 \times 10^{15}/m$
**13.8** (a). No. of cells in series $= 5000$;
(b). No. of cells in parallel $= 2.7 \times 10^9$

## Chapter 14

**14.1** i $= 13.3$ amp

## Chapter 15

**15.1** Change in position $= 0.004$ cm
**15.3** For cornea 41.9 diopters; for lens, min power $= 18.7$ diopters, max power $= 24.4$ diopters
**15.4** $1/f = -0.39$ diopters
**15.5** Focusing power $= \pm 70$ diopters
**15.6** p $= 1.5$ cm
**15.7** (a). Resolution $= 2.67 \times 10^{-4}$ rad;
(b). Resolution $= 6.67 \times 10^{-4}$ rad
**15.8** D $= 20$ m
**15.9** H $= 3 \times 10^{-4}$ cm

## Chapter 16

**16.5** (a). Magnitude of force F $= 0.5 \times 10^{-9}$ N; (b). Q $= 2.4 \times 10^{-19}$ coulomb. This is effectively 1.5 electrons and 1.5 protons facing each other distributed over an area defined by the sampling tip-surface overlap.

## Chapter 18

**18.2** (a). Number ratio of surface to volume atoms $= 0.12$; (b). Number ratio of surface to volume atoms $= 1.2 \times 10^{-6}$
**18.7** No. of particles entering the lung one breath $= 1.15 \times 10^7$

# Index

Note: Page numbers followed by *f* indicate figures, *t* indicate tables, and *np* indicate footnotes.